The Nile

sharing a scarce resource

In memoriam

Paul Howell died in early April 1994 before the publication of the book. He has been associated with the Nile and its countries for nearly sixty years beginning his career in the 1930s in the Sudan and concluding a phase of it in Uganda in the 1950s until 1961. He made an immense impact in several departments of government and yet found time throughout to promote scientific work by others and make some unique contributions himself. Afterwards at Cambridge he was able to sustain his scholarly interests and deploy his administrative flair to develop programmes on the understanding and implementation of social and economic change in the parts of the world with which he was so well acquainted.

His contribution to studies of the Nile and particularly of the Sudd and its peoples and environment is unlikely to be emulated in depth or breadth in that his anthropological work on the law of the Nuer people is an established foundation text and the studies which he supervised as Chairman of the Jonglei Investigation Team provided an environmental evaluation of unprecedented thoroughness. This and his other studies have been inspired in method, sound in detail and priceless in judgement. His work and friendship were greatly valued by friends in Africa where there was much mutual satisfaction in the visits which they made to him in Cambridge and later in Wymondham. Paul Howell left the world a richer place and his friends and fellow authors are grateful to him and to Brigid, his wife, who together enabled such productive times with warm associated memories for so many.

The environmental history of the last three decades has heightened awareness that the volume of water available in the Nile Basin is not sufficient to meet the current levels of water demand reliably, quite apart from rapidly growing needs. The environmental element of the complex matter of evaluating and managing the international water resource of the Nile is addressed in this volume.

The book deals with the global and regional hydrology, examines the scope and emphasis of water resource planning in the Nile Basin over the past century and identifies future options. The material is intended to be relevant to environmental scientists, government officials and water resource professionals in both national and international agencies, as well as those in the academic community concerned with the environment and hydraulic engineering. The economic context of scarce resource development is also examined.

The Nile

sharing a scarce resource

A historical and technical review of water management and of economic and legal issues

Edited by
P. P. HOWELL
Wolfson College, Cambridge
J. A. ALLAN
School of Oriental and African Studies, University of London

CAMBRIDGE UNIVERSITY PRESS

CAMBRIDGE UNIVERSITY PRESS
Cambridge, New York, Melbourne, Madrid, Cape Town, Singapore, São Paulo

Cambridge University Press
The Edinburgh Building, Cambridge CB2 8RU, UK

Published in the United States of America by Cambridge University Press, New York

www.cambridge.org
Information on this title: www.cambridge.org/9780521450409

First published 1994
Reprinted 1996
This digitally printed version 2008

A catalogue record for this publication is available from the British Library

ISBN 978-0-521-45040-9 hardback
ISBN 978-0-521-06555-9 paperback

Contents

Preface

Recent climatic variations and consequent fluctuations in river flows in the Nile Basin suggest that assumptions current in the first half of this century regarding the availability of water may now require drastic readjustment. The mean annual discharge has fallen from 84 cubic kilometres (1900-1959) to 72 cubic kilometres (1977-1987), with an even more dramatic fall in the mean for the years between 1984 and 1987 to less than 52 cubic kilometres.

Historically there have been considerable fluctuations during this century; the annual flow has varied between a maximum of 120 cubic kilometres (1916) and a minimum of 42 cubic kilometres (1984). The Blue Nile, which for most of the century has provided some 86 per cent of the total flow, has shown marked decline in the last few years, with the exception of 1988, following decreases in rainfall over the Ethiopian catchment. On the other hand the White Nile, as the result of exceptionally heavy rainfall over Lake Victoria and most of the East African catchment from the early 1960s, has shown a massive increase in flows over the previous sixty years. These high White Nile flows, especially in the 1960s and early 1970s when flows from the Blue Nile were average or above, have to some extent compensated for the subsequent shortfalls in the Ethiopian tributaries, but inflows from the East African catchment are always masked by the effects of the Sudd swamps. Indeed, as dramatically illustrated in the early 1960's onwards, the greater the inflow the greater the proportional extent of flooding and the higher the percentage loss from evapo-transpiration in that region. To reduce these losses was the purpose of the long mooted but ill-fated Jonglei Canal.

The material assembled in this book is intended to provide a basis for those wishing to understand the environmental circumstances which have shaped the patterns of water availability and water use during the past century and especially during the past thirty years. It is during this last period that the most thoroughgoing measures were taken to control the flow of the Nile for the major Nile waters user, Egypt. The decision of the Egyptian Government to construct the High Dam at Aswan taken in the late 1950s marked the end of a long phase of discussion during which engineers and politicians had agonised over whether to take a basin-wide view of the use of control works or one which concentrated control within Egyptian territory. The decision to build the dam was entirely consistent with the knowledge of Nile flows up to that point and did not seem to contradict the interests of other Basin countries if the principle of prior use was judged to be a reasonable basis for controlling the distribution of its waters. The environmental history of the last three decades has, however, considerably shifted expectations concerning Nile flows and has greatly accelerated the arrival of a widespread awareness that the volume of water available in the system is not sufficient to meet reliably the current levels of water demand, quite apart from rapidly growing needs. The reductions in average flow of twenty per cent in the Ethiopian tributaries in the 1980s confronted the downstream countries with a sharp reminder of the real constraints that the environment could impose on national economic options.

It is this environmental element of the complex matter of evaluating and managing the international water resource of the Nile which is addressed in this volume. The evidence presented enables an appreciation of the latest views on climate change in north-eastern Africa and while it has to be accepted that there is no conclusive evidence of a predictable long term deterioration in the levels of rainfall in the Ethiopian Highlands or those of East Africa, it appears that the relatively benign circumstances of the first half of the twentieth century should not be the basis of long-term water planning and management of either the overall hydrological system or even of the heavily used downstream section. Yet the records of the first half of the century were precisely those on which the decision to construct the High Dam was taken .

The engineering structures which control the Nile are still mainly concentrated in Egypt and the Sudan and their history and current functions are discussed in this volume. It is pointed out that the major sector to which Nile water is allocated is agriculture and it is usual for agricultural priorities to be the deteminant of decisions concerning the management of water at the national level. It is pointed out that the trends in environmentally constrained water availability and those of demographically driven water demand are not in accord and require significant shifts in national and international approaches to the planning and use of management. The options for the downstream states are reviewed in detail and the approach of the country from which over seventy per cent of the current agriculturally accessible water comes are discussed by informed specialists.

As the assessment of the Nile waters in that region has not been evaluated comprehensively in the recent past, this volume includes a discussion of the environmental background to the scientific studies completed in the 1950s with the intent of determining the potential Nile water requirements for agriculture and power in the countries of the East African Highlands. That the modest demands then envisaged have not for the most part been realised is a significant feature and one which suggests that a comprehensive view of current national water plans would be desirable.

As there is a number of scientific and professional interest groups concerned with the Nile and its potential, both in the Nile Basin countries themselves as well as internationally, it was decided to divide the material into two volumes. The first deals with the global and regional hydrology relevant to understanding historical and current water availability and use. It also examines the scope and emphasis of water resource planning in the Nile Basin over the past century as well as identifying future options. The material is intended to be relevant to a wide community of environmental scientists, government officials and water resource professionals in both national and international agencies, as well as those concerned with the environment and hydraulic engineering in the academic community.

A second volume is devoted to the subjects in the socio-economic domain which are much more difficult to quantify. In this second volume the economic context of scarce resource development is examined. It is argued that the adoption of environmentally rational, that is sustainable, economic policies is an essential part of any national or international water allocation policy. Although water is everywhere already scarce, or it will be in a decade or two, the need to view water as one factor in the admixture which enables economic growth is important. It is also important to allocate this scarce factor to productive sectors where it can bring a sustainable economic return, safely and equitably. The problem of scarcity can be approached by seeking to use water more efficiently according to existing water allocative policies or according to new allocative policies based on economic principles. Such innovation has proved difficult to implement because of political inertia and the perceived social costs of a shift in economic emphasis. The problem is also exacerbated by the traditional commitment to providing secure domestic food provision without recognising the non-viability of such a policy because environmental resources will not sustain them. Egypt's food gap represents a water gap amounting to about 25 per cent of its present water use in agriculture. It has already had to substitute other sources of income, including international assistance, to fill this gap and there is no escape from this process in future.

The case of Egypt exemplifies the highly political nature of water at the national and the international level and the second volume addresses the hydropolitical issues which have attended the allocation and use of Nile water. An important element in the hydropolitics of any international river is that of legal precedent and international agreement. The experience on the Nile is similar to that of most such rivers where some agreements have been made and largely observed by signatories, but the agreements have not been sufficiently comprehensive to cope with dynamic economic and political circumstances. As the major supplier of water, Ethiopia, was not a signatory of the 1959 Nile Waters Agreement and does not feel bound by it, is therefore, proposing to embark on a long-term investment programme over the next 50 years which would harness its Nile tributaries for hydropower and agriculture. The preliminaries to reaching an accord which could create a basis for mutually advantageous development are in progress and numerous international meetings, both public and confidential, are in train which will enable some essential preliminary steps to be made to re-install the environmental and hydrological monitoring which is essential if a strong information base is to be created to facilitate discussion and identify where reciprocal benefits can be found and mobilised.

The meeting in 1990 on which the chapters in both volumes are based was part of a long tradition of scientific meetings held in London on the subject of the Nile as well as a significant element in the process of international liaison which gathered pace at the end of the 1980s as the result of the anxieties generated by the accumulating low flows of the middle of that decade. Behind the current debate lie two basic philosophies. The first is the one given expression by the Egyptian initiatives in the late 1950s which led to the construction of the High Dam at Aswan with the intent of securing Egypt's water for the foreseeable future. The other is that which the Egyptian initiative contradicted, namely the basin-wide approach to the management of Nile waters which would take into account the optimum siting of water storage to reduce evaporation and other losses. In the last decade of the twentieth century there will be the chance to discuss and negotiate new options; to make reciprocal gestures using water alone as a currency; to make reciprocal gestures exchanging other resources such as capital investment in return for an assured flow of water. The tension between the local and the basin-wide approaches will be on-going and the debate which will be conducted in many other fora, both generally in public as well as in private, will be at the very highest level. The subject will command the interest not just of specialists in water resources and politics from the region itself. It will also be followed closely by many officials and scientists advising the international bodies which are the only institutions capable of financing the next essential phases of water resource management in the Nile Basin.

The Royal Geographical Society has played a leading role in promoting scientific interest in the Nile for well over a century - first in connection with the exploration of its sources, beginning with Sir Richard Burton in 1856 together with his subsequently much maligned companion John Hanning Speke, who later in 1862 discovered the outlet of Lake Victoria at the Rippon Falls, which he named after the President of the Society of the time. The Society was also supportive of the explorations of Sir Samuel Baker culminating in his discovery of Lake Albert and the falls on the reach of the Nile leading to that lake which he named Murchison, again after the current President of the Society. Later, in 1865, Livingstone was asked by the Society to confirm or otherwise the erroneous theory, advocated by Burton, that there was a further more distant southern source eventually flowing into Lake Albert; and then, when he appeared lost, the Society despatched an expedition to locate him.

From the beginning of this century the Society has turned its attention not only to the geographical aspects of the Nile Valley but to the questions relating to the control of its waters for irrigation and hydro-electric power development, the most notable early contribution being a paper presented by Sir William Garstin entitled *Fifty Years of Nile Exploration and some of the results,* published in the Geographical Journal in 1909. This was not so much concerned with exploration in the sense of the endeavours of the previous century, but with a description of the main hydrological features and the results of his personal expeditions and investigations, especially in the Sudd region. The article was a testimony to the author's mastery of a vast range of specialist knowledge of engineering, water control and conservation, including plans for saving water losses in the swamps by canalisation. Other papers have followed and the Society's interest in the scientific aspects of the Nile Basin has been sustained. In 1982 the Society convened a conference on the *Impact of the Jonglei Canal in the Sudan,* which gave impetus to the idea of a book on the subject which, after further research, was published in 1988. This venture, and the news of unusual changes in Nile flows, prompted the society to convene a conference in which up-to-date views on the re-evaluation of the resource and of future water demand and availability could be discussed.

The School of Oriental and African Studies (SOAS) is an institute of the University of London founded in 1917 late in the period of Britain's imperial presence in Africa. It is since the independence of African countries to which the chapters in this book refer, in the 1950s and 1960s, however, that the major expansion of staff took place, especially in the social and environmental sciences. In 1989 staff of the SOAS Centre of Near and Middle Eastern Studies were preparing to convene a meeting on the economic, hydropolitical and legal issues relating to Nile waters. It was therefore a logical and happy decision that these two institutions should combine to provide a two-day conference for discussion and debate on both recent climatic phenomena and long standing issues of water sharing in a purely scientific and academically orientated forum.

One of the primary concerns was to enable knowledgeable people from all countries of the Nile Basin to attend the conference. The subjects under review were of varying but vital significance to all of them and the subject would have been unrealistic without their participation. We are therefore especially grateful for the very generous response in financial contributions for travel and subsistence for Nile Valley participants from the Overseas Development Administration, the British Council, the World Bank, HRH Prince Muhammad al Faisal and the Research and Publications Committee of SOAS. We are particularly appreciative of the assistance given by Nile Basin governments in supporting the participation of some of their officials who attended the conference and contributed papers and to the debate. The support of Mr H. St John B. Armitage CBE, was immeasurably valuable.

A preliminary publication which was circulated at the conference would not have been possible without the willingness of contributors to prepare material in advance, no easy task for busy and internationally mobile professionals. We owe a debt of gratitude to the staff of British Missions and British Council offices in the Nile Valley capitals for assistance in ensuring that contributions reached us in London in time by Fax and courier. The same staff attended to the myriad of administrative details which enabled observers as well as participants to travel to the conference. Consulting engineering companies in addition to contributing the time of professional staff also provided finance to facilitate the smooth conduct of the conference. We are particularly grateful to the Director of the Royal Geographical Society, Dr John Hemming, and the staff of the Royal Geographical Society for extending their facilities on the first day and to the Director of the School of Oriental and African Studies, Mr Michael McWilliam, and its staff for the second, as well as for the use of services and indispensible technical support. A number of individuals have made very significant contributions to the preparation of the text and diagrams in this volume. Both editors would like to express very warm appreciation to Jon Wild who devoted a great deal of careful attention to improving the presentation of the material. Most of the diagrams had to be redrawn by him with some complicated cartographic work being completed by Catherine Lawrence of the Geography Department of SOAS.

It has been our task to collate the papers presented at the conference, with the addition of further material, into book form. The purpose of the meeting which inspired these volumes was to encourage the exchange at a high official and technical level of information, much of which is controversial and which has been widely discussed elsewhere. It was our hope that the academic circumstances in which specialists, engineers and scientists were enabled to come together on that occasion in London would promote mutual understanding which could lead to some real, if modest, progress with respect to the beneficial development of the waters of the Nile. We trust that this aspiration will be furthered by the publication of this volume.

List of contributors

Dr Zewdie Abate: was General Manager, Ethiopian Valleys Development Studies Authority, PO Box 1086, Addis Ababa, Ethiopia at the time he prepared the material. He has become head of the newly formed Environmental Protection Agency of Ethiopia. His mailing address is PO Box 2509, Addis Ababa, Ethiopia.

Mr Samir Ahmed: 3 Assiouti St, Manchiet El-Bakry, Cairo, Egypt.

Professor J A (Tony) Allan: School of Oriental and African Studies University of London, Thornhaugh Street, London WC1H 0XG.

Professor Robert O Collins: Department of History, University of California at Santa Barbara, California, 93106, USA.

Mr Jeremy Lazenby: Alexander Gibbs and Partners,Earley House,London Road, Earley, Reading RG6.

Mr Peter Chesworth: Mott Macdonald, Demeter House Station Road, Cambridge CB1

Mr Terry Evans: Mott Macdonald Demeter House Station Road, Cambridge CB1

Mr David Knott: Alexander Gibbs and Partners, Earley House, London Road, Earley Reading RG6.

Mr Rodney G M Hewett: Alexander Gibbs and Partners, Earley House London Road, Earley Reading RG6

Dr Paul Howell: Burfield Hall, Wymondham, Norfolk, NR18 9SJ.

Dr Michael Hulme: Climatic Research Unit, University of East Anglia, Norwich, NR4 7TJ.

Mr. Bezazal Kabanda: Commissioner for Water Development, Ministry of Water and Mineral Development, Water Development Department, Box 20026, Kampala, Uganda.

Mr. Patrick Osbert Kahangire: Assistant Commissioner, Ministry of Water and Mineral Development, Water Development Department, Box 20026, Kampala, Uganda.

Dr Michael Lock: Royal Botanical Gardens, C/O The Herbarium, Kew Gardens, Richmond, TW9 3AE.

Dr Chibli Mallat: Department of Law, School of Oriental and African Studies, Thornhaugh Street, London, WC1H 0XG.

Professor Odidi Okidi: Faculty of Earth Sciences, University of Moi, Eldoret, Kenya.

Dr Yvonne Parks: Institute of Hydrology, Wallingford, OX10 8BB.

Prof Rushdi Said: During 1989-90 - Institute of Advanced Studies, Wissenschaftskolleg, Wallootstr 19, d-1000 Berlin 33, W Germany. Tel off-(30) 89001-223, Res-(30) 892 7261.

Dr John Sutcliffe: Heath Barton, Manner Road, Goring on Thames, Reading, Berks, RG8 9EH.

Mr Roy Stoner: Institute of Irrigation Studies, Department of Civil Engineering, University of Southampton, Southampton, SO9 5NH.

Further information: Prof J A Allan, SOAS, Thornhaugh St, London WC1H 0XG. Tel 071 323 6159, Fax 436 3844

Orthography

The spelling of place and tribal names present some difficulty, since the rendering of Arabic names has been inconsistent in literature for a long time. Taking the Bahr al Jabal (the main channel of the White Nile running through the Sudd) as an example, early travellers spoke of this simply as the White Nile, though they referred to the Bahr al Ghazal variously as Bahr el Gasall, Bahr el Gazal, while the Bahr al Zaraf was simply called the Giraffe River. Since then there have been various versions e.g. Egyptian: Behr el Gebel. The correct version is Bahr al Jabal and is now commonly used. However, maps produced during this century and the gazetteer of the Sudan Survey Department consistently refer to the Bahr el Jebel, Bahr el Zeraf, Bahr el Ghazal and are adopted in most of the engineering and hydrological literature to which reference is made. We retain these spellings where appropriate. In Arabic the definite article *al* is assimilated to certain consonants. Bahr al Zaraf, for example in speech is pronounced Bahr az Zaraf, but it is incorrect to spell it so.

Units and conversion factors

Units of Flow

Traditionally, annual Nile flows have been measured in thousand million cubic metres and expressed in "Milliards".

More recent works refer to billion cubic metres per annum (billion m^3) or $m^3 \times 10^9$. Where appropriate we prefer to refer to cubic kilometres per annum = km^3.

Daily flows are expressed in million cubic metres per day =million m^3 / day = 11.6 cubic metres per second or cumecs = 408 cusecs.

Units of Area

1 feddan = 4200 m^2 = 0.42 ha = 1.037 acres
1km^2 = 100 ha = 1,000,000 m^2

Orthography

Introduction

Principles of 'sustainability' have been advocated very insistently during the past decade but they had been eloquently and powerfully enunciated many centuries ago in a saying which in its first part captures with extraordinary simplicity the notion of sustainability and in the second prescribes a moral code which would encourage the principled behaviour necessary for the implementation of sound management of renewable natural resources.

> 'Cultivate your world as if you would live for ever, and prepare for your Hereafter as if you would die tomorrow'
>
> Saying of the Prophet Mohammed, explained by 'Ali Mubarak, (May 1891)

The Nile: an important international resource under pressure

The Nile drains approximately ten per cent of the continent of Africa and includes all or parts of the territories of nine sovereign nation states. The allocation of its waters and the management of them between the competing national and using interests are inevitably complicated and tend to be highly charged with respect to water rights, and the resulting tensions are likely to become of greater significance in future. The present international relations and the pattern of water resource allocation and management are determined partly by the status, that is the volume and quality, of the water resource, and partly by the predictable attitudes of governments and peoples to an annually varying resource, the ownership of which is not clear.

The River Nile is a natural system which moves water, and silt, from mountainous upstream locations to particularly extensive and low-lying downstream tracts, on the way serving many peoples and economies. For the past six or more millennia it has been the unreliability of the flow of water which has been the preoccupying issue for the Nile waters using communities. That the flow could vary from year to year as well as seasonally has been recorded for many thousands of years and the awareness of Egypt's cycles of seven lean years followed by years of plenty were part of the way of life of the peoples residing in the lower Nile valley before the filling of Lake Nasser/Nubia in the 1960s. At the same time millions of people world-wide are aware of the Nile's unreliability through their religious traditions the provenance of which are the monotheistic religions of the Middle East all of which include Egypt's story in their cultural heritage.

Since the 1970s, however, it has been the volume of water in the system which has emerged as the challenge to be faced by engineers and politicians. It is shown in the chapters ahead and it has now been grasped by some of the officials and politicians of Nile country governments that it will only be after major adjustments in international relations have been achieved that the significant investment needed to secure more reliable flows basin-wide will be mobilised to enhance the availability of water in the system. One of the major purposes of the information and analysis contained in the following chapters is to indicate the extent of the hydrological constraints on, and the ecological consequences of, the intensification of water use, as well as to emphasise the inescapable political preliminaries to any improvement in the economic and ecological effectiveness of Nile waters utilisation.

While it has proved to be possible, if controversial, and albeit relatively recent in terms of the Nile Basin's six thousand years of civilisation, to divide up the territory and assert territorial sovereignty, it has been much more difficult to devise an agreed framework, or an underpinning legal arrangement, for the sharing of waters. In terms of recent history, the past 150 years, the division of territory was initially mainly into colonies or dependencies of the British Empire of the late nineteenth century, and subsequently into successor sovereign national entities. Meanwhile the arrangements for the international allocation of Nile waters were and remain extremely rudimentary and have not been helped by the precedents set in the colonial period when political circumstances were very different, with power emanating from what proved to be an unsustainable institution - the British Empire (Chapters 5, 15 and 16).

Ownership of flowing surface water in rivers and of groundwater in subsurface aquifers is much more difficult to establish than ownership of, and sovereignty over, territory. It should not, therefore, be surprising that the international agreements over the sharing of Nile water are rudimentary. Until the past decade they were also untested mainly because it was not until the mid-1970s that the population driven increases in food demands brought water deficits in national water budgets sharply to the attention of professionals concerned with water, as well as to national leaderships and to the international agencies relating to the Nile Basin countries. Also the combination of poor levels of general economic development in the countries of the Nile Basin and the consequent dearth of investment funds made it impossible to mobilise relevant ameliorating technologies and inhibited the increased utilisation of water in the upstream countries. On the other hand selectively applied technology, for example the construction of the High Dam at Aswan in the 1960s had insured for a short time the major user, Egypt, against the consequences of the unreliable annual flow. It was anticipated that the Dam would provide ease from water crisis at least until the end of the century and certainly the new water storage in Lake Nasser/Nubia temporarily addressed Egypt's problem of unreliable annual flow of water in the 1970s and did so spectacularly in the 1980s. The Dam and the new water regime enabled Egypt to cope with the adequacy or not of the annual Nile flood. It also permitted the introduction of perennial irrigation into the remaining tracts of Upper Egypt where the basin system was still operating up to the late

1960s because of the inadequacy of water control in those parts until the High Dam had been completed. Meanwhile the other Nile Basin countries were already using Nile water, especially the Sudan, or were indicating their need to do so.

But by the beginning of the 1990s the ceiling on water availability imposed by nature and the engineering modifications put in place on the Nile was clearly inadequate for Egypt, the major water user. That the ceiling agreed in the 1959 Nile Waters Agreement between Egypt and the Sudan would be further lowered by the increased use of water upstream was a serious political and economic issue for Egypt and these circumstances posed for the Government of Egypt and for its international sponsors a very serious long term problem of economic adjustment. It will be shown that the Egyptian experience will be faced by any Nile country which proposes to use a large volume of water and that economic or political adjustments of some sort will be needed if national economies are to sustain rising populations with an improving standard of living.

The Nile: a hydrological and geomorphological phenomenon with particular management challenges

The Nile is a long river, by some measures the longest in the world. But is comparatively not a big river in terms of the volume of water which it shifts each year from the humid uplands of East Africa and the Horn of Africa to the Mediterranean. The Nile is significant not because of its length and capacity but because it is aligned south to north and therefore crosses a number of climatic, and climate related vegetation, zones. The last tract traversed is at the eastern edge of the biggest desert on Earth to which it brings water to a region which would otherwise be uninhabitable. For millennia it has provided livelihoods for millions of people. In the past, and until recent industrialisation, over ninety per cent of the inhabitants of Egypt depended directly on the waters of the river, and without the river Egypt would have been as empty as the rest of the desert.

The combination of Nile waters with the high prevailing temperatures in central desert locations provided growing conditions which were exceptional in ancient times and still enable Egypt to achieve levels of productivity which place it high in any international league of agricultural performance. Nile waters have been continuously prominent since the dawn of civilisation and the river enabled one of the earliest civilisations to spring up on its banks, and the monuments and architectural relics of Thebes and Memphis bear witness to the length of time over which the Nile has played a major role in human history. The issue of whether the Nile can be assumed to provide the same average flow as in the past is the subject of a chapter by Hulme (Chapter 6). The analysis shows that it is not yet possible to be definitive but it is concluded that the rainfall in the uplands of the Horn of Africa appears to be similar to that which prevailed in the nineteenth century and that the twentieth century up to the 1960s was unusually wet and the Nile flow unusually high compared with long term trends.

The fluctuating volume of water in the system has been a problem throughout history but it has always been the fluctuating seasonal regime of the Nile which has been the major challenge to the people who choose to be sustained by its flow. Its seasonality arises because it is a monsoon river mainly fed by rainfall falling on the uplands of Ethiopia in the summer which floods the channels of the Blue Nile and the other tributaries which rise in Ethiopia. Yet if the Ethiopian waters had been the only source of the Nile it would not have been such an economic and historical phenomenon or a significant stimulus to the development of human civilisation. The virtual total drought which would have marked each winter season would have prevented continuous settlement and the development of sustainable agricultural and urban systems.

Happily for the farmers of Egypt and of the northern part of the Sudan Ethiopian waters are supplemented by a steady flow from the East African highlands which, though less than a fifth of the flow from Ethiopia, is regular throughout the year. This underlying regime provided continuous access to water for the towns and villages to the north of Khartoum even if there were periods of a number of years together when the flow from Ethiopia was below average. In any event most of the water from the Blue Nile system flowed to the sea until regulating structures were built to address the challenge of controlling and utilising a more substantial proportion of the seasonal water. Mohammed Ali's engineers in the first half of the nineteenth century built a Delta Barrage a few kilometres north of Cairo and they extended the area of irrigation progressively. (Ali Mubarak 1306 AH/1889 AD). It was this age old process of land reclamation which had until the twentieth century been the main method of gaining access to more of the water than could be achieved merely by planting crops on the annual natural inundation of the Nile valley silt. It was by constantly devising means of using water at the margins of the annual inundation that the area of cultivated land was extended and more water was retained to transpire productively through crops rather than flooding to the Mediterranean.

The story of controlling the Nile in its desert reaches is told more fully in later chapters and has been very thoroughly recorded by professional engineers since the end of the eighteenth century. The remarkable scientific and engineering contribution of surveyor, engineer and philosopher, Ali Pasha Mubarak (1306AH/1889AD) records the endeavours of engineers and farmers of the first part of the nineteenth century to allocate and manage water and reveals the preoccupation of Egyptian governing and professional society with Nile waters and their potential. The last two decades of the nineteenth century and the early years of the twentieth century were well served by fluent engineers who wrote lucidly about their regulation of the flow of the Nile and of land reclamation and of how they had discovered the realities of reclaiming land, - 'we little knew how difficult it was to reclaim land', (Willcocks and Craig 1913) - and emerged wiser and respectful of the capacity of Egyptian farmers and officials to manage water and agriculture.

The mighty hydrological system of the Nile proved a compelling subject for generations of Egyptian and British engineers and scientists (Garstin 1905, Hurst, Black and Simaika 1931-1966, Jonglei Investigation Team 1946-1953, Howell 1953, Howell *et al* 1988, Morrice and Allan 1959) and subsequently for those in the Sudan and more recently for the engineers of Ethiopia and the East Africa countries of the upper Nile. Despite its scale the Nile is a comparatively well understood system which has hydrological records going back for over two thousand years as described in Chapter 2. Adding to this body of knowledge and making contributions to a fuller understanding the phenomenon of the Nile has attracted an extraordinary volume of scientific attention (Collins 1990).

The approach taken by those conducting studies of the Nile has been determined by where they have come from. Engineers and scientists from Britain tended to take an approach founded on the notion that it was the securing of a flow of water to Egypt which was the main priority. During the first half of the twentieth century Britain had political influence over approximately three quarters of the area of the Nile Basin, and a determining political influence over the engineering of Nile waters, and as a result the vision of British engineers was of great significance. The 1929 Nile Waters Agreement was a product of this situation and the agreement placed extremely small Nile waters quotas on the Sudan and forbade the use of water by the colonies in the uplands of East Africa (Chapter 4). The interests of Egypt were also taken into account in a comprehensive Nile Basin approach, enunciated especially by Hurst, (Hurst and Black 1955) which argued for the storage of water in cool upland locations in the south of the basin in order to avoid evapotranspiration losses. The approach was hydrologically sound but was politically unacceptable to Egypt in the late 1950s and Egyptian engineers and political leaders chose the more strategically secure option of building the High Dam at Aswan close to the border with the Sudan to control a volume of water approximately three times the annual flow of the river for the exclusive use of Egypt. In reaching this decision Egypt also had to accept the loss of about sixteen per cent of the annual flow through the evaporation losses at the Lake, a position which becomes progressively less tenable as the water deficit of the country mounts.

Egyptian engineers and diplomats have always been extremely clear about Egypt's national interest and adopted and still advocate a water allocation and management strategy based on the High Dam. At the same time they argue very strongly indeed for the creation of a comprehensive and reliable hydro-meteorological data monitoring system for the whole Nile Basin. That such data would demonstrate the wisdom of allocating and managing water according to a strategy which would downgrade the storage capacity at Aswan and justify the increase of surface storage upstream is a contradiction which has not yet become widely obvious because the upstream states make the provision and sharing of information a hydropolitical issue. All hydrologists, both those from the Basin countries and international professionals, urge very strongly that a monitoring system should be operational (Chapter 7) but unfortunately only Egypt and the Sudan have sufficiently reliable data for incorporation in a comprehensive optimising system as a basis for the allocation and management of the increasingly scarce Nile water. (And even the records of the Sudan have become less reliable in the past decade through the political disturbances in the south of the country.) The absence of agreement on how to proceed in the area of monitoring and simulation significantly inhibits the development of a basin-wide system to provide decision support at the international level.

The upstream countries are much less enthusiastic than Egypt to put a comprehensive monitoring system in place. This is mainly because little Nile water is at present used in the upstream countries and in any event they are unwilling to cede ownership of the water rising or passing through their territories without some reciprocal benefit being achieved. Impeding the establishment of a reliable data-base which could be the basis of agreements alienating permanently a proportion of water is a sensible approach from an upstream riparians point of view.

The chapters in the first part of the book provide an analysis of the scientific information available on the hydrology of the river over the past two millennia and of how the countries of the basin have approached the management of the shares of water which have naturally fallen to them in the past century. There are also detailed analyses of the environment of the Sudd in the context of the availability of water for development *in situ*, a strategy which is shown to be environmentally and politically hazardous. The utilisation of new water from the Sudd for use in the northern part of the Sudan and in Egypt according to the sharing of investments and benefits as arranged in the 1959 Nile Waters Agreement is also examined and it is concluded that a limited volume of water could be beneficially engineered from the Sudd marshes without significant environmental impact. Moreover, it is relevant to point out that the international position on wetlands, of which ecotype the Sudd is a major global exemplar, has changed greatly in the past decade and as a result it is less likely that a Sudd drainage scheme would receive development backing of the international agencies in future and as a consequence it may be unlikely that further international money will be made available for its completion.

There is also a review of the position of Ethiopia and its approach to the development of water during the coming decades. The impression conveyed by the author of the Ethiopian material, Zewdie Abate, is that the professionals responsible for allocating and managing water in Ethiopia are fully aware of the advantages of managing the Ethiopian Nile tributaries according to a basin-wide and comprehensively planned scheme taking into account the interests of all riparians (Chapter 10). But in order to make the essential inputs of data to the information management system there would have to be a shift in the position of the signatories to the 1959 Nile Waters Agreement - Egypt and the Sudan - such that the rights of Ethiopia to use water are recognised. Ethiopian aspirations to develop water will, however, be unrealised as long as the internal conflicts which beset the 1980s continue. Ethiopian institutions will not be strong enough nor have the capital resources to create the physical infrastructure to control the Blue Nile tributaries as long as instability continues nor will international agencies and major national sponsors make available the essential finance without some assurance that the political climate has changed in Ethiopia.

The Lake Basin countries of the East African Highlands have few plans as yet to utilise Nile water and even if they did it would not have the same impact as would the utilisation of Blue Nile water by Ethiopia. Evaporative losses in the Sudd reduce by fifty per cent the impact of upstream use in the White Nile system. It is clear from the contribution of the Ugandan authors, Kabanda and Kahanghire, (Chapter 10) that power generation is the main interest in Uganda and at the same time there is strong recognition that there is an urgent need to rehabilitate the hydrometeorological network and to strengthen natural resource inventory activity and planning so that a water master plan can be developed for the coming decades. Insecurity and political instability have also had a very serious negative effect on the Ugandan economy during the 1970s and the 1980s and it is anticipated that the 1990s will be marked by a stabilisation of the political economy which will enable development including the development of Nile water to take place.

Taking the theme of sustainability as a concluding element in this cursory introductory review of some of the special features of the hydrology and geomorphology of the Nile Basin it is important to emphasise some important characteristics of the Nile's geomorphology which make it a particularly valuable resource to the agricultural communities of Egypt and the Sudan. The Nile valley has been built up by the annual deposition of Ethiopian silt on either side of the river channel. The resulting profile has proved to be fertile for irrigated agriculture, but its most important quality is its drainage qualities which means that the soils of the Nile valley are generally robust in the face of repeated irrigation. The soils are also good in that they do not seriously reduce the quality of the water as it passes through the profile into the groundwater table. By the time the control works had been built on the Egyptian Nile in the 1930s any Nile water which reached the Mediterranean had passed through a number of Nile silt profiles on its way from Aswan. This pattern of use has been intensified since the construction of the High Dam and as a result the water use efficiency of Egyptian irrigated farming is very high indeed and compares very favourably internationally. Copious high quality water and robust soil profiles made the Nile and its generous geomorphology seem a boundless source of sustenance for its major users, the farmers of Egypt. Now that there is no certainty that there will be sufficient water in the medium term even to service the existing irrigated tracts of Egypt this, the major user, is rapidly adjusting its expectations concerning the availability of Nile waters.

The Egyptian experience is important for many reasons and one of them is the example it demonstrates for other riparians. Egypt, like them, has allocated and managed Nile water until very recently without consideration of the ecological and economic efficiency of such use. In recognising the limited share of Nile water to be available in the longer term, Egyptian institutions will have to make a number of adjustments and recognise principles such as ecological sustainability and returns to water in developing water policies and then implement them. The Sudan will be the first to follow this path and Ethiopia is likely to be the next. Meanwhile it is to be hoped that the principles will be adopted before the shortages of water have serious impacts on the using sectors of the respective economies. It is possible that this could be the case in that professional engineers have already begun to write about the necessity of deploying ecological and economic principles as a basis for renewable natural resource allocation and management (Abate 1993).

The Nile as an economic resource

The Nile has throughout history been the key economic resource of the communities living and gaining livelihoods in the desert part of the basin in Egypt and the Sudan. Nile waters have been little used in the other seven countries for agriculture as there have been sufficient rains in their humid uplands to support rainfed farming. Uganda has since the 1950s been able to use the water for power generation and the Owen Falls hydropower station provides a substantial proportion of Uganda's power needs as well as making a contribution to those of Kenya. The study confirms that Nile water will remain a crucial economic resource in the desert countries and it will become of great importance in Ethiopia, for both agriculture and power, and to a lesser extent in the countries of East Africa.

The nature of the economic significance of the Nile waters will change, however, in line with the rate of transformation of the economies of the Nile Basin countries. The agricultural sector has always been the dominant user in the past except in Uganda, accounting for about eighty per cent of the total use, but the part played by water will change as the relative place of agriculture in the national economies declines. Water will remain important in the agricultural sectors and the demand for water in agriculture will be sustained not least because of the demographic pressures which affect all countries of the basin. Egypt is the country which has demonstrated the pattern which will be followed by the other countries of the basin. The share of agriculture as a contributor to Egypt's GDP has progressively fallen in past decades and by the early 1990s it only accounted for about seventeen per cent. In these circumstances when water has become a scarce resource the question has to be asked whether water brings the best possible economic return when allocated to agriculture. As the answer is clearly no, then it is to other sectors that the water will be allocated in future.

By the beginning of the 1990s the ceiling on water availability imposed by nature and the engineering modifications put in place on the Nile were clearly inadequate to meet the needs of the major Nile water user, Egypt. That the volume of water agreed in the 1959 Nile Waters Agreement between Egypt and the Sudan would be further lowered by the increased use of water upstream was a serious political and economic issue for Egypt and these circumstances posed for its government, and for its international sponsors, a very serious long term problem of economic adjustment. That the problem was seen as one of resource inadequacy rather than one of economic and social adjustment has been unfortunate for the development of a realistic economic strategy for Egypt. Such a strategy must include the building of the capacity to import food by strengthening the sectors which generate foreign exchange together with policies which reduce the growth in population in order to reduce food consumption since food production is the major user of Egypt's scarce water. The issue of increasing population is only addressed briefly in the chapters ahead. It is, however, concluded that population increase is the major problem for the country already coping with a very serious water deficit, Egypt, and is a major general problem in all the countries of the basin.

The Nile flow had been revealed by the early 1980s to be progressively inadequate even if the hydrological performance of the two tributary catchments, the Blue and the White Nile, had accorded with the long-term average of the previous eighty years - the period of data which was the basis of the planning assumptions for the High Dam design. That the 1980s yielded flows only about eighty per cent of the long-term average aggravated the alarm of the governments and national water managing agencies in the downstream countries and accelerated by at least a decade widespread awareness of the crisis in water availability. It still has not, however, been widely recognised that it is the economic policy of Egypt, as the major water user, which determines the significance or not of the inadequate water supply in the Nile system.

Egypt has a long standing strategy aimed at the achievement of food self-sufficiency for reasons of national security. This policy is unattainable but while in place it distorts Egypt's hydropolitical behaviour. Such a policy has a determining affect on the expectations and decisions of those who allocate and manage water at all levels in Egypt from the highest policy maker to the individual farmer. Meanwhile the growing food gap of Egypt in the 1970s and the 1980s, represented by its food imports, amounted almost exactly to the increase in the national water gap. The economic expression of this food/water gap almost exactly matches the annual international assistance received by Egypt, mainly from the United States. Of the Nile Basin countries it was only Egypt that had by the early 1990s moved to a position in which its water demands seriously exceeded its water supplies.

Within Egypt itself the pursuit of the food self-sufficiency goal has also had a very distorting effect on the sectoral national water allocation, especially as the performance of the irrigated sector has not been subject to economic scrutiny. That agricultural sectors all over the world, whether in countries of the developing or the industrialised worlds, escape such scrutiny does not make it any easier for Egypt to cope with its particularly extreme version of the national food gap. And as agricultural output is not rated at its real value to Egypt's national economy it is particularly difficult to identify and implement economically efficient practices. Because the other Nile Basin countries are not yet facing water deficits their food self-sufficiency policies have not to date had serious impacts on water allocation practice at the individual country level. But such national policies will in due course prove to be inimical to the development of sound ecologically and economically based water allocation and management practice at the international level, just as they have had serious consequences for the Egyptian economy through distorting the allocation of water between sectors within its national economy.

In the chapters which deal with the Nile as an economic resource it is pointed out that the river has not been a unifying factor in terms of international trade. There are few complementarities between the national economies of the Nile Basin which might be the basis of effective trade. And those that there are are not worth converting into commercial transactions as the transport costs would be prohibitive between sources of production and markets in the very large Nile Basin. The river itself is not without its complications in terms of transport. The main trading partners - Egypt and the Sudan - did not have an uninterrupted water transport link even before the construction of the High Dam and once in place this structure merely emphasised the problems of river transport in a channel crossed by basalt and other volcanic rocks which create the well known cataracts of the Nile valley.

The weakness of all of the economies of all of the Nile Basin countries is another severe impediment to trade. So scarce are the additional factors of production with which to mobilise the economically effective use of natural resources in all the economies of the Nile Basin (with the possible exception of the externally supported economy of Egypt) that all of them face severe problems in implementing sound natural resource managing policies. Water is not scarce in all nine countries of the basin, but the major user, Egypt, has entered an irreversible phase of shortage if its water allocation policies are not changed; and the Sudan, the second ranking user could quickly reach a scarcity position if its political economy could be stabilised and its potential for further development set in motion again. That the upstream countries use scarcely any water and do not regard the Nile as a constraint is a potentially seriously destabilising scenario in that it is only when a resource is perceived to be scarce by those who manage it that sustainable principles even get on the agenda. Meanwhile the interests of peoples and economies beyond the sovereignty of the upstream states can be given little consideration by the farmers and water managing officials of such economically constrained entities as the upstream countries of the Nile Basin.

The Nile in international relations: scarcity for the moment has been accommodated

The division of territory is almost always attended by predictable sensitivities on the part of centrally organised national entities even if these national political systems lack the essential government institutions of an effective nation state. The governments of these national entities do, however, all possess some military potential and can count on the willingness of their peoples to support the deployment of military force if they can establish that there has been a territorial transgression by a neighbour. In practice there have been few territorial disputes within the Nile Basin, and none of them has so far been serious. International tension has come mainly as the result of environmental, political and economic dislocations reflected in the movement of refugees across borders. Such alienated communities become the focus of organised opposition which can provide support to the forces confronting the governments from which they have fled. The reciprocal sponsorship of opposition groups by neighbour governments has been the major source of international tension which complicates relationships and makes the negotiation of other issues, such as water, difficult.

The absence of tension over territory, in contrast with that over the movable resources of people and water, is mainly explained by the location of the majority of the international boundaries in remote and uninhabited or scarcely inhabited regions which have not as yet proved to be potentially mineral resource rich. Also the countries of the Nile Basin cannot be considered to be short of land. Some of the Nile Basin countries are amongst the largest in the world. With over half of the catchment, however, being located in zones enduring conditions ranging from the semi-arid to the extremely arid the importance of water is disproportionately significant to the peoples of these dry downstream tracts. Attitudes towards water have, therefore, for millennia been different from those towards land and territory.

Coping with the unreliability of the flow of the river was the main requirement for the six millennia of Nile water use until the very recent past. While anxiety about upstream use was real from the moment that the colonial government in the Sudan began to plan seriously to irrigate the clay plains in the 1920s, the interests of the major user were explicitly taken into account in the 1929 Nile Waters Agreement between Egypt and the Sudan. The Sudan was only to receive annually less than five per cent (4.76%) of the long term average flow. The Egyptian interests were shared by the British who were at once the initiators of the Gezira irrigation scheme and the water control works on the Blue Nile which ensured the reliable flow of water to it.

The chapters relating to Nile hydropolitics reflect the points of view of the authors and the countries from which they come. Officials and professionals from upstream countries argue for the reappraisal of agreements and do not see the urgency of contributing to comprehensive basin-wide environmental information systems nor determining precisely the amount of water crossing borders. Officials from downstream countries on the other hand argue strongly for precedent and the observation of existing agreements and are very keen to establish international environmental management information systems.

In the following chapters the changing status of the Nile water resource is examined, as well as the increasing demands placed upon its waters by the rising populations of all the Nile Basin countries who will continue to generate increased demands for food. The Nile countries also face an urgent challenge to adjust their strategies of renewable resource management, with water being by far the most important of these natural resources. This adjustment of strategy would be part of an overall shift in emphasis first in Egypt and ultimately in other countries of the basin, though not necessarily in the near future, away from dependence on indigenous natural resources towards the production of agricultural and industrial commodities which enable the individual economies to participate in world trade. The Nile is not a sufficient resource for the populations of the future which live in the countries which depend on it.

The material contained in this book embraces both environmental history and environmental science as well as the contemporary socio-economic and institutional contexts of the Nile Basin and its political entities. Some might have preferred an apparently more objective editorial approach which set the sections dealing with the planning, allocation and management of Nile water in a consistent framework, moderating the opinions of contributors according to a conceptualised hydro-economic policy. But this would have been unrealistic and would have led to unhelpful discussions of the merely hypothetical. The alternative adopted has been to assemble the chapters contributed by scientists and officials from a wide variety of backgrounds whose approaches reflect their national concerns and these concerns are sometimes reflected in their scientific partiality or at least in the selection of a preferred interpretation of uncertainty. The editors also felt that it was important to record the emphasis given to arguments presented by authors from the respective riparians to convey some of the flavour of the basin's hydropolitics, even where the evidence presented is arguable scientifically, although in such cases an editorial comment has been introduced. It is also clear that there are many environmental issues on which a definitive scientific statement still has to be made and this situation makes possible conflicting scientific interpretations fuelling polarised opinion which in turn sustains the political positions adopted by different interest groups. For example the issue of evaporation from the Sudd marshes and the significance of this atmospheric moisture in the regional climate of the Horn of Africa remains to be definitively resolved, and this is significant in Nile hydropolitics even if it is a minor matter on the agenda of those struggling to understand global climate. Until there is a definitive and unarguable scientific statement on the issue, Ethiopian, and even southern Sudanese interests will argue persuasively that there should be no further interventions in the hydrology of the Sudd region and similar environments in the area.

The wish to provide a review which contained insights into the real current and future tensions over Nile water was by no means fully realised, however, mainly because the chapters are in the public domain and authors from the Basin countries understandably in general took a non-conflictual line. The chapter by Abate, for example is strong on the need for cooperation and for international initiatives to optimise the allocation and management of Nile water but cautious and unspecific on the issue which concerns downstream riparians, namely the advantageous geography of Ethiopia and the volumes of Nile water likely to be directed for irrigation and power generation within its boundaries. This in itself is a statement on current hydropolitics and reflects the reality that such public statements cannot be made irresponsibly by officials as they could easily become part of the evidence used in negotiations between riparians, and even more important in litigation, should at any point the disputes over the scarce Nile water be taken to international arbitration or to an international court.

References

Abate, Z. (1993). *Water Resources Development in Ethiopia an Evaluation of Present Experience and Future Planning Concepts* ., [in press]

Ali Mubarak (Ali Pasha Mubarak) (1306 AH/1889 AD). *Al-Khitat al Taufiqiya al-Jadida,* (The New Tawfik Survey of Egypt and its towns and villages), 20 vols, Bulaq, published between 1304-1306 AH/ 1886-1889.

Ali Mubarak, (1891). 'Sharh al-hadith a-nabawi: ihrith li dunyak ka'annaka ta'ish abadan, wa i'mal li akhiratik ka'annaka tamut ghaddan', *Al-Ahzar,* Vol. 4, no. 10, May 1891, pp 309-315.

Collins, R. O. (1991). *The waters of the Nile: an annotated bibliography.* London, Hans Zell Publishers.

Garstin, Sir W., 1905, *Some problems of the Upper Nile, Nineteenth Century and after.* London, HMSO.

Howell, P.P. (1953). The Equatorial Nile Project and its impact on the Sudan, *Geographical Journal*, Vol. **119**, pp. 33-52.
Howell, P.,Lock., M. and Cobb, S. (1988). *The Jonglei Canal: impact and opportunity,* Cambridge, Cambridge University Press.
Hurst, H.E. and Black, R. P.(??). *Report on a hydrological investigation on how the maximum volume of Nile water may be made available for development in Egypt and the Sudan.* Cairo, The Sadd el ali Authority, Republic of Egypt, Misr Press.
Hurst, H.E. Black, R.P. and Simaika, Y.M. (1931-1966). *The Nile Basin,* 10 Vols. plus supplements, Cairo, Ministry of Public Works (Vols. 1-9). Egyptian Ministry of Irrigation (Vol 10).
Jonglei Investigation Team (1946-1954). *Reports of the Jonglei Investigation Team,* Khartoum, The Sudan Government.
Morrice, H.A.W. and Allan, W.M., (1959). Planning for the ultimate hydraulic development of the Nile Valley. London, Paper No. 6372, *Proceedings of the Institute of Civil Engineers*, 14 October 1959, pp 101-156.
Willcocks, Wm. and Craig J.I., (1913). *Egyptian Irrigation,* 3rd edition, London, Spon.

Part I

Environmental history of the Nile and its management

1

Origin and evolution of the river Nile

R. SAID

Introduction

In this chapter a brief outline is given of the evolution of the River Nile. The river seems to have developed as a result of the interconnection of several independent basins and rivers. The Egyptian Nile is shown to have undergone great changes since it started excavating its channel in the late Miocene time. The shape and regimen of the extant river developed only about 10,000 years ago during the last wet phase which affected Africa after the retreat of the last glacial.

The shape of the modern Nile is a very recent development; it is but the last stage of a continuously evolving river which changed its face many times before it assumed its present configuration. The River Nile spans 5646 kilometres from Lake Victoria to the Mediterranean Sea (figure 1). The slope of the river shows marked changes passing through several gently sloping 'landings' which are connected by steeply-sloping rivers. Figure 2 is a longitudinal section of the river from the equatorial lakes to the sea. It shows five 'landings' which are from south to north: Lake Victoria, Lake Kioga and the stretches from Lake Albert to Nimule, from Juba to Khartoum and from Wadi Halfa to the Mediterranean. The stretches of the river which connect these 'landings' are extremely steep, obstructed by waterfalls and cataracts and youthful in appearance and age. Before the river assumed its present-day course, these different 'landings' seem to have formed independent basins which were disconnected. Each of these basins had its own peculiarities with regard to size, cross section, the amount of water it held and, more importantly, its geological history and evolution.

The four southern 'landings' seem to have formed interiorly for a long time. Their access to the sea throughout most of their history was tenuous indeed. At times of high rainfall they swelled, assumed enormous dimensions and overflowed their shores to other basins; at times of low rainfall they shrank or dried up completely. The three southernmost 'landings' belong to the hilly Lake Plateau region which has a high rainfall (1200 mm/year). The fourth 'landing', which stretches from Juba to Khartoum, occupies a large part of the Sudan. This basin is drained today by the river as it flows across the Nubian swell into Egypt and the Mediterranean by way of a series of cataracts.

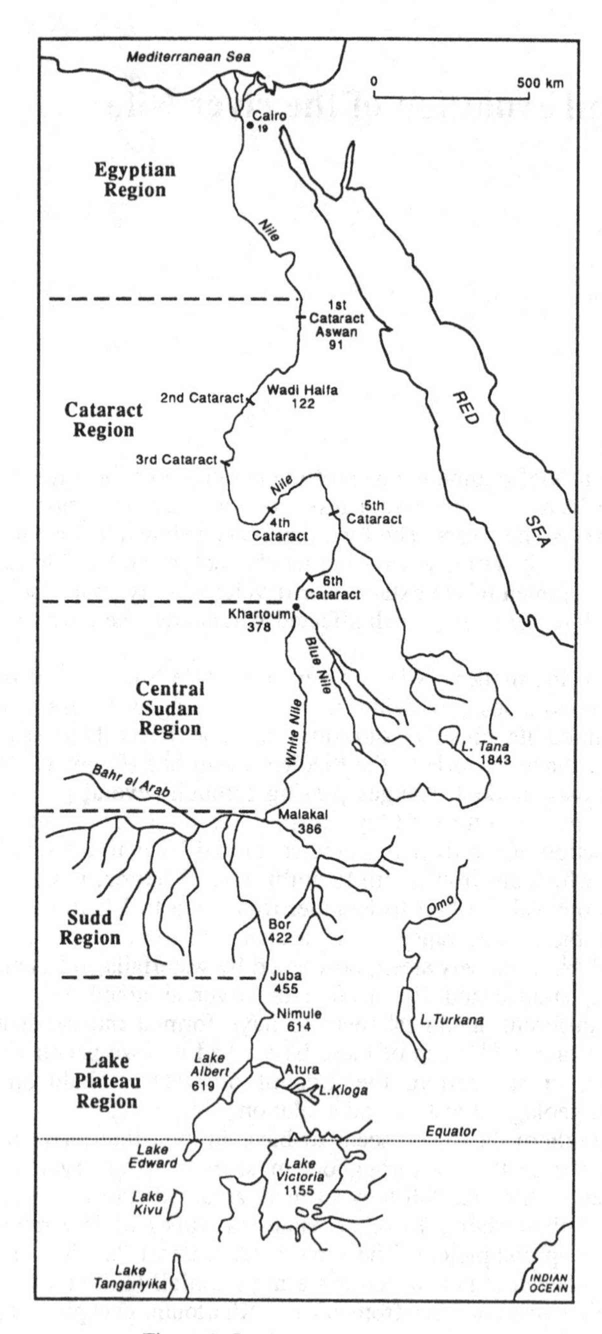

Figure 1: Sectional view of the Nile

The interconnection of these different basins and their integration into one drainage system is a relatively recent phenomenon, for the Nile is not one river; it is in reality a collection of basins and rivers which were connected together to form the present-day system at a very late date. There is now good evidence that the River Nile evolved to its present form as recently as 10,000 years ago.

The origin of the numerous basins which constitute the modern river is closely tied to the history of the African continent. The Sudd basin, for example, is one of the old basins which evolved, like many other interiorly-drained basins of that continent, as a result of the extended history of erosion which affected the elevated lands of Africa. While this and other basins of Africa were able to get access to the sea, some are still interiorly-drained such as the Chad and Elari. The River Nile, which has a more complex history, has a north-south direction, drains more than one basin and spans more than 35^0 of latitude. It drains an area of close to 3 million square kilometres and connects regions which are different from one another in relief, climate and geologic structure. The main sources of the present-day Nile are the Equatorial Lake Plateau which constitutes the southern swell bordering the Sudan basin and the Ethiopian Highland which forms part of the east African coalescing series of plateaus traversed by the great African Rift.

The Egyptian Nile

The Egyptian Nile, through which the drainage of the Nile basins reaches the Mediterranean, has a unique history. From its inception in late Miocene time and up to the early Pleistocene it drained the elevated lands of north-east Africa and had little if any connection with Equatorial Africa. The beginnings of the modern valley can be traced to the late Miocene when the river started incising its course in its present tectonically-controlled valley to a great depth in response to the lowered Mediterranean base level, which occurred at that time as a result of the severing of the Mediterranean from the world oceanic system. When the Mediterranean sea-level rose again in early Pliocene time, the excavated canyon of the Nile was inundated by the rising waters of the sea up to the latitude of Aswan. It was later converted into an estuary and then into a veritable river. The result was the filling up of the canyon and the assumption of a gradient comparable to that of the modern Nile.

The Pre-Nile: The first African connection

The first African connection occurred probably some 700,000 years ago. Table 1 summarizes the developmental history of the Nile since it established an African connection. The new river, here termed the Pre-Nile, was the result of the new drainage pattern which developed after the relief of Ethiopia and the Lake Plateau had approached their present-day shape as a result of the great earth movements of that age. These earth movements resulted in the development of Lake Tana and the Main Ethiopian Rift and also in the appearance of Lake Victoria. The new river brought an enormous quantity of coarse sands which piled up in the Valley and the Delta expanse, contributing to a landscape which started to take its present day shape.

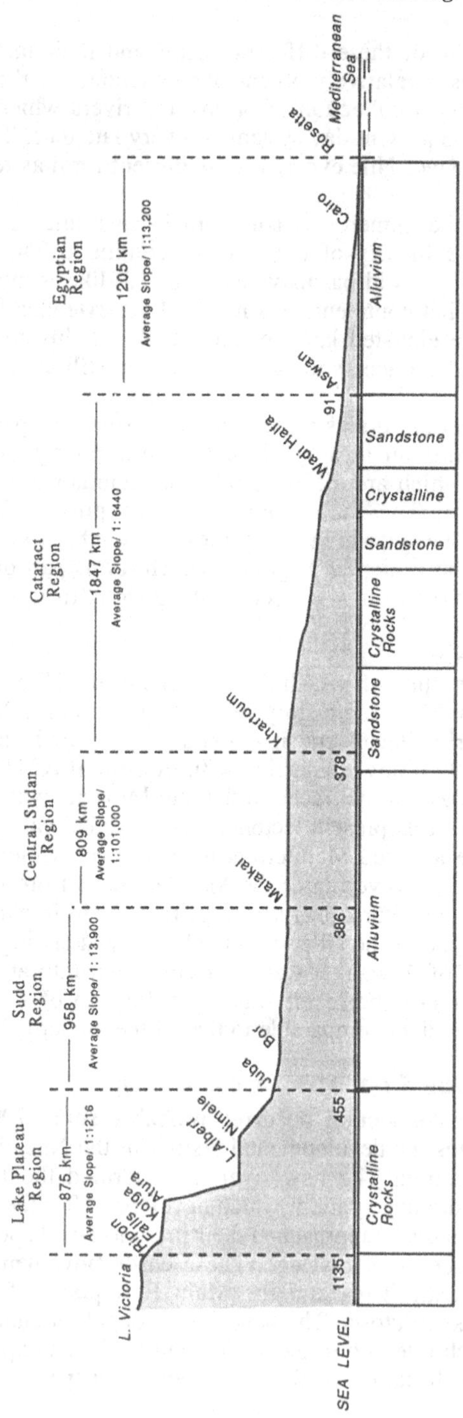

Figure 2. Cross sectional view of the Nile

The Neo-Nile: The modern regimen established
A new phase in the evolution of the river occurred approximately 300,000 years ago when the Ethiopian connection became tenuous. When it was severed the river became ephemeral, deriving its supply from local wadis activated during pluvials which brought enormous quantities of gravel. At other times the connection was resumed but the sediment brought in was different from that of the Pre-Nile and became mostly silt, very similar to that of the modern Nile. There were four main siltation episodes (alpha, beta, gamma and delta Neo-Niles). The alpha/beta Neo-Nile interval had a relatively longer duration than the intervals which separated the other siltation phases. It was an interval of low Niles and high local rainfall (Abbassian and Saharan Pluvials). With the exception of the silts of the low Niles of the alpha/beta interval, all the other silts were deposited during periods of high aridity in Egypt.

The last three siltation episodes occurred within the span of the last 70,000 years. They were separated by episodes of downcutting. The silts of the older two of these phases (the beta and gamma Neo-Niles) were essentially formed under similar conditions by rivers which seem to have had a similar regimen, while the silts of the third and extant phase were formed by a river with a regimen which ushered in the modern Nile.

Firm dates exist for the terminal part of the beta Neo-Nile (26,000 B.P.), for the gamma Neo-Nile (20-12,000 B.P.) and for the beginning of the modern or delta Neo-Nile (10,000 B.P.). The beta, gamma and delta Neo-Niles developed during the last glacial and post-glacial time, becoming seasonal during times of lower temperatures and lower rainfall and flowing during times of rising temperatures and greater rainfall.

Evidence at hand shows that the two older rivers were formed during a period of lesser rains; they were seasonal in nature approximating in their regimen to that of the river Atbara which rises in pulses during flood time and almost dries up during the dry season. The newest river, on the other hand, formed during a wet interval, the so-called Nabtian (Neolithic) Wet Phase, which allowed it to flow all year round. The modern Nile is indeed the child of that phase.

For most of their duration the beta and gamma Neo-Niles were contemporaneous with the Würm or the last glacial age when ice sheets covered large areas of Eurasia and North America. During this age the climate of Egypt was colder and many of the peaks of the mountains of equatorial Africa were glaciated. There is ample geomorphic, faunal and floral evidence that the last glacial was marked in Africa by a dry episode.
During that period the area of the headwaters of the Nile received considerably less rainfall than today. The pollen spectra of this age from the Lake Plateau region indicate assemblages dominated by grasses; the African rain forest had shrunk. It did not come back until around 12,500 years before present when forest trees replaced the grasses. Lakes Victoria and Albert, both important sources of the White Nile, were closed basins before 12,500 years ago. The Sudd region became considerably drier, receiving hardly any rain and was reduced to a series of saline lakes. The channel of the White Nile was partially or completely blocked by dunes south of Khartoum until perhaps

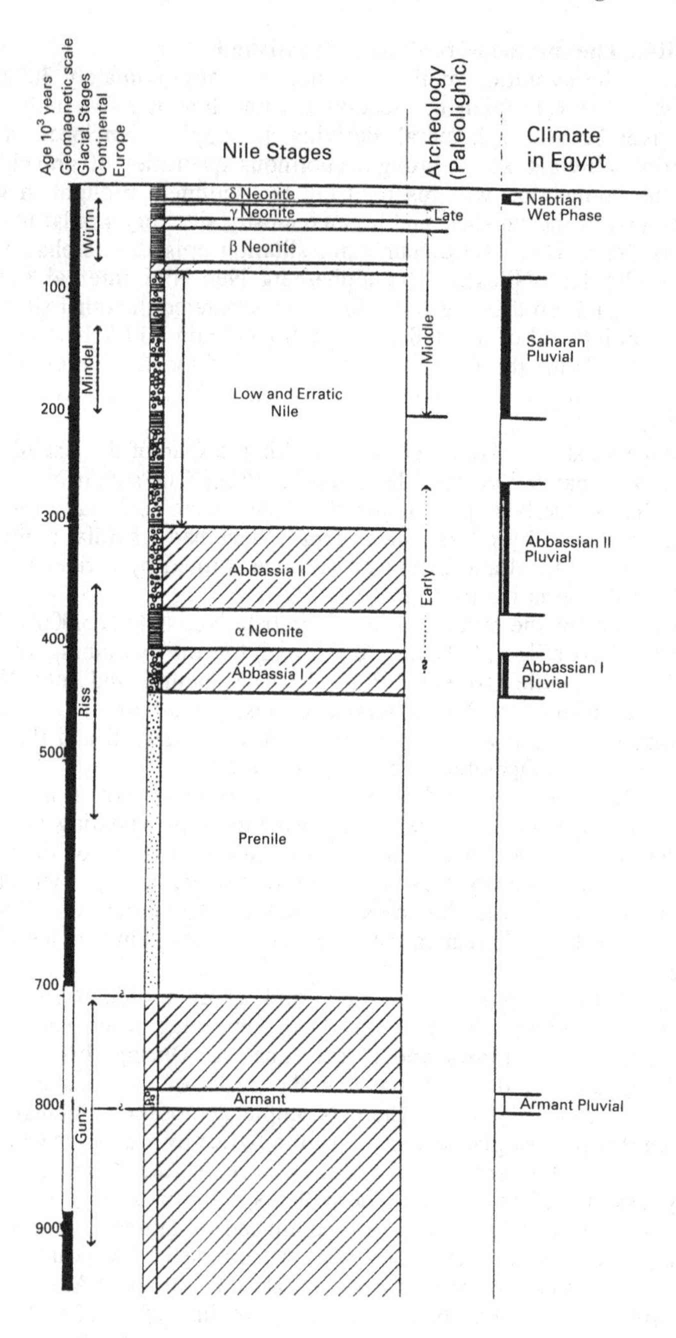

Figure 3: Nile Time Scale

12,500 years ago when it was overwhelmed by the early Holocene transgression. It seems the White Nile did not contribute any appreciable quantity of water to the flow of the beta and gamma Neo-Niles. These earlier rivers derived their waters almost entirely from the Ethiopian Highland; the White Nile was choked by sand dunes and made little if any contribution to the river's discharge. This must have resulted in rivers that dried up almost completely during winter time.

The Ethiopian Highland also received less rain during the last glacial age but the evidence here is not as conclusive.

The Present Nile

The modern Neo-Nile came into Egypt about 10,000 years ago after a transitional period which lasted for at least 2,000 years.

This transitional period, which covers the span between 12,000 and 10,000 years before present, witnessed dramatic climatic events over the Lake Plateau, events which instigated an enormous surge in the volume of the waters of the then waning gamma Neo-Nile, raising its levels to heights never reached before or since that time. After this short period of great and exceptional floods which were followed by a period of incision, the modern Nile with a regimen similar to the one we know today came into being. In contrast to the earlier seasonal rivers, the modern river flowed freely all year round, albeit with different intensities during the different seasons. It received its waters from the Ethiopian Highland as well as from the Equatorial Plateau via the White Nile which by now was cleared from the dunes that had blocked its way. The runoff from the two plateaus plays very different roles in the regimen of the modern Nile. The Equatorial Plateau contributes a small but regular amount to the Nile in Egypt throughout the year; and if this source of supply were to be cut off, as happened with the two previous rivers, it can hardly be doubted that the river would run dry in the spring months. The rainfall on the Ethiopian Highland, on the other hand, is seasonal and comes in the form of a flood. The flood brings to Egypt more than 80 per cent of the discharge of the river and also great quantities of suspended matter which it deposits on its flood plains.

The retreat of the ice, which started immediately after the last glacial had reached its maximum some 15,000 years B.P., was complete around 3,500 years later. There was a rapid rise of the surface temperature of the oceans between 13,500 and 11,500 years B.P. In the headwaters of the Nile this warming trend was reflected by a retreat of the mountain glaciers from their earlier maxima; by 14,750 years B.P. Mt. Ruwenzori was already ice-free.

With the retreat of the ice a period of increased rainfall followed over the Lake Plateau. The beginning of this episode of increased rainfall, which seems to have been considerable, causing the rise of some of the lake levels to more than 100 metres, is not well dated. There is ample evidence, however, that a dramatic change of climate took place at about 12,500 years B.P. There were major vegetational changes in the East African Highland around this time. Pollen analyses of cores raised from boreholes drilled in the equatorial lakes show that the forest tree pollen replaces the grass pollen which dominates the earlier spectra. Lakes Victoria and Albert overflowed into the Nile some

12,500 years B.P. and the Kabarega (Murchison) Falls were active about that time. The overflowing waters of the equatorial lakes cleared the White Nile from the dunes which had choked its channel during the earlier drier periods and reached Egypt in large quantities. These waters together with the supply of the rivers of the Ethiopian Highland swelled the river in Egypt and brought large quantities of silt which had been eroded from the weathered and parched lands of the headwaters of the Nile. These silts reached Egypt during the last phases of the gamma Neo-Nile, the last of the seasonal rivers, where they were piled over the earlier silts. Their level marks the highest level ever reached by any of the rivers of the Neo-Nile. The silts are recorded at 27 metres above the floodplain in Aswan and 6 metres in Qena. These were indeed times of wild floods.

The early equatorial rains did not last for long; they declined after 12,000 years B.P. and were followed at about 10,000-9,500 years ago by an interval of rains, the so-called Nabtian (Neolithic) Wet Phase. This phase affected large parts of the African continent. Lake Turkana, which was very low after 35,000 years B.P., rose at about 9,500 years ago to a height that allowed it to overflow and pass into the Nile by way of the River Sobat. The levels of Lakes Kivu and Tanganyika, which had been at -300 metres during the last glacial, rose around that time to a height of 100 metres above their present level. In Egypt the interval was marked by the birth of the modern river, the oldest sediment of which is dated 10,500 years B.P.

The Neolithic Wet Phase (Nabtian Pluvial) did not only affect the headwaters of the Nile but also the northern Sudan, Nubia and, indeed, the entire Sahara including the desert expanses of Egypt. There is ample evidence that during this interval the desert formed a great stretch of grassland or steppe over which roamed nomadic families of hunters following the rainfall. Much of the evidence for this wet phase is archeological, such as artifacts found in desert areas where man cannot live today, rock drawings of animal species that require at least a savanna type of vegetation, and fossil roots and tree stumps in wadi bottoms and the low desert where no trees grow today. The Nabtian (Neolithic) Wet Phase continued with intensity until its decline around 2,350 B.C., toward the end of the reign of Dynasty V of Ancient Egypt. In the deserts of southern Egypt and the northern Sudan the phase is reported to have lasted from about 10,000 to 5,400 years B.P. during which time it was interrupted by at least two short periods of lesser rains between 6,500 and 6,200 and between 5,900 and 5,700 years B.P.

With the onset of the new pulse of rains around 10,000 years ago the river assumed its new regimen. Its silt started to build up to the north of Aswan forming the famous agricultural layer of the fertile land of Egypt. In Nubia, however, the river continued, as had been the case with the older rivers, to incise its channel. The oldest of the modern silts, dated 10,600 years B.P., occurs at 12 metres above the floodplain in Wadi Halfa. Following this maximum, the river downcut its bed in Nubia to a level of five metres above its modern floodplain during Predynastic times, then went further down in an oscillatory manner to about three metres in early Dynastic times, and then to the present level by the time of the New Kingdom some 3,000 years ago. There has been very little downcutting in Nubia since that time.

References

Ball, John (1939). *Contributions to the Geography of Egypt.* Cairo, Survey and Mines Dept.

Butzer, K. W., Isaac, G.L., Richardson, J.L. & Washbourne-Kaman C.K. (1972). Radiocarbon dating of East African lakes. *Science*, **175,** pp. 1069-76.

Butzer, K. W. & Hansen, C.L. (1968). *Desert and river in Nubia*, Univ. Wisconsin Press.

Degens, E. T. & Hecky, R.E. (1974). Paleoclimatic reconstruction of late Pleistocene and Holocene based on a tropical African lake. In: *Colloques Internationaux du C.N.R.S.*, **219**, pp.13-24.

Doornkamp, J. C. & Temple, P.H. (1966). Surface, drainage and tectonic instability in part of southern Uganda. *Geograph. J.*, **132**, pp.238-252.

Fairbridge, R. (1963). Nile sedimentation above Wadi Halfa during the last 20,000 years. *Kush*, **11**, pp.96-107.

Gasse, F., Rognon, P. Street, F.A. (1980). Quaternary history of the Afar and Ethiopian rift lakes. In: *The Sahara and the Nile*, Williams, M.A.J. & Faure, H. (eds). pp. 361-400, Balkema.

Heinzelin, J. de (1968). Geological history of the Nile Valley in Nubia. In *The Prehistory of Nubia*, Wendorf, F. (ed) pp.19-55, Southern Methodist Univ., Dallas, Texas.

Livingstone, D. A. (1976). Paleolimnology of headwaters. In: *The Nile, biology of an ancient river*, J. (ed) pp.21-30. The Hague, Junk.

Livingstone, D. A. (1980). Environmental changes in the Nile headwaters. In: *The Sahara and the Nile*, Williams, M. A. J. & Faure, H. (eds) pp.339-360. Balkema.

Nyamweru, Celia (1989). New evidence for the former extent of the Nile drainage system. *Geograph. J.*, **155**, pp.179-188.

Paulissen, E. & Vermeersch, P.M. (1989). Les comportement des grands fleuves allogènes: l'exemple du Nil saharien au Quaternaire supérieur. *Bull. Soc. gól. France*, 1989(8). t. V:pp.73-83.

Said, Rushdi (1981). The geological evolution of the River Nile. Springer.

Said, Rushdi (1983). A proposed classification of the Quaternary of Egypt. *J. Afr. Earth Sci.*, **1**, pp.41-45.

Stanley, D.J. (1988). Subsidence in the northeastern Nile Delta: Rapid rates, possible causes, and consequences. *Science*, **240,** pp.497-500.

Wayland, E.J (1921). *A general account of the geology of Uganda.* Ann. Rep. Geol. Surv. Uganda, (1920), **14**.

Wendorf, F. & Schild, R. (1976). *Prehistory of Wadi Kubbaniya III*. New York, Academic Press.

Wendorf, F. & Schild, R. (1980). *The Sahara and the Nile*. New York Academic Press.

Wendorf, F. & Schild, R. (Assemblers), & Close, A.E. (ed) (1989). *Prehistory of Wadi Kubbaniya*. Dallas, Texas. Southern Methodist Univ. Press.

Williams, M.A.J. & Faure, H. (1980). *The Sahara and the Nile*. Balkema.

Williams, M.A.J. & Williams, F.M. (1980). Evolution of the Nile Basin. In

The Sahara and the Nile, Williams, M. A. J. & H. Faure (eds), pp.207-224. Balkema.

Williams, M.A.J. & Adamson, D.A. (1980). Late Quaternary depositional history of the Blue & White Nile rivers in central Sudan. In: *The Sahara and the Nile*, Williams, M.A..J & H.Faure (eds) pp.281-304. Balkema.

Williams, M.A.J. & Adamson, D.A. (1982). *A land between two Niles*. Balkema.

2

History of Nile flows

T. EVANS

Introduction

The drought which has ravaged most Sahelian countries during the 1970 and 1980 decades is unprecedented in recent records. The objective of this chapter is to compare the recent sequence of low Nile floods with past historical evidence of droughts in Egypt; an attempt to put this century's flows in an historical context.

The major question facing engineers, planners and politicians alike is whether this drought will continue; whether this persistence of low rainfall years in the Sahel is just further evidence of the natural 'Hurst Phenomena', whereby sequences of low Nile flows and large Nile floods tend to occur together, or whether it is man-made, possibly a manifestation of global warming representing a permanent or semi-permanent change in the general circulation of the atmosphere influencing the continental climate of Africa.

The Sahel is a region with steep rainfall gradients with annual rainfall decreasing rapidly from 600 mm to less than 200 mm in a few hundred kilometres. Slight shifts in the circulation can cause major problems which would go unnoticed in other climatic zones.

The Nile is the birth place of hydrology. No other river provides such a wealth of information. Available records reach back to before 3000 BC. The heavy dependence of Egyptian civilisation on the size of Nile floods, leading to years of famine or plenty, and the ability of Egyptian dynastic society to record evidence for posterity provides a unique opportunity to investigate historical river flows.

Man's historical fascination with the Nile

The Nile has fascinated philosophers, geographers, historians, engineers and politicians of all creeds and races over many centuries, since man first set eyes on its waters. Four thousand years ago three major civilisations flourished: the Egyptians in the Nile valley; the Sumerians in Mesopotamia; and the Harappans in the Indus valley. The emergence of the sophisticated Egyptian civilisation at the threshold of history with its unique dependence on the rich annual flood from an unknown source has mesmerised scholars through the ages. Herodotus, 'the father of history' in 450 BC described Egypt as 'an acquired country, a gift of the River Nile'. Greek philosophers were so intrigued with the Nile that they believed its origin was not like that of other rivers but it had been created along with the world.

The mystery of the Nile lay not only in its source, but in the predictability of the rise and fall of its flood. Thales of Miletos, the chief of the seven wise men of ancient Greece, believed that the northerly Etesian winds with their constant blows held back the Nile before releasing the pent up giant to enter the Mediterranean. Herodotus rejected this theory as did Seneca, the Spaniard, owing to the lack of coincidence between the onset and dying of the wind and the Nile flood.

Two centuries after Herodotus, Eratosthenes (276-194 BC) described the Nile source much more accurately, separating the White and Blue Niles and describing its source as lakes fed by summer rains. However, this was not before early Greek geographers, such as Hecatios of Miletos, pictured the world surrounded by Oceanus as the source of the Nile, and the sea into which discharged other major rivers such as the Euphrates, the Tigris and the Indus. (see Figure 1). Other theories for the source of the Nile abounded, including locations in Libya and a connection with the Niger River in West Africa.

Ptolemy, the Roman astronomer and geographer who resided in Alexandria in the second century AD, prepared a remarkable map of the Nile basin (Figure 2) showing the three lakes Tana, Victoria and Albert with the main source of the White Nile in the snow capped mountains of the Ruwenzori range, the Mountains of the Moon. Aeschylus, 500 years BC, had talked of Egypt nurtured by the snows.

The exploration of the Nile's source is an epic story which has captured the imagination of the world and will continue to do so for many years to come.

In spite of all the mysteries surrounding the Nile, no other river provides such extensive information of river level data. Available records date back to before 3000 BC, but before briefly discussing this information it is appropriate to have a look at Palaeoclimatic evidence which has been increasing rapidly in the past few decades.

Palaeoclimatic evidence

In mid-Tertiary times prior to the major faulting of the Rift Valleys the modern Nile Basin is thought to have been divided into five separate zones (Butzer and Hansen, 1968). The Lake Victoria complex drained into the Congo basin; the basin below Khartoum as far as the Ethiopian and Ugandan borders, including the Sudd, drained internally, whilst the reach between Wadi Halfa and the Atbara River (the Howar basin) discharged eastwards into the Red Sea via Wadi Odib. Only the area above the third cataract near Wadi Halfa flowed towards the Mediterranean Sea.

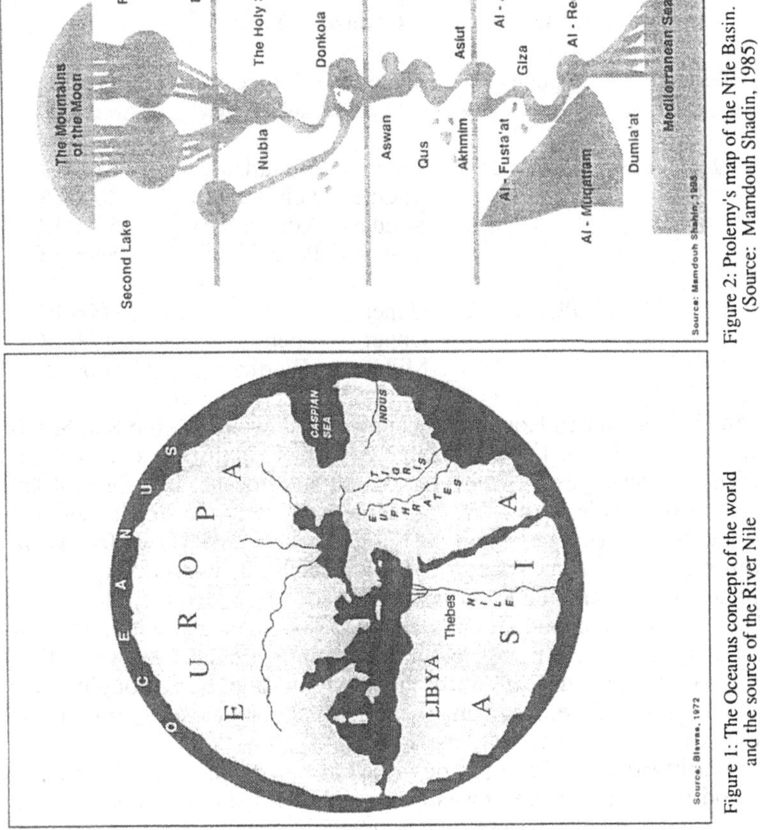

Figure 1: The Oceanus concept of the world and the source of the River Nile
(Source: Biswas, 1972)

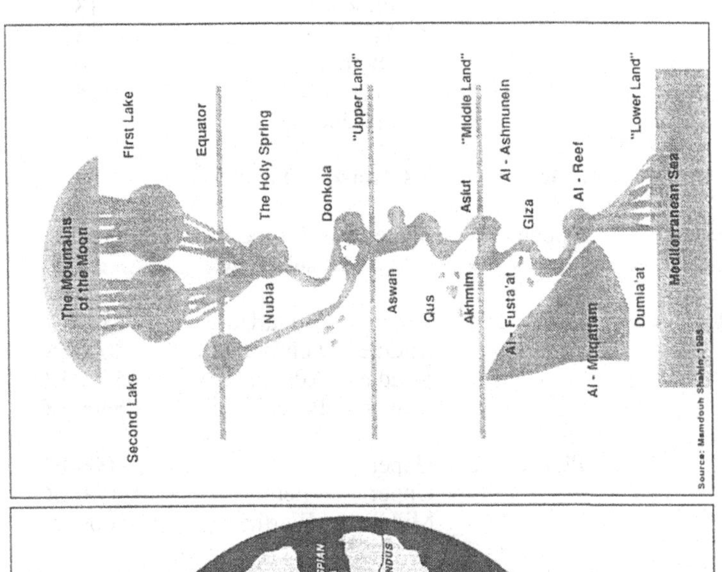

Figure 2: Ptolemy's map of the Nile Basin.
(Source: Mamdouh Shadin, 1985)

Table 1 Geological Time Scale

Era	Period	Epoch	Beginning dates (BP) (million years)
Cenozoic	Quaternary	Holocene	(10,000 years)
		Pleistocene	2.75
	Tertiary	Pliocene	13
		Miocene	25
		Oligocene	36
		Eocene	58
		Palaeocene	63

Table 2 Late Cenozoic Time Scale

Period	Epoch	Stage	European sub-stage	Beginning dates
Quaternary	Holocene	Upper	Sub-atlantic	800 BC
		Middle	Sub-boreal	3000 BC
		Middle	Atlantic	5600 BC
		Lower	Boreal	8300 BC
	Pleistocene	Upper	Würm	70,000 BC
		Upper	Eem	90,000 BC
		Middle	Warthe	120,000 BC

Both the Ethiopian and Howar basins probably drained into the Red Sea as late as the mid-Pleistocene times, although a river of significant size emptied into the Mediterranean west of Cairo during the late Eocene. It is thought that the White and Blue Nile basins did not merge until early Pleistocene and there is evidence that the summer flood was not established before 25,000 years BP. Therefore, although the present hydrological basin is relatively recent, the Egyptian and Nubian Nile River is ancient.

The last ice age ended abruptly about 18,000 years ago with a temperature rise of some 9°C reaching the 'climatic optimum' around 7000 years BP (see Figure 3). A slow descent into the next ice age was thought likely to commence at that time, following a cycle which may have occurred twenty or more times in the last two million years.

The early Holocene has been described as the 'golden age' (Roberts, 1989), 'A period when soils were unweathered and uneroded and Mesolithic people lived off the fruits of the land without physical toil or effort'. In the first half of the Holocene nature remade itself, recovering from the burden of the ice age or glacial drought. For the past 5000 years man has been increasingly taking charge of the planet and if this time scale is thought of in terms of less than 200 generations the transformation is staggering, however one views man's achievements.

Figure 3: Change in temperature and Carbon Dioxide from 17,000 BP. (Based on ice core samples.) (Source : Oesihger and Langway, 1989)

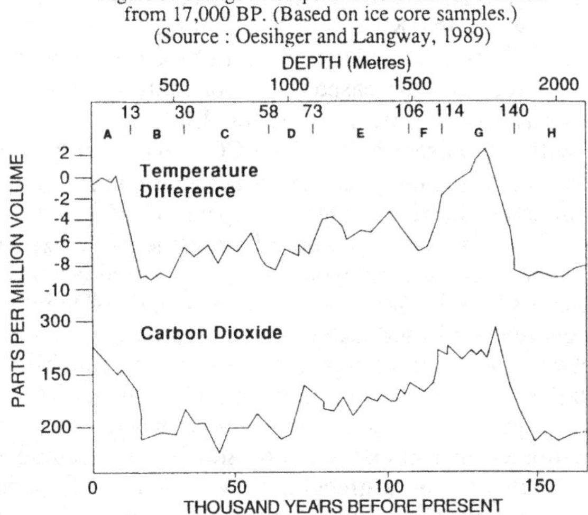

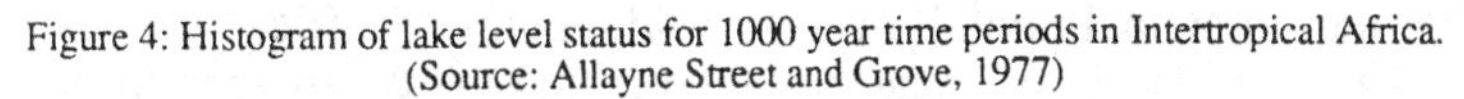
Figure 4: Histogram of lake level status for 1000 year time periods in Intertropical Africa. (Source: Allayne Street and Grove, 1977)

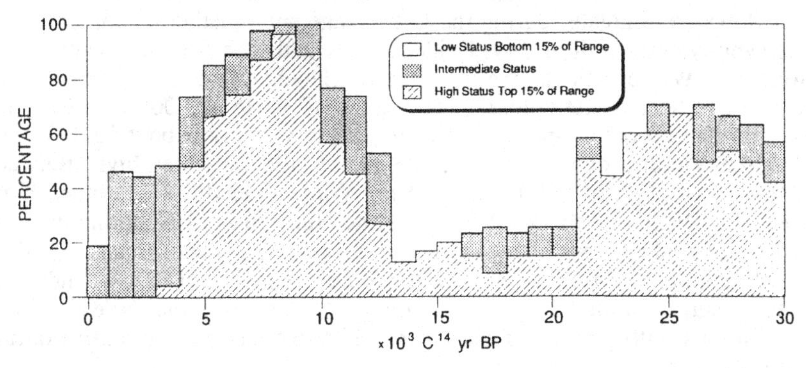

Figure 5: Calibrated C_{14} dates for increased lake volumes and/or stream discharge in the Sahara and East Africa since 5000 B.C. (Source: Butzer, 1976)

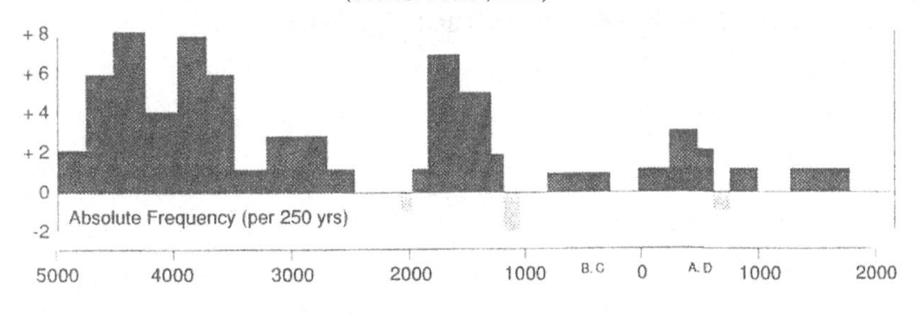

The term 'Little Ice Age' has been used to describe the cooling of the earth which occurred between the Middle Ages, around 1200 to 1300 AD, and ending in the 19th century, documented in detail by Grove, J.M. (1988).

It can be speculated that increased CO_2 concentrations and other man-induced changes may have helped to bring the 'Little Ice Age' to a premature end. Moreover, with the present increases in CO_2 levels going unchecked and probably uncheckable in the next century, concentrations could double by the year 2030 with an estimated rise in global temperature of 3 to 5°C. As can be seen from Figure 3 such conditions would lie outside those experienced over the last 150,000 years and the earth would lie in uncharted waters.

Although a history of the earth's climate and periods of high humidity and aridity have been revealed, and dated, there is still controversy as to the relationships between the earth's temperature and overall rainfall. In simplistic terms lower temperature leads to lower evaporation which should result in lower rainfall and vice versa. Budiko, the famous Russian climatologist, for example, has postulated that global warming should be encouraged, based on the most recent analogue for a greenhouse world, the 'Climatic Optimum (7000 year BP)', a period when most climatologists believe that few deserts existed in Africa and rainfall abounded. As a result he has predicted increases of 300 mm in annual rainfall over the arid regions of Africa.

In 1974 Williams and Adamson highlighted the controversy over whether Nile flows were greater during the last ice age and whether or not glaciation was synonymous with aridity. Evidence was gathered from prehistoric sites along the White Nile above Khartoum. Geochemical and palynological evidence indicated that before 14,500 years BP up to 12,000 years BP Lake Victoria had no outlet and the regional climate was dry. About 12,000 years BP Lake Victoria overflowed and the White Nile flow was high. Between 12,000 years BP and 8000 years BP the White Nile flows continued high. After 8000 years BP the level of the White Nile fell until Neolithic man (an unsubstantiated and extravagant assumption by Williams and Adamson, 1974) and climatic desiccation accelerated the trend towards semi-desert conditions. These results support the contention that aridity in the Nile Basin is synonymous with periods of glaciation whilst wetter sequences arise during periods of higher temperature.

These conclusions are supported by Alayne Street and Grove, A.T. (1979) in a detailed review and collation of scientific evidence on lake level fluctuations since 30,000 years BP. A summary of some of their results relating to Africa is included in Table 3 and Figure 4.

Table 3 Lake Level Fluctuation in Africa
(after Alayne Street and Grove, A.T. 1979)

Period	Pluvial conditions in Africa
21,000 to 20,000 years BP	Onset of dry conditions in intertropical Africa
18,000 to 17,000 years BP	Aridity spreads across Africa except north-west Sahara
15,000 to 14,000 years BP	Aridity more intense
12,000 to 11,000 years BP	Lake levels rising including Chad and equatorial lakes
9000 to 8000 years BP	Lake levels rose dramatically in a coherent pattern after 10,000 years BP. Although north-west Sahara remained anomalously dry.
6000 to 5000 years BP	Lakes began to dry out at around 7000 years BP although Ethiopia subject to wetter influences.
4000 to 1000 years BP	After 4500 years BP strong drying trend began in tropical Africa. High lake levels recede towards equator with minor reversal between 3500 and 3000 years BP.
1000 years BP to present	Present millennium appears to be the most arid of the late Quaternary.

In a study of 'Early Hydraulic Civilization of Egypt', Butzer (1976) looked at available environmental parameters influencing the dynastic periods of Egypt. Figure 5 illustrates the frequency of records for high and low lake levels. It shows a fluctuating wet period from 5000 BC which ended around 2700 BC with the submaxima centred at 4500, 3750 and 3000 BC. After 2700 BC lake levels and flows reduced considerably until wetter conditions returned around 1850 BC. Lake Rudolf was estimated to be 70 m deeper than at present and overflowed into the Nile basin and the White Nile was 2 to 3 m higher than the modern river and it has been speculated that flows may have been 5 to 10 times greater (Williams and Adamson 1974). The termination of this wet phase occurred around 1200 BC with the total desiccation of Lake Naivasha. Later moist intervals are relatively unimportant and limited to the Chad basin and Ethiopian lowlands, occurring between 100 to 1000 AD at a time when Lake Rudolf was some 30 m higher than at present.

What is quite clear is that during any long term trend, such as the atmospheric warming of some 9°C which took place between the end of the recent ice age, 18,000 years BP and 7000 years BP (see Figure 3), there are numerous reversals of trends, oscillations, persistences and no doubt short and long term cycles in climate which are significant in relation to long-term changes which occur over many millennia.

Current research into global warming using general circulation models and investigations into the complex circulation pattern of the oceans, which play such a dominating role in determining climate, should throw up answers to many of the outstanding questions. Calibration of such models will depend on palaeoclimatic evidence amongst which the most exciting is the vast cellar of information of past climates locked away in the ice sheets of Antarctica and Greenland.

Apart from the palaeoclimatic evidence, the foundation of the Egyptian State in 3050 BC produced another unique source of records. With their absolute dependence on the Nile, Ancient Egyptians have regularly measured the height of the maximum flood and recorded it in their royal annals. These are discussed briefly in the following section.

Dynastic Nile records

No other river provides such a wealth of recorded data. Maximum annual flood records were engraved on a large stone stela dating from 2500 BC back to the beginning of the First Dynasty about 3000 BC. Unfortunately only small fragments of the pillar have been found, the most important of which is the Palermo stone, so named after the Palermo museum where it is housed. Figure 6 shows a plot of annual maxima abstracted from the Palermo stone and other ancient fragments. A fixed stable zero of the gauge has been assumed, which is very unlikely but if it were, then a fall in Nile flow is indicated. This is consistent with known Lake volumes (see Figure 5). The scatter during the First Dynasty is also fairly typical of recorded data, though somewhat less than in the late 19th and 20th century which is thought to be atypical.

Much other valuable information exists in the form of flood level marks and inscriptions on buildings and on cliffs. One such series of inscriptions recorded between 1840 BC and 1770 BC is located at Semna near the Second Cataract.

Table 4 Egyptian Chronological Framework

Date	Civilisation
4500 BC	**Late Palaeolithic** (Stone Age)
	Predynastic (Badarian and Nagada cultures)
3050 BC	**Foundation of Egyptian State** (3050 BC)
	Early Dynastic Period (2920 to 2575)
	1st to 3rd Dynasties
2575 BC	**Old Kingdom** (2575 to 2134)
	4th to 6th Dynasties
	1st Intermediate Period (2134 to 2040)
	9th to 11th Dynasties
2040 BC	**Middle Kingdom** (2040 to 1640)
	11th to 13th Dynasties
	2nd Intermediate Period (1640 to 1532)
	15th to 17th Dynasties
1500 BC	**New Kingdom** (1550 to 1070)
	18th to 20th Dynasties
1070 BC	**3rd Intermediate Period** (1070 to 712)
	21st to 25th Dynasties
712 BC	**Late Period** (712 to 332)
	25th to 30th Dynasties including 1st and

	2nd Persian Periods
332 BC	**Greco-Roman Period** (332 BC to 395 AD)
	Macedonian and Ptolemaic (332 to 30 BC)
	Roman Emperors (30 BC to 395 AD)
395 AD	**Byzantine Period** (395 to 640 AD)

Other important data include marks on temples, foundation levels of buildings, water steps leading down to the Nile from forts constructed on the river banks. The analysis of this information has been primarily the preserve of archaeologists and historians. The heavy dependence of Egyptian civilisations on the magnitude of the Nile floods and sequences of low and high flows is of crucial importance in the interpretation of the rise and fall of Egyptian dynasties. The 'First Dark Age' in Egypt which marked the end of the prosperous Dynasty VI and the collapse of the Old Kingdom around 2150 BC is likely to have been the result of a failure of Nile floods which probably lasted only for about 25 years.

Bell (1970, 1971 and 1975) has compiled considerable evidence to postulate that the 'First Dark Age', which began around 2150 BC and the Second Dark Age (1200 BC), after which Egypt went into a prolonged decline, was the result of short but intense sequences of drought years which would stand comparison with those of the 1980s (see also Figure 5).

Bell (1975) has prepared extensive documentation to demonstrate with confidence that the period during the Middle Kingdom and Second Intermediate Period from 1991 to 1570 BC was one of high floods. Of particular importance are the 18 flood levels at Semna above the 2nd Cataract which, because of their high levels, some 8 to 11 m above those of present floods, have been difficult to accommodate (see Figure 7). In the past these have generally been explained by erosion of the bed and particularly erosion of the enclosing cliffs which apparently are of a softer sandstone (Jarvis, 1935). Bell rejects this conclusion. Using evidence of flood damage at levels of 155 m above mean sea level and recent geological information on the slow rate of erosion of the diorite bedrock, she concludes that the flood level was produced by a short-lived climatic fluctuation of a few decades duration. A peak discharge of 3.2 km^3 per day was estimated by extending the rating curve for Kajnarty. This size of flood was stated to be about 3 to 4 times the maximum flood recorded towards the end of the 19th century. However, most other evidence points towards flows similar to those recorded in the 19th and early 20th centuries. It would appear unlikely from our present knowledge of fluctuations of the Nile that a sequence of high floods of such magnitude could have occurred. These levels are more likely to have resulted from a compromise scenario of a significant measure of erosion and a sequence of lower but still high floods.

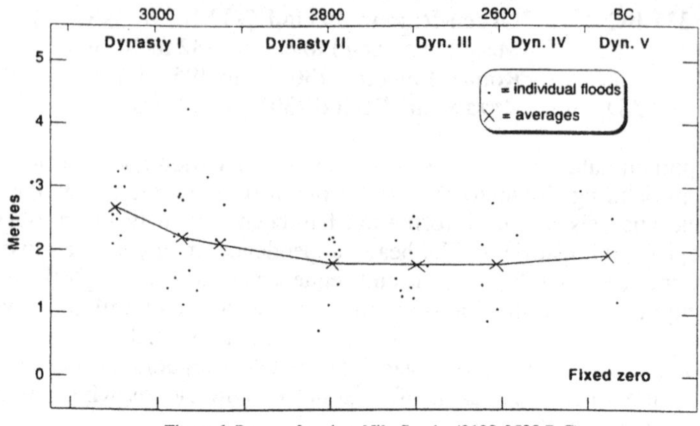

Figure 6: Range of ancient Nile floods. (3100-2500 B.C)
(Source: Bell, 1970)

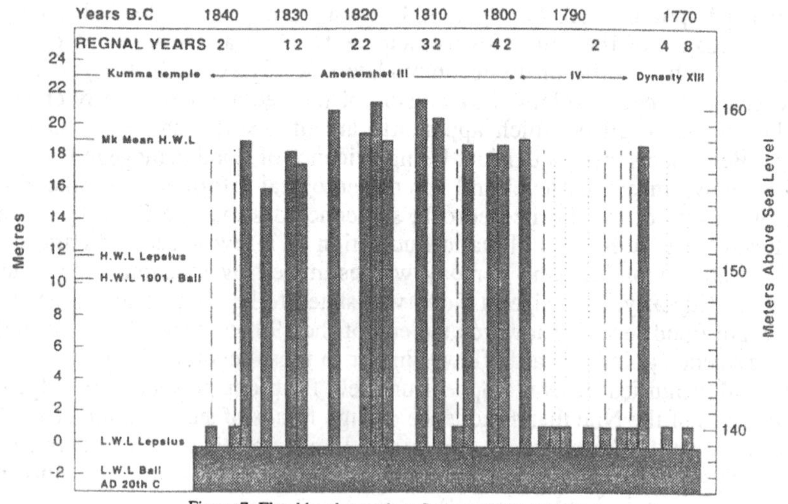

Figure 7: Flood level records at Semna. (1840-1770 B.C)
(Source: Bell, 1970)

According to Lyons (1906), 'At Karnak in 1895 M. LeGrain found a series of 40 high Nile flood levels marked on the quay walls of the great temple. They date from about 800 BC and the mean altitude given by them for a high Nile is 74.25 m above sea level, while that of today is 76.93 (in 1906), showing a rise of the river bed of 2.60 m in 2800 years or 10 cm per century.' These levels indicate a flood regime very similar to those of more recent years.

Although much of the flood information is speculative and cannot be related to present day conditions, it is quite apparent that short-term sequences of drought and floods were a common feature of the Nile basin in the Dynastic period of the Egyptian civilisation and that these played a major role in affecting the success of individual dynasties.

Roda Nilometer

Description

The day when the Nile rose and irrigation of the rich fertile floodplain began was the most important in the Egyptian calendar. The ancient Egyptian's very existence depended on the Nile flood and on its peak reaching 16 cubits; a level above which food sufficient to feed the nation for a year would be forthcoming. To record the event some structures housing level gauges were built which were described by the ancient Greeks as Nilometers. Several were constructed but the most famous, and the one with the longest continuous record, was located on the southern end of Roda Island, opposite Old Cairo.

Several Nilometers were in use when the Arabs conquered Egypt in 641 AD. However, it was 861 AD before authentic details were available. The present gauge is thought to have been built or reconstructed at this date. Some earlier records from different gauges on Roda Island date back to 641 AD and 621 AD.

The gauge, consisting of an octagonal shaft of white marble graduated into cubits (or pics) and fingers, was housed in a well which was connected by conduits to the River Nile (see Figures 8 and 9). Arabic descriptions of the gauge date back to Ahmad Ibn Muhammad al-Hâsib in 861 AD, but generally lack precise details. Not until the time of the Napoleonic Expedition (1798 to 1801 AD) was a complete investigation made.

Reliability of records

The principal sources of data are provided by Ibn Taghri Birdi and Ibn al-Hijazi, who supply similar but not identical values for almost all the 828 years between 641 to 1469 AD. Statistics for the Roda gauge fall into three distinct periods: 641 to 1522 AD; 1523 to 1860 AD; and 1860 to 1890 AD (Popper, 1951). Most of the statistical and hydrological analyses of these old records have been based on those translated and published by Prince Omar Toussoun (1925), and most investigations assume that the same graduated scale and zero of the scale as described in Napoleonic times applied throughout the whole period. As with modern records of river stage, a major problem arises when

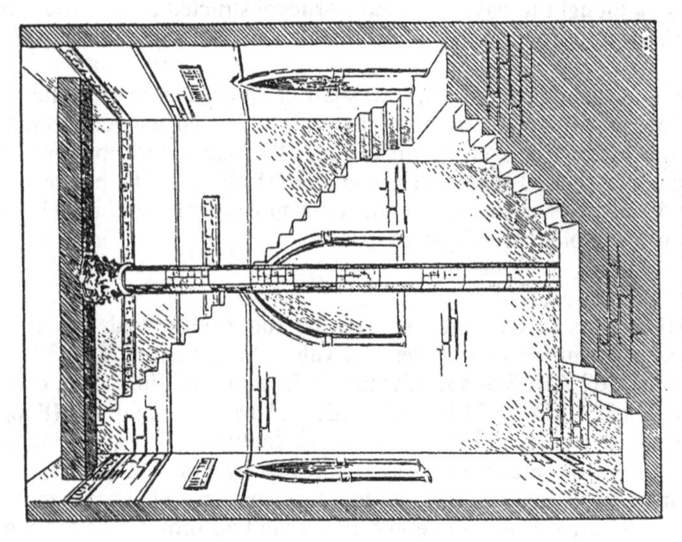

Figure 8: The Roda Nilometer.
(Source: Biswas, 1972)

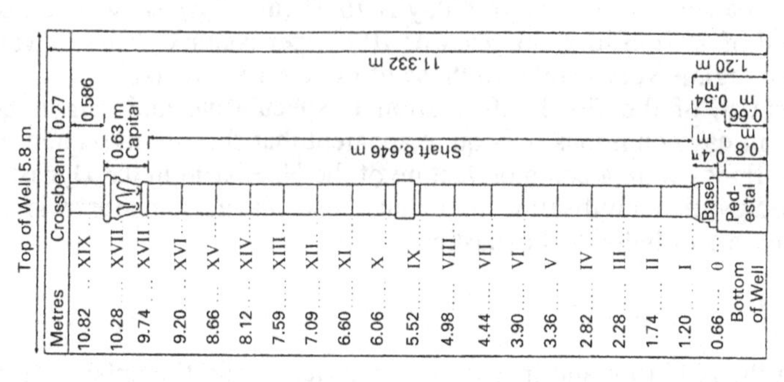

Figure 9: The Roda gauge after repairs in 1800.
(Source: Popper, 1951)

there is an unknown change in zero position of the gauge, resulting in inhomogeneous records. Such records, unless the changes can be identified, significantly reduce their value. There is little doubt that the Roda gauge has been frequently reconstructed (Willcocks, 1913). K. D. Ghaleb in a discussion on the paper by Jarvis (1935) suggested that during the Napoleonic Expedition the base had been badly repaired with the lowest cubit losing 23 cm in height. He also suggested that the French investigators mistook the top cubit as the sixteenth whereas it was really the nineteenth, so that the large increase in flood levels during the nineteenth century were primarily due to this cause.

However, without doubt the most exhaustive study of the Roda Nilometer has been undertaken by Popper (1951). He concluded that three different scales had been used for the three periods quoted above. In the early record up to 1522 AD he recommended using a composite scale with a large cubit of 28 fingers (0.539 m) combined with a small cubit of 24 fingers (0.462 m). From 1522 to 1860 a composite scale had been used but with a small cubit of 24 fingers, equivalent to 0.361 m, above the 9th cubit; whilst after 1860 the top cubits were half the dimension of the large cubit. Moreover the zero of the gauge was estimated to have been raised from 0.0 m during the period 641 to 1522 AD to 1.62 m in 1523 AD and dropped to 0.66 m in 1860 AD. Popper (1951) incorporated all these changes into a new record, which was based on an in-depth study of original arabic scripts as well as the results of all the available past investigations. Unfortunately, although the new data were reworked statistically to compare results with previous analyses, the full revised data set was not published with his studies.

Although the various problems concerning the accuracy of the records as discussed above could impair seriously the analysis of the full record as a single unit, it is suspected that individual parts of the records over shorter term spans are likely to be accurate, and possibly more so than much data collected world-wide today. The onset of the Nile flood was an occasion of great celebrations; with a 16 cubit level reached, the *wafa* festivities proclaimed that lowlands could be flooded and Egypt would be free of famine for the coming year. When the level reached 18 cubits, flooding of the 'middle land' commenced and the festivities of *Neirouz* inaugurated 'New Years Day of the Agricultural Year'. It was followed by the *Saleeb* festival if a 20 cubits level was reached and the 'high land' could be irrigated. Because of the religious importance, the prosperity of the state and even the survival of its people, were so dependent and intricately woven into the pattern of the rise and fall of the Nile, the actual readings are likely to have been taken and recorded to a high degree of accuracy. The main objective in this paper in looking at past records, is to check persistences and relatively short term fluctuations to compare with the extreme events of the past 120 years. In spite of the inherent problems with ancient records the Roda gauge data is able to provide unique evidence in this regard.

Analysis of the Roda records

Siltation of the Nile bed

With the high sediment load carried by the Blue Nile from soil erosion over the Ethiopian Plateau, the Nile Delta actively grew up to the time of the construction of the High Aswan dam. Over the centuries slow accretion of the bed of the Nile and the agricultural lands has been a regular feature of the river's regime. Toussoun (1925) estimated a rise of 1.3 m over a 1000-year period (722 - 822 AD to 1722 - 1821 AD) and arrived at an average 100 year increase of 13 cm; Willcocks (1913) assumed a rise of 12 cm per 100 years, whilst many other authors have assumed 10 cm per 100 years.

It is interesting to speculate on the other factors involved in the siltation process. During this century sea levels have been rising at a rate of at least 10 cm per 100 years. It is probable, but I have seen no records to verify it, that during the 'Little Ice Age' (1400 AD to 1750 AD) (Grove, 1988) sea levels fell, so reducing the siltation rate, all other factors being equal.

Willcocks (1913), however, using J. I. Craig's Roda data set demonstrated that the range of the flood decreased from 6.5 m in the 7th century to 6.1 m in the 13th century before increasing to 7.3 m in the 19th century. He equated this change with the migration of the Head of the Delta at the bifurcation of the Nile into its Damietta and Rosetta branches, with the head moving toward Cairo up to the 13th century and then retreating. However, without a major change in the flow regime such conditions are difficult to visualise. Relating such a scenario to sea level changes would require high sea levels in the 13th century falling until the 19th century, not completely at odds with the 'Little Ice Age'.

A high siltation rate of 29 cm per 100 years is estimated by Popper (1951) based on the analysis of his own adjusted data sets for maximum floods. A large measure of this increase is due to Popper's assumed increase in the gauge zero by 0.66 m at some date between 1522 and 1587 AD. As a result a siltation rate of 116 cm per year is estimated for the period 1587 to 1630 AD and 56 cm for the period 1630 to 1900 AD. This can be compared with an average of 15 cm for the remaining eight centuries. These anomalous results throw some doubt on the validity of the adjusted datasets prepared by Popper (1951).

Periodicities

In 1927 C.E.P. Brooks published a study of periodicities in the Nile floods based on Roda gauge records prepared by J. I. Craig. A plot of annual maxima is reproduced in Figure 10. As is inevitable in such a study, a large number of cycles with different periodicities was identified. However, all the periodicities only exhibited small amplitudes, the maximum being 16.9 cm for a 76.8 year cycle. This can be compared with a standard deviation of 56 cm for all the data. The periodicities therefore, whether real or not, contribute little to the annual fluctuations in levels and cannot be of real significant value in predicting flows in the Nile.

Frequency of low Nile flood

In a contribution to the publication, *Nile Control,* by Sir Murdoch MacDonald (1921), Hurst analysed departures of Nile flood levels from the mean and concluded that the 1913 low level of 17.17 m above sea level was 2.36 m below the mean for the period 1737 to 1920. Two floods had been recorded of a similar low level in the period 641 to 1451 AD and one extreme value,

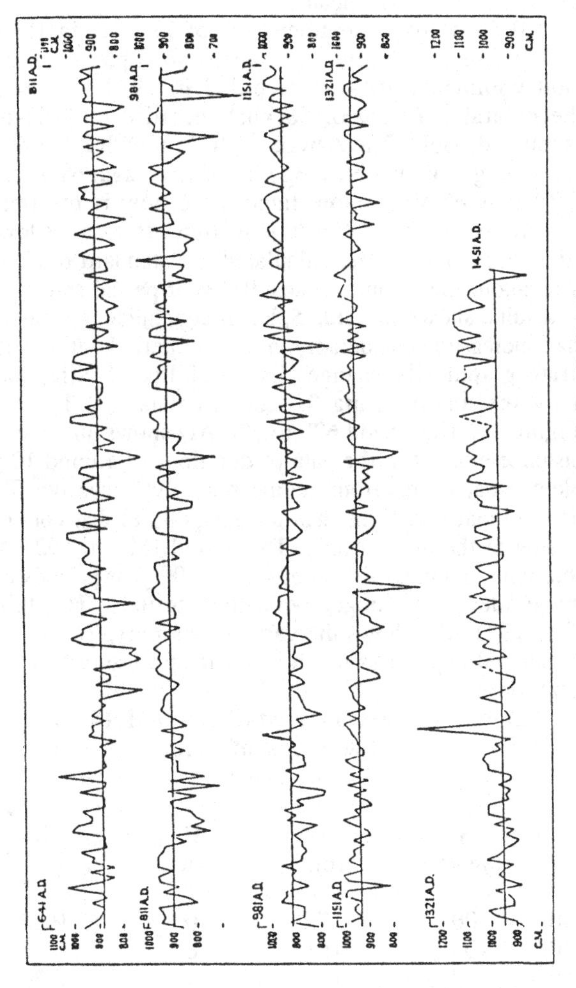

Figure 10: Variations in the height of the Nile flood.
(Source: Brooks, 1927)

presumably in 967 AD, was a further 1 metre lower. However, if one allows for sedimentation and rising land and bed levels, then 1913 can be demonstrated to be the smallest flood within the 1350 year record. Although the 967 AD flood was over 3 m below the 641 to 1451 AD mean, it was only 1.89 m below the 922 to 1021 AD mean.

Willcocks (1913) does give an extreme low of 13.14 m in 1809, but this appears to have been ignored by Hurst and others so is presumably incorrect. The annual flood volume of 46 km^3 recorded in 1913 was considerably smaller than the naturalized flow of 58 km^3 recorded in 1984 during the height of the Sahelian drought. The Aswan High Dam with its large over-year storage (century storage) now gives Egypt security against such extreme annual events. What is of much more importance now is the frequency of sequences of low years or signs of a definite trend towards a lower runoff regime within the Nile Basin. Hurst calculated the standard deviation of the Roda gauge levels about the mean for each 100-year period starting from 641 AD. From these results, shown in Table 5, Hurst concluded that fluctuations in the Nile flood had increased considerably in the last period 1866 to 1946.

To demonstrate graphically changes in variability of Nile floods, plots were prepared by the author using Toussoun's data set. Three plots are reproduced in Figure 11. The period 620 to 720 AD shows an even variability with some persistence evident in the data of durations of around 10 years. No trends are visible and departures from the mean are not excessive. This period does exhibit wider annual variations than all others up to that commencing in 1741 and continuing to the present time. The period 1822 to 1921 AD shows the largest fluctuations in annual floods of all the 100 year periods and is also accompanied by evidence of longer persistent periods. The third period illustrated, 1222 to 1321 AD, shows the variation more typical of the bulk of the datasets with annual departures much lower than those experienced in the 19th and 20th centuries.

Table 5 Roda Gauge Statistics 641 to 1946
(1080 years of Data)
(after Hurst)

Period	**N** (years)	σ (m)	**R** (m)	**K**
641 to 740	100	0.72	10	0.68
741 to 840	100	0.61	8	0.65
841 to 940	100	0.51	9	0.74
941 to 1041	100	0.38	5	0.65
1042 to 1142	100	0.44	11	0.82
1143 to 1242	100	0.48	14	0.86
1243 to 1344	100	0.44	7	0.72
1345 to 1445	100	0.66	14	0.78
1446 to 1741	100	0.73	11	0.69
1741 to 1866	100	0.77	18	0.84
1867 to 1946	80	0.86	22	0.74

Figure 11: Roda gauge departures from the mean for annual flood levels for selected periods.

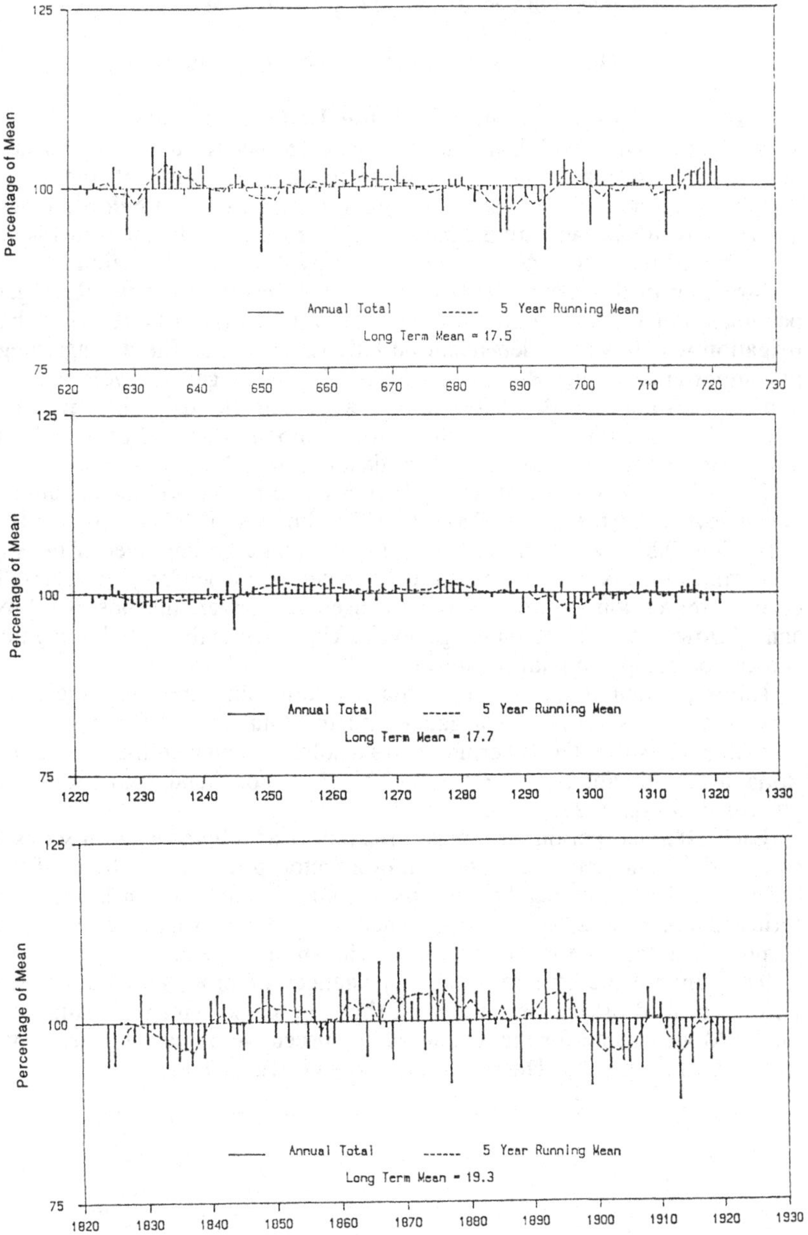

Where: N = number of years recorded
σ = standard deviation of annual maximum Nile levels (metres)
R = maximum accumulated departure of peak level from mean (metres)
K = Hursts 'K' exponent (See following section)

Variation in Nile floods based on the Roda gauge levels

To relate the Nile flood data collected since 1871 with Roda gauge data, the latter series needs to be brought up-to-date. Because of the gradual implementation of regulation and controls in Egypt of both Nile flood levels and flow, which began with the construction of the Delta barrages in 1842 and 1861 by Mohammed Ali, Viceroy of Egypt, and culminated with the construction of the Aswan High Dam in 1963, records from the Roda gauge becomes much less significant. Perennial irrigation has replaced basin irrigation which was so dependent on Nile flood levels. The downgrading of the importance of the Roda is reflected by Roda gauge levels not being published in the Nile Basin Reports. However, as the main objective of this chapter is to place the flows of this century into an historical context, it is of interest to extend the Toussoun (1925) data beyond 1921.

Recorded annual flows at Aswan have been correlated with maximum Roda gauge levels for the period 1871 to 1921. For the 51 values a correlation coefficient 0.83 was obtained. This no doubt could be improved if the water year values for discharge had been used rather than calendar year values. The derived regression equation was then used to convert the post 1921 Nile annual flows into peak Roda gauge levels. The result is illustrated in Figure 21 and can be compared with Figure 11.

Having calculated an approximate relationship between the annual maximum levels at the Roda gauge and the total annual flood volume as measured at Aswan, the opportunity was available to convert the long series of Roda levels to annual Nile flows. This was done and some results are presented in Figure 22.

Each 100-year period was treated separately as a homogeneous series and corrected for siltation by the addition of a factor to remove the trend of rising bed levels. The difference between the recorded mean 100-year levels in each series and the 1822 to 1921 mean maximum level, was selected for this purpose. The factors for each period are shown in Table 6.

To compare the extremes and persistences of droughts existing in the series, a ranked list has been prepared showing the percentage departure from each 100-year mean for the annual, and consecutive 5-year, 9-year, 15-year, and 19-year sequences. The results are shown in Table 7.

Table 6 Correction factors to remove the trend of bed accretion

Period	Mean maximum annual flood level at Roda (amsl)	Correction factor added to levels (m)
1822-1921	19.32	0
1722-1921	19.12	+0.20
1622-1721	18.72	+0.60
1522-1621	18.79	+0.53
1422-1521	18.21	+1.11
1322-1421	17.98	+1.34
1222-1321	17.66	+1.66
1122-1221	17.69	+1.53
1022-1121	17.66	+1.66
922-1021	17.47	+1.85
822-921	17.47	+1.85
722-821	17.41	+1.91
622-721	17.50	+1.82

This approach of assuming that the trend of increasing Nile levels is all accounted for by bed accretion cannot be substantiated, but over the whole series is probably a reasonable assumption. However, actual variation in mean flood level between the individual 100-year periods will be masked by this approach. Although this is the case, as our main objective is to compare fluctuations and persistences of low flow years, this assumption is considered reasonable. The mean annual flood as calculated for each 100-year period varied only between 92 and 93 km^3.

Figure 22 shows annual departures and 5-year running means for Nile flows over three periods, 622 to 721 AD, 1722 to 1821 AD and 1822 to 1987 AD. The period from 1470 up to 1719 AD generally has insufficient data to provide meaningful running mean values. The earliest recorded period, 622 to 721 AD, shows stronger persistence and larger annual departures from the mean than in most of the 100-year periods up to 1521. However, departures from the mean are much more extreme from 1722.

Table 7 Lowest sequences of Nile flows for individual years and for 5, 9, 15 and 19 consecutive years.

Single Year			5 year			9 Year			15 Year			19 Year		
Rank	Year	Flow Km^3	Rank	Year	Flow Km^3	Rank	Year	Flow Km^3	Rank	Year	Flow Km^3	Rank	Year	Flow Km^3
1	1913	45.5	1	1985	68.4	1	1983	73.8	1	1906	81.1	1	1919	81.9
2	1984	58.2	2	1913	70.5	2	1903	79.3	2	1918	81.3	2	1978	82.0
3	1200	58.3	3	1837	73.8	3	1923	80.2	3	1980	81.6	3	1906	82.6
4	1987	60.8	4	1942	76.9	4	1911	80.6	4	1789	83.0	4	1935	84.1
5	1647	61.2	5	1784	78.3	5	1941	81.2	5	1946	83.3	5	1790	84.5
6	967	61.5	6	1926	78.9	6	1831	84.4	6	1831	84.4	6	839	86.0
7	903	64.5	7	1901	80.0	7	690	83.0	7	835	85.1	7	1633	86.9
8	694	66.6	8	965	81.2	8	1786	84.6	8	688	86.6	8	1332	88.5
9	650	66.6	9	832	82.5	9	833	85.1	9	946	87.3	9	1209	88.6
10	1902	68.9	10	1144	83.8	10	967	86.2	10	769	88.0	10	945	88.8
11	1907	69.0	11	801	85.7	11	1141	87.2	11	1205	88.0	11	1065	89.0
12	1928	69.2	12	1451	86.8	12	800	87.4	12	1332	88.1	12	958	89.1
13	841	69.6	13		-	13	1337	87.8	13	1068	89.0	13	1136	89.3

What is immediately apparent from the results in Table 7 is that in terms of low flow sequences, the 1900s have been exceptional. For each series analysed, from the single year up to 19 consecutive years, the two lowest flow sequences have always fallen in the 1900s. In the nine consecutive years series the lowest five events have occurred during this century.

It should be stressed that the analysis of Roda gauge series presented here is superficial in many respects. Although this is so, the indications are clear that both annual variability persistences and possible trends are much more pronounced in 20th century records if the Roda records are accepted as having even a low level of accuracy between successive yearly levels.

Lake Nasser storage from Roda Gauge records

Hurst's lifelong association with the hydrology of the Nile from 1906 to the 1970s and the monumental contributions he made to the science of hydrology as a result, culminated in his analysis of the storage requirements needed to provide full regulation of the Nile; to remove finally, the vagaries of drought and floods, of famine and plenty, the extremes that played such an important role in the fortunes of the Dynasties of Egypt. As noted previously, the fall of stable societies into anarchy in the 'Dark Ages'(c. 2200 and 1200 BC) was probably caused by relatively short sequences of dry years (Bell, 1970, 1971 and 1975).

Hurst's interest in long term storage on the Nile began arround 1920 when he worked on data for assessing the works proposed by Sir Murdoch MacDonald, adviser to the Ministry of Public Works (*Nile Control,* 1920).

An analysis of 1080 years of Nile flood levels recorded at the Roda gauge indicated that their frequency followed the normal Gaussian probability.

However, he also found that although the individual flood events followed normal probability laws, sequences of low flow and high flow tend to be grouped together. Probably as a result of the discussion paper by Jarvis, Hurst analysed over 120 long term sequences of natural events, ranging from river flows, river levels and lake levels, varves, temperatures, air pressures, sunspots and tree rings (see Table 8).

Table 8 Properties of Hurst's 'K' from natural phenomena

Parameter	Range of N years	Number of data sets	Mean
Hydro-Meteorology			
River discharges	10 - 100	39	0.72
Roda gauge	90 - 1080	1	0.77
River and lake levels	44 - 176	4	0.71
Rainfall	24 - 211	39	0.70
Lake Varves			
- Lake Saki	50 - 2000	1	0.69
- Moen and Tamiskaming	50 - 1200	2	0.77
- Corintos and Haileybury	50 - 650	2	0.77
Climate			
- Temperatures	29 - 60	18	0.68
- Pressures	29 - 96	8	0.63
- Sunspots	38 - 190	1	0.75
- Tree rings and spruce index	50 - 900	5	0.79
- Totals and means of sections			
- Hydrometeorological statistics		83	0.72
- Lake varves		5	0.74
- Climate and tree rings		32	0.72
Grand Totals and Means	10 - 2000	120	0.726

For a normal Gaussian distribution with events independent from each other, the relationship between differences behind cumulative departures from the mean in a sequence, the standard deviation measure of the individual departures and the number of events, is expressed by the formula:

$$R = 1.25\,(N)^{K}.\sigma$$

Where R = maximum difference between departures from mean (see Fig.12)
σ = standard deviation of events

N = number of events
K = exponent = 0.5

However, Hurst found that the records for the Roda gauge and for all his series of natural phenomena gave an exponent larger than 0.5. The mean value was 0.72 and a mean general equation of $R = 0.61\ (N)^{0.72}\sigma$

Therefore, for example, for N = 500 years, the storage required on the Nile to fully regulate flows over this period would be estimated as 53σ rather than 28σ which would have been the case had each annual flood been independent.The statistics for the Roda gauges for the departures from the mean, as calculated by Hurst (1966), are as shown previously in Table 5. By analysing records in 100 year subsets Hurst minimised the problem of changes in datum of the Roda gauge.

The Nile Waters Agreement (1959) was based on the average natural river yield at Aswan in the years of the first part of this century. It was estimated at about 84 km^3 per year. The relevant analyses to confirm the suitability of Lake Nasser storage to supply this yield were undertaken by Hurst *et al.* (1966).

Table 9 Statistics of Nile Flows used by Hurst to Estimate Lake Nasser storage

Number of years	Period	Mean flow (km^3)	Standard deviation (km^3)	Hurst's 'K' coefficient	Storage 'R' (km^3)
30	1870-1899	110.0	17.1	0.65	99
69	1900-1959	84.5	13.5	0.50	79
90	1870-1959	92.6	19.8	0.90	608

For N = 100 years, R = 90 km^3 of reservoir storage and K = 0.72 (Hurst's average value from full Nile records and various time series of natural phenomena), a standard deviation of 5.38 km^3 in Nile flows was required to ensure complete regulation, a figure well below that recorded for the 1870 to 1959 period. However, for a K value of 0.5 (as calculated from 1900 to 1959 flows) a 90 km^3 reservoir capacity would be sufficient to regulate all flows.

Figure 12 (after Hurst, 1965) shows a plot of cumulative flows and clearly demonstrates the striking difference between the flow regimes between 1870 to 1898 and 1899 to 1957 and the wide variation in storage requirements dependent on which of the three periods is selected. If pre-1899 flows were reduced by 8 per cent (Hurst and Phillips, 1933) then the storage requirement for full Nile regulation for the period 1899 to 1957 would be reduced from 500 km^3 to 307 km^3.

Hurst in his analysis assumed evaporation losses of 10 per cent of the reservoir losses at the year end and assumed Sudanese abstractions taken at Aswan. For a draft of less than the mean river flow Hurst derived the

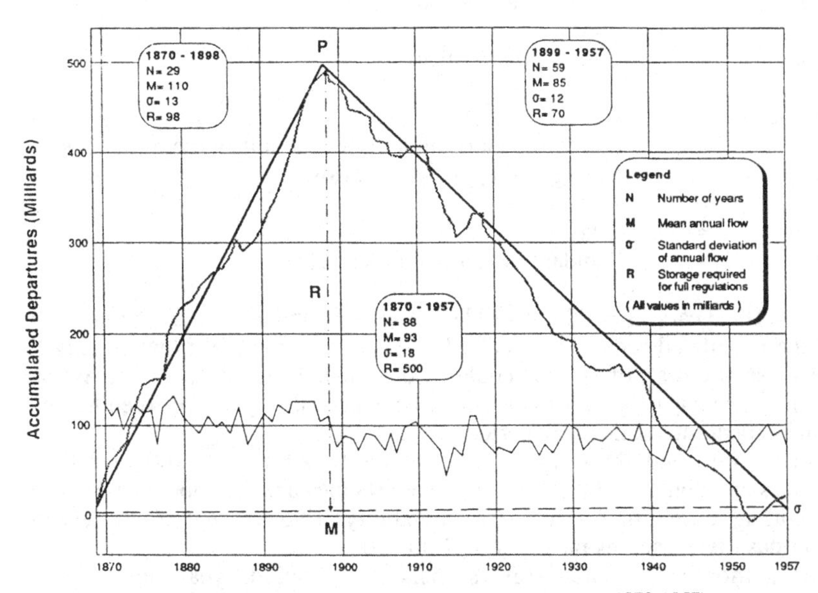

Figure 12: Nile discharges at Aswan, cumulative values (1870-1957).
(Source: Hurst, Black and Slamalka, 1966)

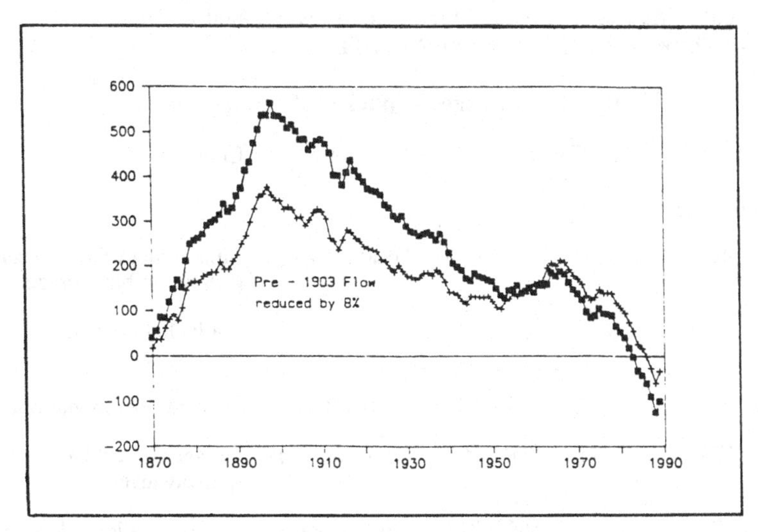

Figure 13: Annual accumulated departures (1870-1989).

following equation from 38 examples of time series of natural phenomena (implicit in the relationship is a K value of 0.72)

$$\text{Log } S/R = -0.08 - 1.05\,(M-B)/\sigma$$

where:-

S	=	Storage to maintain a draft of less than the mean flow
R	=	Storage to regulate mean flow
M	=	Mean flow
B	=	Draft
σ	=	Standard deviation of mean flows

For N = 100 years, S = 90, M = 92.6, σ = 19.8, and K = 0.72 (giving R = 331) the calculated secure draft is 82 km^3. For 1900 to 1959 flows (giving R = 140) a secure draft of 90 km^3 could be obtained. Thus a reservoir at Aswan with an active capacity of 90 km^3 might be expected to yield between 82 and 90 km^3 depending on the flow period selected.

Hurst confirmed these values by generating 20 sets of 100-year flow sequences and concluded that a reservoir of 90 km^3 capacity should be capable of providing a yield of 84 km^3 with a reliability of 96 per cent. The reservoir as built has the capacities presented in Table 10.

The author has updated Hurst's analyses to include years up to 1989 (Figure 13). The influence of the Sahelian drought has increased the full regulation storage from Hurst's original figure of 500 km^3 to 600 km^3, though its full effect is tempered by the high discharges which were present in the late 1950s and 1960s. Taking an effective storage of Lake Nasser as 90 billion cubic metres and using Hurst's analyses updated to 1989, the reliable yield would be reduced from 82 km^3 to 78 km^3.

Table 10 Storage Capacity of Lake Nasser

Level (m) (amsl)	Capacity (km^3)	Description	Comment
87 - 146	0 - 30 = 30	Dead storage	Minimum level for power generation but storage available for use in emergency
146 - 175	30 - 121 = 91	Active storage	Planned operational range
175 - 178	121 - 137 = 16	Flood storage	Storage to Toshka spillway level
178 - 182	137 - 169 = 32	Flood storage	Maximum allowable level
196	169 - 240 =71	-	Dam crest

The Hurst analysis and its update are more of academic than practical interest. If a climatic change has occurred, which is in part or wholly due to man's activity predominantly during this century, then a radical new approach to the estimation of the yield of the Nile is required. However, the Hurst approach was an original attempt in reservoir yield analysis to take account of the anomalous sequences of low or high values found in time series of natural phenomena. The method (making use of K = 0.72) should therefore account for the more random intra- and extra-terrestrial influences, including such biophysical short-term feedback processes as reduced vegetation cover and increased albedo, increased dust storms associated with drought, and natural persistence resulting from such variables as soil moisture (Walker and Rowntree, 1977).

The Hurst equation could certainly account for the 'Joseph' and 'Noah' occurrences set off by such trigger mechanisms as volcanic eruptions, fluctuations in ocean currents such as the El Ñino and extra-terrestrial processes including solar flares, cyclic long period tides, etc: events which may in part account for the 'butterfly effect', a central feature in the theory of chaos.

Any new approach to reassess the yield of the Nile Basin should take account of the distinctive and different circulation and rainfall regimes dominating the catchments of the White and Blue Niles.

All previous yield studies have tended to analyse River Nile flows at Aswan. However, flow records over the past two decades provide the evidence, if any is needed, of the importance of treating the Blue and White Niles separately. The following section looks at recent flows and this problem.

Recent Nile data

Flow records

River Nile flows at Aswan (1871 to 1982) have been published in the Nile Basin Volumes back to 1871. Figure 14 shows annual departures from the mean for naturalised flows, that is flows corrected to allow for abstractions, reservoir losses, etc. The record is unusual in that three distinct periods are clearly visible. From 1871 to 1898 the mean discharge was 102 km^3 falling to 88 km^3 from 1899 to 1971 before descending to 77 km^3 over the latest period 1972 to 1986. Between 1954 and 1967 there was a sequence of above average flows averaging 98 km^3.

Before 1902, flows were based on a gauge at Aswan calibrated by infrequent float measurements. After the construction of the First Aswan Dam sluice discharges, calibrated by discharging into large downstream tanks, were used. In *The Nile Basin* Volume IV (1933) it is stated that evidence from comparisons between the Aswan and Wadi Halfa gauges before and after 1902 shows that the pre-1902 Aswan flows were over-estimated by about 8 per cent. Todini and O'Connell (1979) agree with this conclusion and the pre-1902 flows shown in Figure 14 include these reductions.

Since the completion of the construction of the Aswan High Dam (AHD) in 1963, river discharge based on flow measurements at Dongola have been generally used to represent flows entering Egypt. However, water balances of Lake Nasser enable valuable checks to be made of these flows. Some evidence

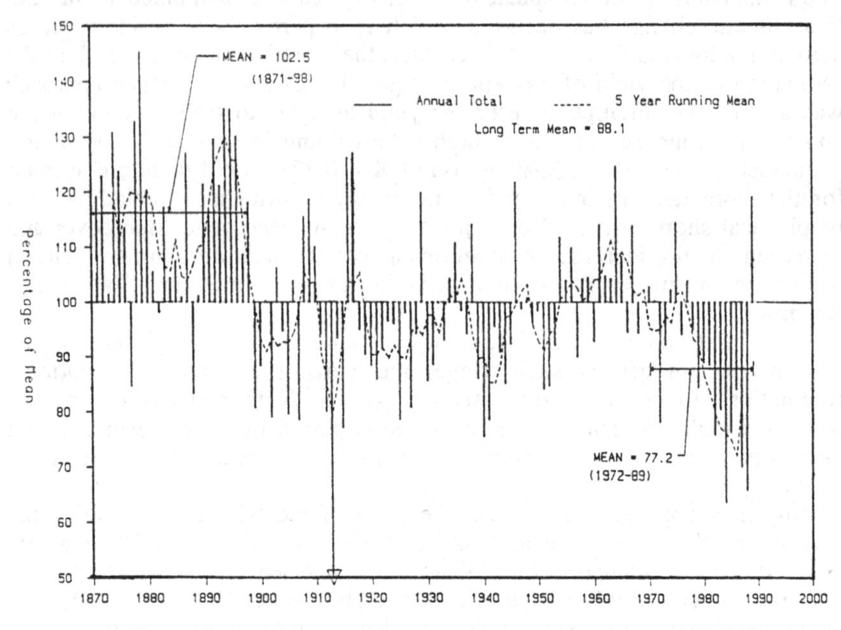

Figure 14: Departures from the mean annual flows at Aswan (1871-1989).

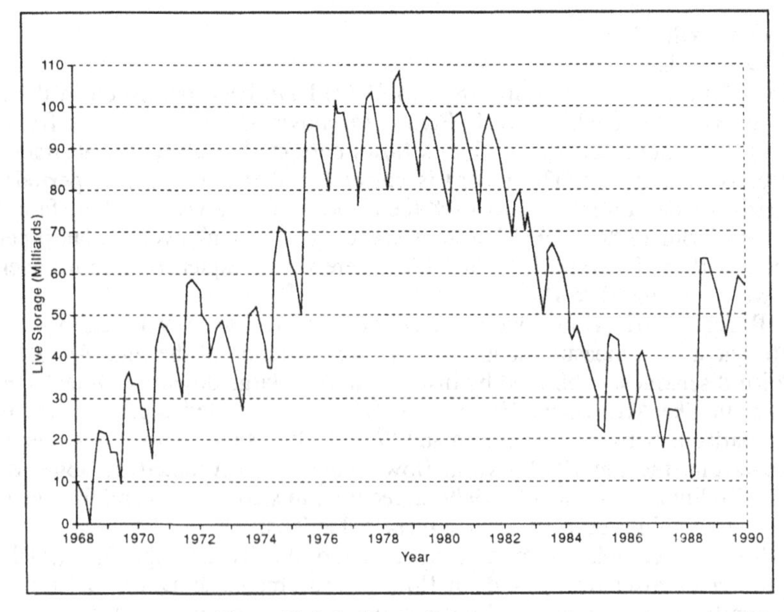

Figure 15: Lake Nasser live storage.

exists that the use of a standard rating curve for Dongola may provide more accurate estimates of flow than individual annual rating based on flow measurements taken each year (Sir M.MacDonald & Ptrs/UNDP/IBRD, 1988).

Lake Nasser started to fill in 1963 following the construction of the Aswan High Dam. Its construction provided Egypt with the complete regulation of annual Nile flows and gave sufficient over-year storage to supply its share of 55km^3 provided under the Nile Waters Agreement between Egypt and the Sudan, with a very high level of reliability. From 1963 the reservoir filled steadily and seepage and evaporation were as anticipated. By 1978 Lake Nasser was almost full with a storage capacity of 134 km^3. Since 1978 Lake Nasser levels have fallen (see Figure 15). The very steep reduction in volume since 1981 is quite marked and the whole sequence, in view of the accepted impression that the drought commenced in the late 1960s or early 1970s, requires some explanation. However, the influence on lake levels of the Sahelian drought in 1984 and 1972, associated with the failure of crops and starvation in Ethiopia and other Sahelian countries, is clearly visible.

Flows in Blue and White Niles

Previous Nile basin yield studies have tended to analyse River Nile flows at Aswan. However, the flow records over the past two decades demonstrate clearly to everyone that the rainfall regimes over the Blue Nile and White Nile catchments differ significantly. The importance of this fact on Nile flows at Aswan is in normal times somewhat diminished as the Blue Nile and River Atbara contribute on average about 70 per cent of the discharge at Aswan or 84 per cent if the River Sobat is included. Although this is the case, the influence of the different rainfall regimes can clearly be seen by comparing Figures 16 and 17. The White Nile flows between 1962 and 1985 have been increased by 32 per cent or 8 km^3 above the 1912 to 1961 mean. This has occurred at a time when Blue Nile flows have been decreasing and for the period 1965 to 1986 were 16 per cent or 8 km^3 below their 1912 to 1964 mean. Over this latter period the White Nile contributed 44 per cent of the Main Nile flows at Khartoum, a significant increase.

Two conclusions can be drawn from a perusal of Figures 17 and 18: firstly, high levels in Lake Victoria have helped cushion Egypt from the more severe influences of the Sahelian drought; and, secondly, lake outflows are falling to values closer to their long-term mean and more in keeping with those seen during the first 60 years of this century. The beneficial effect of the high Lake Victoria levels, therefore, may no longer be available to supplement the falling discharges of the Blue Nile in the future.In the light of these two factors future estimates of yields from Lake Nasser should be based on separate White and Blue Nile flow inputs.

Blue Nile flows

The Blue Nile rises on the Ethiopian Plateau as the River Abbai at elevations up to 3000 m and has a catchment area of 324,500 km^2. When compared with the White Nile, the Blue Nile's catchment is relatively small. The Bahr el Ghazal alone has a basin of 526,000 km^2. However, because of the high rainfall of up to 1,500 mm, mainly falling between July and September, the

Blue Nile is the main contributor to the Main Nile flows. When the rivers Sobat and Atbara are included with the Blue Nile, the importance of fluctuations in rainfall over Ethiopia become apparent. For this chapter interest is focused only on the discharge of the Blue Nile and its variations over the years. Figure 16 shows how discharges have fallen fairly consistently since the high flow regime of the mid-1950s to mid-1960s. What is of concern is the apparent intensification of the drought up to 1988.

Blue Nile flows have been shown to be closely mirroring Sahelian rainfall. It is clear that they are controlled by shifts in general global circulation patterns which have been the dominant feature of climate and runoff in the Sahelian region this century. This close relationship is somewhat surprising as the Ethiopian Plateau with its pronounced orographical influences has led many climatologists to exclude Ethiopia from their studies of the largely homogeneous Sahel region, which spans the African continent.

White Nile flows

Recently a notable contribution to the hydrology of Lake Victoria has been produced by the Institute of Hydrology (IH) (1984 and 1985) and summarised by Piper *et al.* (1986). In the IH report (1984) it was demonstrated that the steep rise in levels between 1961 and 1964 could quite effectively be reproduced, as could subsequent and previous levels, by simply increasing lake rainfall by 7.5 per cent. Using a simple soil moisture balance procedure, catchment runoff was estimated for the period of ungauged flows and a realistic sequence of Lake Victoria levels was produced for the period 1925 to 1978. It was concluded that the 1961 to 1964 increase in lake levels was not unique and that its occurrence could be reproduced by a rainfall/runoff model based on 1925 to 1979 records. The results of this study have been used by the author to investigate White Nile flows.

A simple monthly water balance model has been developed by the author for Lake Victoria. The analysis was based on published data for inflows and outflows since 1900, updated to mid-1986. Separate estimates of lake rainfalls are only available from 1925 to 1978 from which predicted inflows have been determined (IH, 1985). For the earlier period 1900 to 1924 and post-1978 end of month lake levels at Jinja were used.

Central to the modelling exercise was the determination of outflows at the Owen Falls Dam. For this purpose the 'agreed curve', was assumed to have operated over the full 87-year period.

In addition to monthly outflows, storage in the lake is depleted by losses due to evaporation which are offset by rainfall over the Lake. As in the later IH studies (IH, 1985) potential evaporation has been taken as constant at 1595 mm a year. The model assesses the evaporation losses and rainfall gains as a product of the current surface area of the Lake, whilst volume, area, elevation curves for the lake were taken from the WMO study (WMO, 1981). For each month in the sequence an initial estimate of overall losses was obtained using the surface area at the beginning of the month followed by two further iterations using successive refinements of the storage/area for that month.

After very satisfactory calibration, the model was used to examine the sensitivity of the Lake to changes in starting level at Jinja The results of

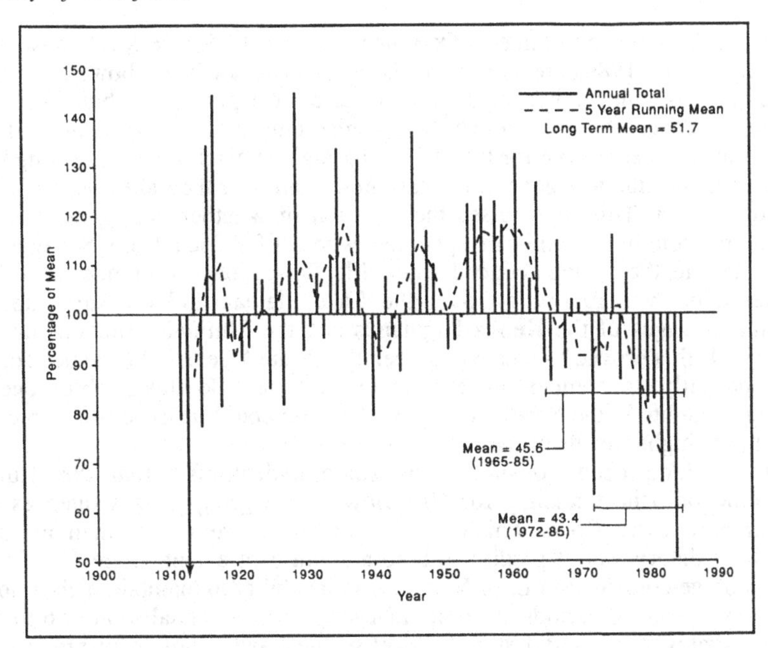

Figure 16. Departures from mean of naturalised annual Blue Nile flow at Khartoum (1912-1986)

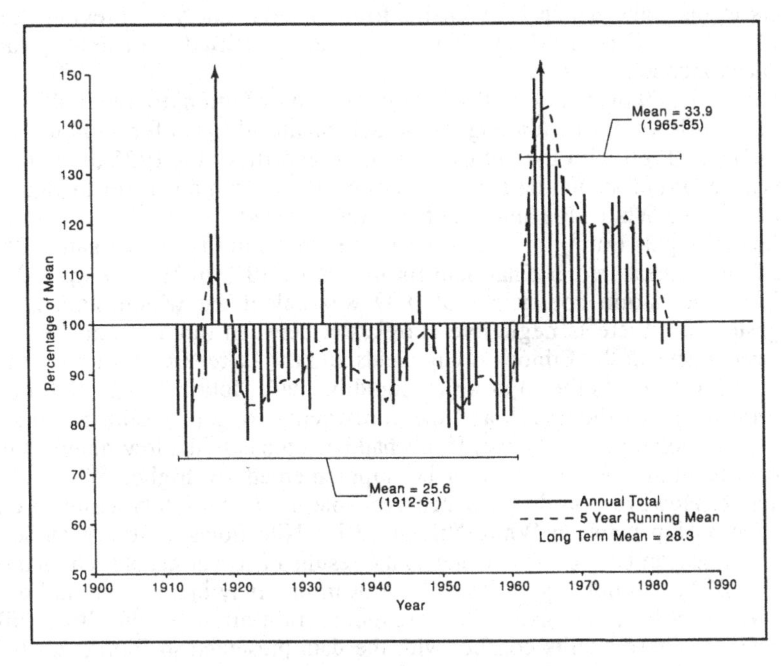

Figure 17. Departures from mean of naturalised annual White Nile flow at Mogren (1912-1985)

lowering the starting volume in December 1964 to 11.5 m (a level consistent with December 1986) are shown in Figure 19. This analysis showed that the Lake level is independent of starting level after a period of about 6 to 12 years. Moreover, it indicated that the extreme jump in levels between 1961 to 1964 was not responsible for the continuing high levels in the Lake during the 1970s but was due to a general increase in rainfall over the Lake and the Lake Victoria basin. This then raises the question of whether during periods of Sahelian drought, which can span the whole of Africa from Senegal to Ethiopia and the Sudan reduced Blue Nile flows, may be compensated by higher than average rainfall over the Lake Plateau and the surrounding mountain ranges of the Kenya Highlands and the Mufumbiro mountains in Uganda. If this proved to be the case then the White Nile could be expected, at least partially, to compensate for the lower Blue Nile flows which occur during droughts in the Sahel. Also Lake Victoria could be expected to return now to the higher levels of the 1960s and 1970s.

It has long been considered by some hydrologists that the 'Hurst Phenomenon' (the tendency for Nile flows to be grouped in sequences of higher and lower flows which are longer and more severe than normal statistical theory would predict) was to a significant extent the result of the large storage contained in Lake Victoria and its ability to maintain high or low outflows over long periods. The fact that levels tend to equalise after 6 to 12 years, whatever the initial starting level of the Lake, tends to disprove this hypothesis.

The water balance calculations for Lake Victoria provided a series of inputs to the Lake which is amenable to time series analysis. Previous time series analyses, Kite (1981 and 1982), have been restricted to Lake levels and produced inconclusive results.

This series comprises both the rainfall contribution to the Lake and tributary inflows. By combining the model calculated flows for 1900 to 1924 and 1977 to 1986 with the published rainfall and flows for 1925 to 1976, an 86-year record of annual lake input, for comparison with downstream gauging stations on the White Nile and Blue Nile, was obtained.

Annual inputs (km^3) to Lake Victoria are plotted in Figure 20 against Blue Nile flow as measured at Khartoum for the period 1912 to 1985 (74 years). A positive correlation coefficient of 0.09 was calculated which statistically suggested that there is negligible correlation between runoff from the Blue Nile catchment in the Ethiopian Highlands and the Lake Victoria Basin. This result is important as the supposition could be made, following the high Lake Victoria levels of the 1970s and the concurrent drought in Ethiopia, that a negative correlation might exist. If this had been the case the low runoff in the Blue Nile could be expected to be supplemented by higher White Nile discharge. However, on the evidence available, it can be safely assumed that the relationship between White Nile and Blue Nile flows is almost random. However, as can be seen in Chapter 1, the results of recent general circulation model (GCM) studies suggest that global warming may increase rainfall over the White Nile catchment whilst reducing rainfall over the Blue Nile Catchment. These results conflict with the data presented in Figure 20. It is possible that the effect of global warming has only recently, over the past two

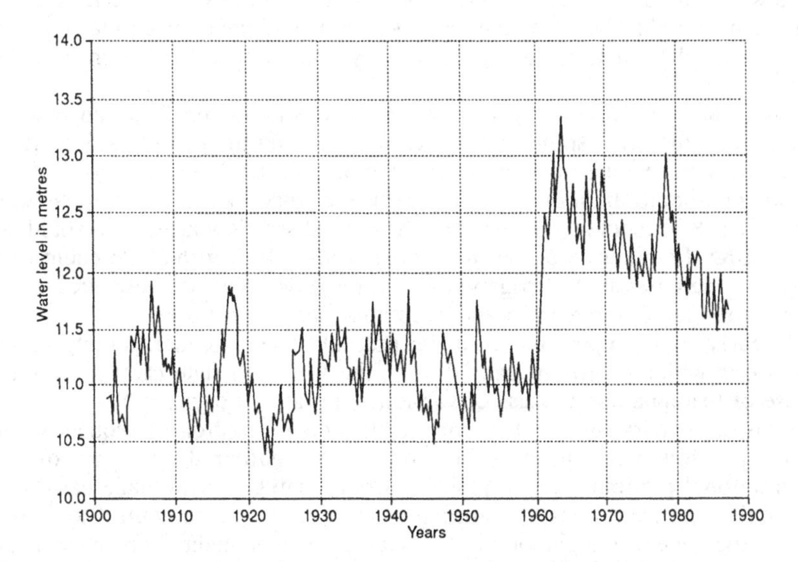

Figure 18. Lake Victoria levels at Jinja (1900-1988)

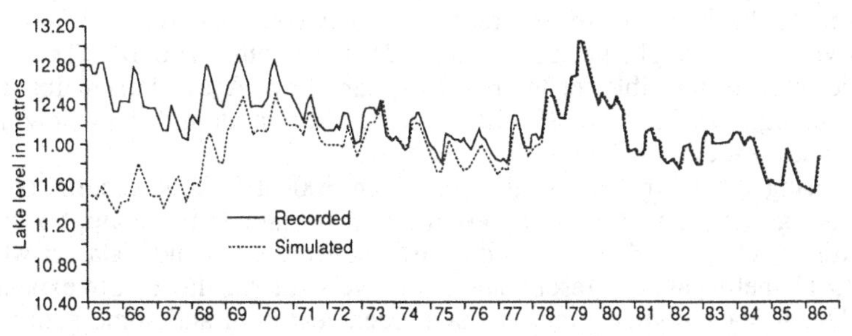

Figure 19. Lake Victoria levels - simulation with starting level at 11.5m (Jinja Datum)

decades, started to influence the rainfall regimes of the region.

To investigate what effect high Lake Victoria levels had in safeguarding Nile water in the Sudan and Egypt from the most serious consequences of the Sahelian drought, the high lake level period was removed from the sequence and replaced by a flow series statistically similar to those which occurred before 1963.

The results of this analysis indicated that should the Sahelian drought continue at the same strength as experienced during the period 1968 to 1987, then the naturalised flow at Aswan would fall from 84 km^3 to 77 km^3. However, should the more severe drought sequence, which occurred between 1978 to 1987 become the norm, then the naturalised flow at Aswan would fall to 72 km^3. In either case such reductions in the yield of the Nile could have serious repercussions for irrigation within the Nile Basin and measures would be required to conserve water and to optimise its efficient use.

Clearly, it is impossible with the present state of understanding of the processes which determine rainfall over the basin to state categorically the cause of the apparent increase of variability in rainfall over the Nile Basin in the past 200 years and in particular for the increase in drought sequences this century. However, information now being gathered, as part of the international programme into global warming, suggest man-made causes. In particular, different rates of warming of the oceans, which control to a large extent the general circulation of the atmosphere, is thought to be responsible (Folland *et al.*, 1986). This, of course, is one manifestation of the increase in 'greenhouse gases', primarily carbon dioxide.

Comments and Conclusions

1. Between the last ice age 18,000 years BP and the climatic optimum 7000 years B there has been a number of rapid changes of climate (Figure 3). These are the result of naturally occurring phenomena, such as feedback process, extraterrestrial process (variations in Earth's motion) or random events such as volcanic eruptions, but not man-made.
2. Lake volumes and river flows in Africa reached a peak around 8000 BP prior to aridity during and after the last ice age (Figure 4). On a larger scale (Figure 5) lake levels were generally high between 5000 BC and 3000 BC with two short drought periods arround 2150 BC and 1200 BC. These are claimed to be responsible for the two Egyptian 'Dark Ages'. Palaeoclimatic evidence suggests that from 100 BC extremes in rainfall and river flow have been modest except for the last two centuries.
3. During the Egyptian Dynastic era from 3000 BC flood records on the Nile have generally tied in with present day expectation. There are, however, anomalies such as the Semna floods marks above the Second Cataract which are 8 to 11 metres above present day flood levels and are difficult to explain.
4. The Roda Nilometer records have been reviewed and some additional analysis undertaken. The Roda gauge levels have been converted to discharge and drought sequences studied. It would appear that Nile flows have since 1720 become much more variable and high flow sequences and low flow sequences are much more acute and persistent. The 1900s have been characterised by exceptional droughts which can be explained by differential

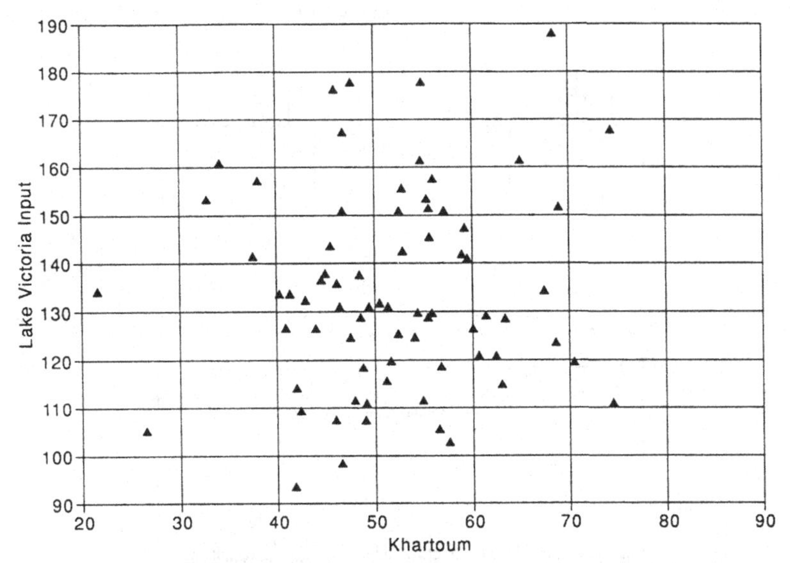

Figure 20: Scatter diagram relating annual Lake Victoria inflow to Blue Nile discharge at Khartoum.

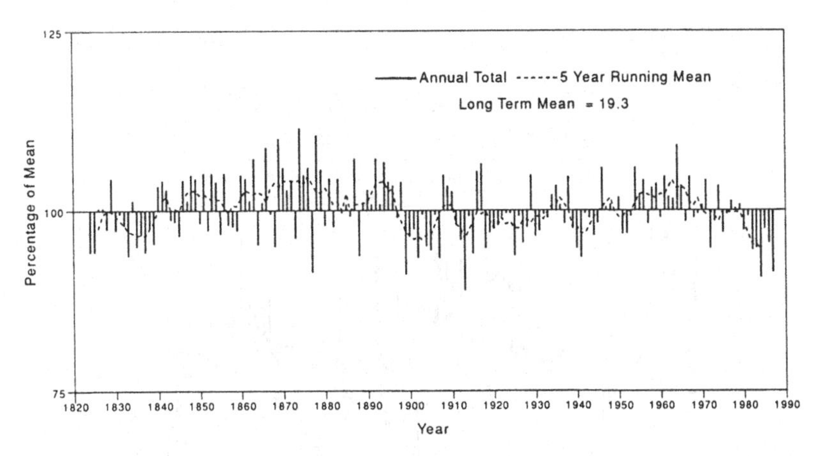

Figure 21: Departures from mean annual maximum levels at Roda Gauge (1822-1987 AD).

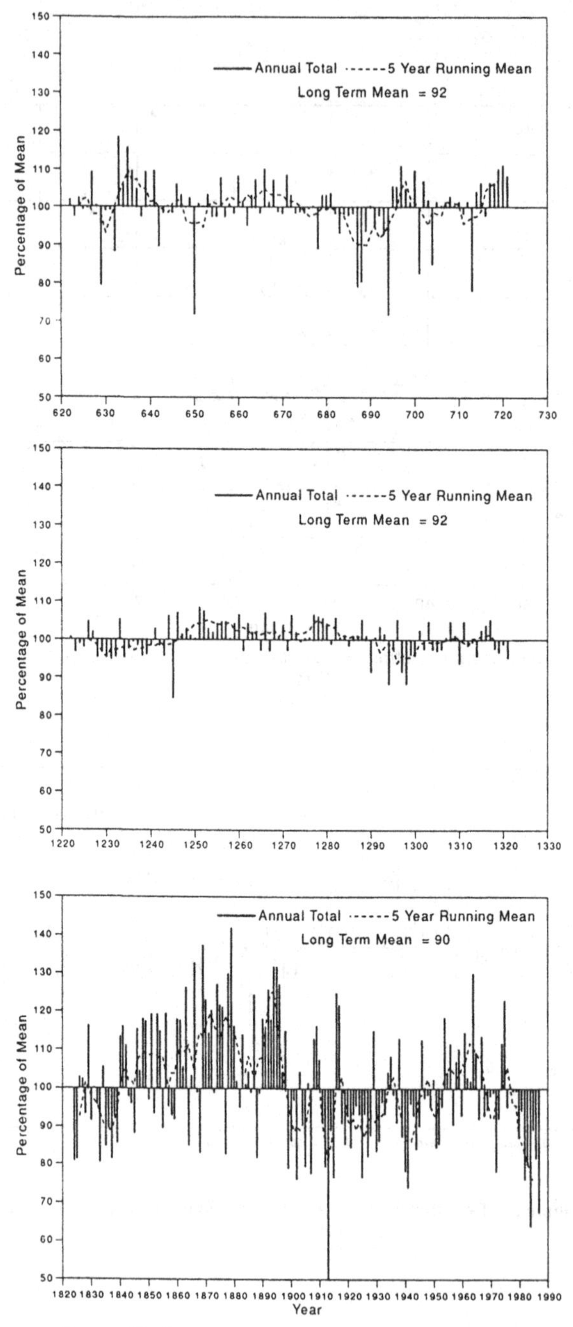

Figure 22: Departures from mean annual Nile flow (milliards) for periods 622-721, 1222-1321, 1822-1987 AD.

increases in sea temperatures between Northern and Southern hemispheres. The strong correlation between temperature and carbon dioxide obtained from the ice cores suggests that a significant increase in the Earth's air temperature should be anticipated with a doubling of atmosphere carbon dioxide.

5. The Roda gauge records present a good opportunity for research. One avenue to explore could be to compare short-term climatic changes obtained from ice cores with the Roda gauge flow series. It would also be valuable to prepare a more definitive series of Roda gauge levels incorporating the research of Popper (1951) and to convert these levels into Nile flows.

6. An analysis of recent Blue and White Nile flows are presented. To a large extent Egypt has remained unscathed from drought due to the large over-year storage present in Lake Nasser. However, other factors have contributed to reducing the droughts impact. One of these has been the exceptionally high levels in Lake Victoria which have helped to maintain higher White Nile flows. The higher levels have resulted from the heavy rainfall between 1961 and 1963 and from above average rainfall since. The higher White Nile flows have helped to compensate for the lower discharges in the Blue Nile which have been adversely affected by the drought in Ethiopia. There has been an intensification of the Sahelian drought since 1978 (apart from the flood in 1988) which has reduced storage in Lake Nasser significantly. The reduction in Blue Nile flows has been accompanied by a fall in Lake Victoria levels, whose regime has returned to the pre-1963 conditions. As a result annual White Nile flows at Mogren have fallen to their pre-1963 level.

7. A study of the hydrology of the Upper Nile Catchments indicates that the rainfall of the Upper Blue Nile Catchment has negligible correlation with rainfall contributing to the inflow of Lake Victoria. Consequently it is fortuitous that Lake Victoria has been supplying above average flows during the Sahelian drought of the 1970s and early 1980s. (This conclusion is contrary to those obtained from the latest ECM studies)

8. Because of possible intensification of the drought it is imperative that water conservation measures are implemented to conserve reservoirs and water management techniques improved to alleviate the more damaging effects of reduced Nile discharges.

References

Alanye Street, F. and Grove, A.T. (1979). Global maps of lake level fluctuations since 30,000 yr. BP, *Quaternary Research* , **12**, pp. 83-118.

Baines, J. and Malek, J. (1980). *Atlas of Ancient Greece*. Oxford, Phaidon.

Bell, B. (1970). The oldest records of the Nile floods, *Geographical Journal*, **136**, pp. 569-73.

Bell, B. (1971). The Dark Ages in Ancient History: the First Dark Age in Egypt, *American Journal of Archaeology*, **75**, pp. 1-36

Biswas, A.K. (1972). *History of Hydrology*. North Holland, Amsterdam and London.

Brooks, C.E.P. (1928). Periodicities in the Nile floods, *Mem's of the Royal*

Meteorological Society, **11**, Nr. 12.

Butzer, K.W. (1976). *Early Hydraulic civilisation in Egypt*. University of Chicago press, Chicago and London.

Butzer, K.W. and Hansen, C.L. (1968). *Desert and rivers in Nubia*. University of Wisconsin Press, Madison, Milwaukee and London.

Folland , C.K., Palmer, T.N. and Parker, D.E. (1986). Sahel rainfall and worldwide sea temperatures, 1901-85. *Nature*, 320, **17** April 1986.

Grove, J.M. (1988). *The Little Ice Age*, London and New York, Methuen.

Hurst, H.E. (1952). *The Nile*. London, Constable.

Hurst, H.E. (1965). *Long Term Storage*. London, Constable.

Hurst, H.E. *et al.* (1966). The major Nile projects, **X**, *The Nile Basin*. Cairo, Physical Dept.,Ministry of Waters, Government Press.

Hurst, H.E. and Phillips (1933). *The Nile Basin*, **II**. Cairo, Physical Dept., Ministry of Works, Government Press.

Institute of Hydrology (IH) (1984). *A review of the hydrology of Lake Victoria*. London, ODA and Wallingford (IH).

Institute of Hydrology (IH) (1985). *Further review of the hydrology of Lake Victoria*, London, ODA and Wallingford (IH).

Jarvis, C.S. (1935). Flood stage records of the River Nile. *Transactions of the American Society of Civil Engineering*, Nr 1944, August 1935, pp.1013-1071.

Kite, E.W. (1981). Recent changes in the level of Lake Victoria. *Hydrological Sciences Journal,* **26**(3), pp.233-243.

Kite,E.W. (1982). Analysis of Lake Victoria levels. *Hydrological Sciences Journal*, **26**(3), pp.233-243.

Lyons, H.G. (1906). *The physiography of the Nile River and its Basin*. Cairo.

MacDonald, Sir M. (1921). *Nile Control*. Cairo, Ministry of Public Works, Government Press.

MacDonald Sir M. and Partners (1988). *Rehabilitation and improvement of water delivery in Old Lands*. IBRD/UNDP Cairo, Ministry of Public Works and Water Resources.

Oeschger, H. and Langway (1989). *The environmental record in glaciers and ice sheets*.Wiley, p.404.

Piper, B.S. *et al.* (1986). The water balance of Lake Victoria, *Journal of Hydrological Science*, **31**(1), 3/1986.

Popper, W. (1951). *The Cairo Nileometer*. Berkley and Los Angles University of California Press.

Shanin, M. (1985). *Hydrology of the Nile Basin*. Amsterdam, Oxford, New York, Tokyo. Elsevier.

Todini, E. and O'Connell, P.E. (1979). *Hydrological simulation of Lake Nasser*. IBM/IH UK (also WMP Technical Report 15, Cairo).

Toussoun, Prince Omar. (1925). Memoire sur L'histoire du Nil, MIE, **IX**.

Walker, J. and Rowntree, P.R. (1977). The effect of soil moisture on circulation and rainfall in a tropical model. *Quarterly Journal of the Royal Meteorological Society*, **103**, pp. 29-46.

Willcocks, W. and Craig, J.I. (1913). *Egyptian Irrigation*. London and New

York, Spon.

Williams, M.A.J. and Adamson, D.A. (1974). Late Pleistocene desiccation along the White Nile. *Nature*, **248** (April 1974).

3

The history of water use in the Sudan and Egypt

P. M. CHESWORTH

Introduction

The Nile has provided the basis of agricultural development in Egypt and the Sudan since the start of agriculture in the area about 7000 years ago. Artificial irrigation began about 5000 years ago and continued largely unchanged until the early years of the nineteenth century when use of Nile water on a significant scale started. By the end of the nineteenth century further agricultural expansion was limited by the availability of water during the low river season. This led to the provision of annual storage with construction of the first Aswan Dam in 1903 and its subsequent heightening in 1912 and 1934. Further storage was provided by dams at Sennar (1925), Jebel Aulia (1937), Roseires (1966) and Khasm el Girba (1966). In more recent years the total availability of Nile Waters rather than its seasonal availability became the constraint and the main reason for the construction of the Aswan High Dam (1963). The enormous storage provided by the Aswan High Dam enables the variations of the Nile flows from year to year to be evened out and the potential of the Nile to be utilised. Until recent years the use of Nile Water has gradually increased, keeping pace with the increase in population. Egypt and the Sudan are now, however, faced with the prospect of continuing population increase but with only limited further Nile water available for agricultural expansion.

It is estimated that the amount of water that would flow from the Nile to the sea, without the interference of man, is some 80 billion cubic metres/year (km^3). This is equivalent to a run-off of less than 3 mm per year over the whole of the Nile catchment of 2.9 million km^2 (one tenth of the surface area of Africa). In these circumstances, the variability of Nile flows is not surprising since relatively small changes in rainfall will result in much larger changes to run-off. The rainfall that does not flow to the sea is accounted for by evapo-transpiration from vegetation and evaporation from lakes and river surfaces. Water use results mainly from irrigation. Other uses are industrial and domestic and additional evaporation from man-made reservoirs and channels. The present flow of water to the sea amounts to some 17 billion cubic metres(km^3). If water use is defined as the reduction of natural flow to the sea as a result of the intervention of man the total water use amounts to some 60 km^3 or three-quarters of the natural flow.

Discussion in this chapter is limited to water use in the Sudan and Egypt as these countries have been by far the major users of Nile water and use by other countries to date has been negligible.

Until the completion of the Aswan High Dam the main interest in accounting for Nile Water was for so-called 'timely water', which was defined as the water required during the period February to July when natural river flows were insufficient to meet demands and water had to be drawn from storage. Following the completion of the Aswan High Dam, with its enormous storage capacity, the concept of timely water was no longer relevant and closer attention was paid to total annual water use. Thus the figures for total annual water use prior to 1966 are only broad estimates based on cropped areas and population data, whereas after 1966 'official' data of total water use are available.

Egypt

Water development in early times

In ancient times the Nile in Egypt at times of flood overflowed its banks and inundated large areas of adjacent land. When the flood abated the river returned to its main channel leaving behind it a layer of silt on the now uncovered land. The first agricultural activity in Egypt took place by the population sowing seeds on this land, watered and fertilised by the natural floods. Archaeological evidence suggests that this first agriculture started in about 5200 BC.(Goldman, McEvoy, Richardson, 1973)

The first agricultural revolution in Egypt came with the start of artificial irrigation, including deliberate flooding and draining by sluice gates and water contained by longitudinal and transverse dikes. Such basin irrigation was established by the first Dynasty (3050 BC). The first recorded evidence of such irrigation is found on the mace head of the so-called Scorpion King, which has been dated at 3100 BC (Figure 1). The mace head shows the Scorpion King (so called because the King is always shown preceded by a scorpion) cutting an irrigation channel that then bifurcates and appears to feed an irrigated field, which is surmounted by an unmistakable palm tree.

Although this control of the flood waters was an improvement on total dependence on the vagaries of the annual Nile flood, the variations in flood level from year to year were critical and no irrigation was possible except at time of flood (winter crops only). The second agricultural revolution came with the introduction of lift irrigation. The first mechanised irrigation by the introduction of the *shadoof* (Figure 2) took place during the 18th Dynasty (1550-1307 BC) and the more sophisticated persian wheel or *saqia*, able to lift substantial quantities of water, was introduced in early Ptolemaic times (323-30 BC). These lifting devices permitted increased reliability in years of low flood and the introduction of limited summer crops. Summer crops were, however, mainly confined to horticultural practice, as much as anything due to the need for fertiliser for summer crops in the absence of the natural fertiliser provided by the Nile flood. The system of irrigation continued largely unchanged until the middle of the nineteenth century.

Estimates of the population in Egypt in early times are given in Table 1 and illustrated in Figure 3. The peak population in the first century AD was in response to exploitative, labour intensive agriculture designed to supply Rome with food. In later Roman, Byzantine and early Islamic times the population of Egypt declined rapidly as a result of plagues, civil strife, floods and famine, and was largely unchanged until early in the 19th century. The cultivable area in Egypt in ancient times has been estimated at between 4 million and 5 million feddans. The area cultivated each year varied greatly depending on flood levels, food demands and labour available. It is thought on average about 1 feddan of area was cultivated per head of population.

Table 1 Early growth of Egyptian population
(millions)

4000 BC	0.35	820 AD	2.37
3000 BC	0.87	869 AD	2.64
2500 BC	1.61	884 AD	2.37
1800 BC	1.96	975 AD	1.76
1250 BC	2.89	1090 AD	1.68
150 BC	4.92	1189 AD	2.35
50 BC	4.40	1298 AD	4.20
300 AD	3.20	1315 AD	4.20
450 AD	3.50	1380 AD	3.50
680 AD	2.70	1430 AD	3.00
743 AD	2.20		

Source: (Butzer, 1976 and Russell, 1966)

First modern development

The impetus for modern development was provided by an Albanian soldier of fortune, Mohammed Ali Pasha, who rose to become the ruler of all Egypt in 1805. Mohammed Ali realised that the revenue of the country was to be attained mainly from agriculture and he decreed that cultivable land be distributed among the people and introduced more valuable crops such as sugar cane, vegetables, fruit and especially cotton. The production of cotton necessitated a radical change in the irrigation system since it needed to be planted before the natural rise of the flood, required regular watering and needed to be protected from inundation during the flood. For the first time controlled irrigation was required but natural variation of water level in the river system caused great problems. Mohammed Ali therefore called upon his engineers to take measures to solve this problem and the outcome was the first man-made structures (The Delta Barrages) across the Nile. These barrages were constructed at the head of the Delta across the main Damietta and Rosetta branches to hold up the low summer water levels and to enable it to flow into the higher flood level canals. The construction of the Delta Barrages was started in 1843 but various engineering difficulties hindered the progress of the works and it was not until 1861 that they were completed. Fortunately the engineers were able to dissuade Mohammad Ali from his suggestion that stone from the pyramids be used for the construction (Hussein Sirry Pasha, 1937).

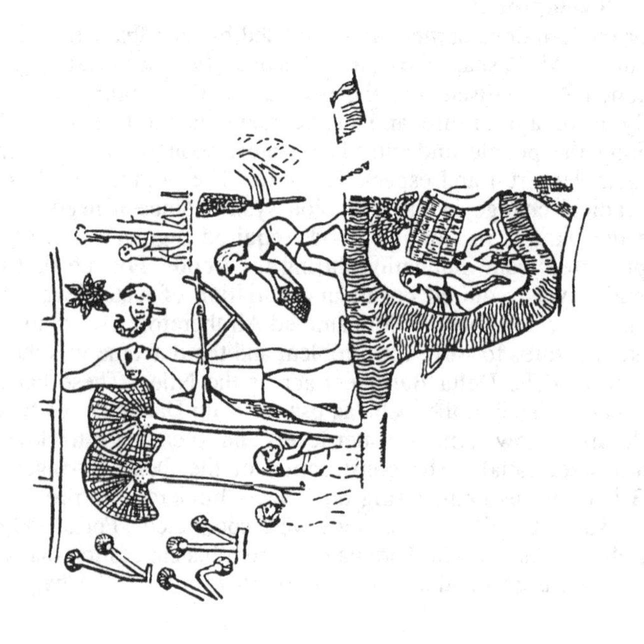

Figure 1: The Scorpian King inaugurating an irrigation network.
(Source: Butzer, 1976)

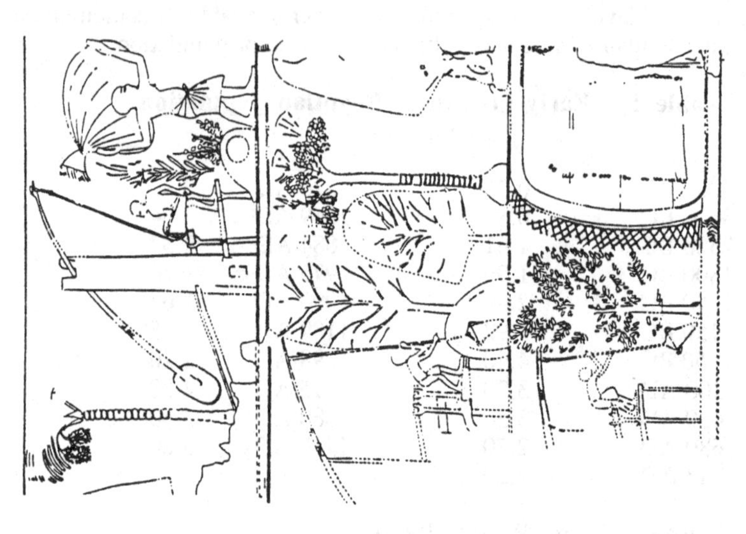

Figure 2: Shadufs of the Amarna period from the tomb of Nefer Hoteb at Thebes.
(Source: Butzer, 1976)

Mohammad Ali died in 1848 before completion of the barrages, but although their hydraulic functions were taken over by replacement barrages in 1939, the original Delta Barrages are still used as road bridges and stand as an elegant monument to the founder of modern irrigation in Egypt. Following completion of the first barrage, remodelling of canal systems was undertaken to give the canal system largely as it is today.

As well as the improvement in irrigation supplies, a great impetus to the cultivation of cotton was provided by the American Civil War which resulted in high prices for Egyptian Cotton. Production increased from 600,000 to 2,000,00 kantars (45 Kg) from 1860 to 1864 and by 1900 reached 6,440,000 kantars.

By the end of the nineteenth century, however, agricultural production was constrained by another factor. The natural low flow was only sufficient for 1.5 million feddans in a low year and this shortage of water outside the flood season for late season and summer crops was to lead to the first storage works on the Nile, the Aswan Dam.

First half 20th century

The first half of the 20th century saw tremendous improvements in the Egyptian irrigation systems. The first Aswan Dam was completed in 1902 with a storage capacity of 1 km^3 and it proved so successful that it was heightened in 1912 and further heightened in 1934 to increase the storage capacity to 5.1 km^3.

Associated with the development of the Aswan Dam, was the extension of the perennial irrigation areas with construction of further barrages at Assiut (1902, remodelled 1938), Zifta (1902) on the Damietta Branch, Esna (1908, remodelled 1947 and now being replaced), Nag Hammadi (1930) and Edfina (1951) on the Rosetta Branch to limit discharge of excess water to the sea.

Second half 20th century

The completion of the Aswan High Dam in 1963 constituted the most recent revolution in Egyptian agriculture. The enormous storage in the reservoir formed by the Aswan High Dam (total storage 162 km^3, live storage 107 km^3; see Chapter 2 Figure 15) is sufficient to make Egypt virtually independent of the vagaries of the year to year variations in the annual Nile flood. After nearly 7000 years during which Egyptian farmers regularly suffered from the effects of annual droughts or floods, the impact of the dam on Egyptian agriculture was nothing less than revolutionary and brought immense benefits from increased irrigated areas, increased cropping intensities and increased yields.

The estimated population, cultivated areas and cropped areas in Egypt from early in the nineteenth century until the present day are given in Table 2 and illustrated in Figure 4.

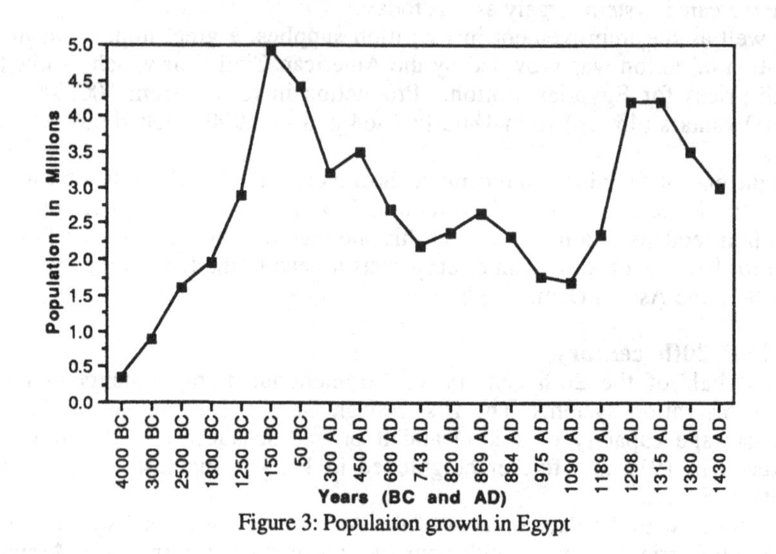

Figure 3: Populaiton growth in Egypt

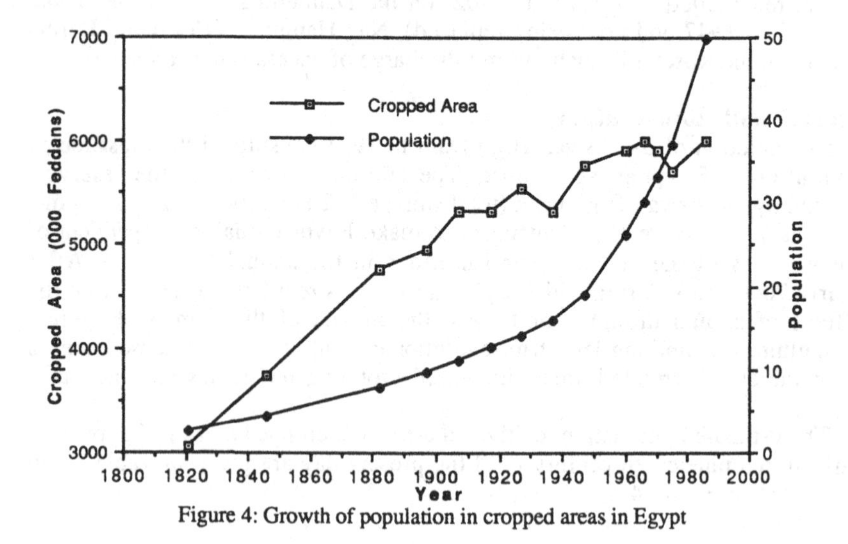

Figure 4: Growth of population in cropped areas in Egypt

Table 2 Growth of Egyptian Population and Cultivated and Cropped Land

	Estimated Population (million)	Cultivable Area (000 feddans)	Cropped Area (000 feddans)	Cropping Intensity (%)
1821	2.51 to 4.23	3053	3053	100
1846	4.50 to 5.29	3746	3746	100
1882	7.93	4758	5754	121
1897	9.72	4943	6725	136
1907	11.19	5374	7595	141
1917	12.72	5309	7729	146
1927	14.18	5544	8522	154
1937	15.92	5312	8302	156
1947	18.97	5761	9133	159
1960	26.09	5900	10200	173
1966	30.08	6000	10400	173
1970	33.20	5900	10900	185
1975	37.00	5700	10700	188
1986	49.70	6000	11400	190

Sources: (Waterbury, 1979) for population and cropping data up to 1975; (Sir M.MacDonald & Partners Ltd., 1988) for 1986.

The Sudan

Irrigation in the Sudan started in the north of the country at much the same time and in the same form as in Egypt, with the first basin irrigation starting about 3000 BC. Basin irrigation together with lift irrigation using the *shadoof* and *saqia* continued virtually unchanged until early in the 20th century. The first modern irrigation development was at Zeidab in 1906, where cotton was cultivated using pumped water.

In 1910 a group of English weaving companies formed the Sudan Plantation Syndicate who started their first irrigation by a pump scheme at Taiba. Their initial success encouraged the company to extend its irrigation area in 1914 by 6000 feddans at Barakat, followed by a further 19,500 feddans at Hag Abdalla and 30,000 feddans at Wad el Nau in 1921. The major landmark in irrigation development in the Sudan was, however, the completion in 1925 of the Sennar Dam on the Blue Nile which irrigated an area of 300,000 feddans in the Gezira. The Gezira scheme gradually expanded, reaching at the time of independence in 1955 an area of 1 million feddans (Republic of Sudan, 1955 and 1957). Soon after independence the Government of Sudan embarked on the Managil extension which brought the total area of the Gezira scheme, still fed by a single headworks, to the present total of over 2 million feddans. In parallel with the development of the Gezira scheme, pump schemes were developed on both the Blue and White Nile and now serve an estimated total area of almost 1 million feddans. Other major schemes implemented in recent years are New Halfa (400,000 feddans) in the 1960's and Rahad (3,000,000 feddans) in the 1970s. About 40,000 feddans of basin irrigation is still practised in the northern Sudan on the main Nile with areas irrigated varying from year to year depending on flood levels.

The expansion of the Gezira scheme and the implementation of the Rahad scheme were based on the availability of low season stored water from the Roseires reservoir which was formed by the completion of a dam on the Blue Nile in 1966. Similarly, the New Halfa scheme depends on the reservoir formed by the Khasm el Girba dam which was also completed in 1966 on the Atbara river. The original storage of Khasm el Girba was 1.3 km^3 but this has been reducing by some 40 million m^3 per year due to siltation (Sir M. MacDonald & Partners, 1978 and 1979). The estimated population, cultivated areas and cropped areas (using Nile waters) in the Sudan from early in this century until the present day are given in Table 3 and illustrated in Figure 5.

A dam was also built in 1937 at Jebel Aulia on the White Nile just upstream of its confluence with the Blue Nile. This dam formed a reservoir with an original storage capacity of 3.5 km^3 which was used to provide 'timely water' for summer cropping in Egypt (See Chapter 9, p.)

Table 3 Growth of Sudanese Population and Cultivated and Cropped Land

	Estimated Population (million)	Cultivable Area (000 feddans)	Cropped Area (000 feddans)	Cropping Intensity (%)
1905	3 ?	100	?	?
1930	5 ?	400	?	?
1940	6 ?	1100	?	?
1957	10.3	2245	1216	54
1970	14.3	3218	2348	73
1980	19.2	4385	3600	82
1986	22.6	4155	2772	67

Source: Various

Reservoir evaporation

The development of reservoirs to store water inevitably results in water use by evaporation from the reservoir surface. The present average annual water use by evaporation from reservoirs is approximately as given in Table 4.

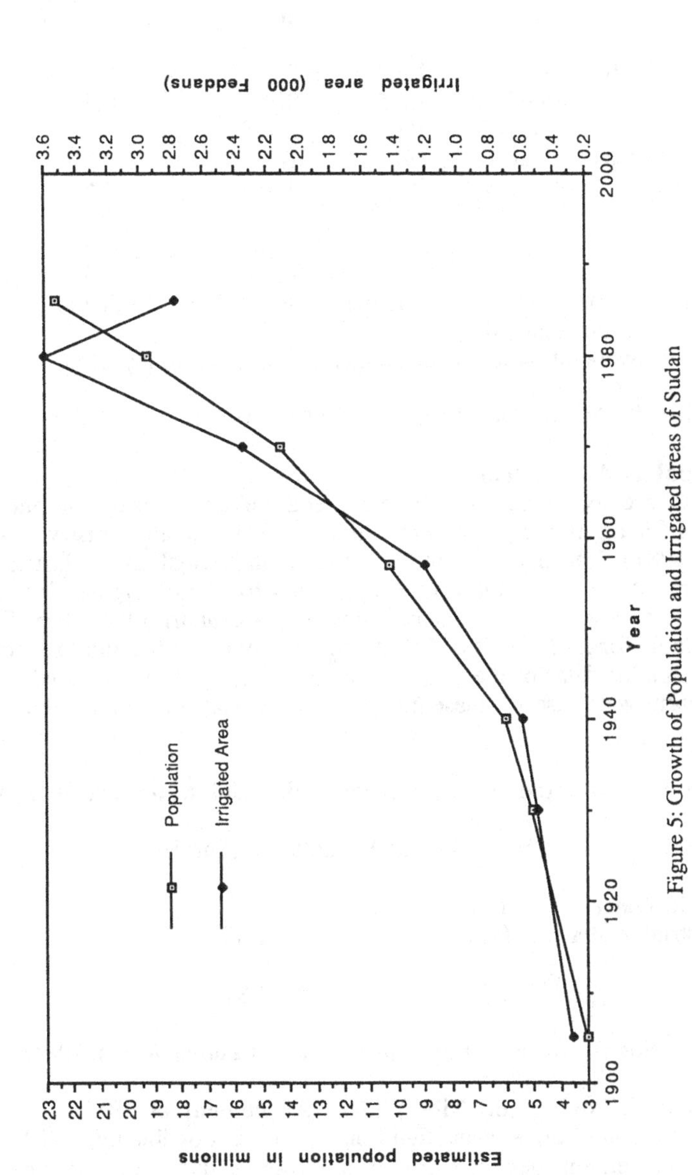

Figure 5: Growth of Population and Irrigated areas of Sudan

Table 4 Existing reservoirs in Egypt and the Sudan

Reservoir	River	Year	Live Storage (km^3)	Annual Evaporation Loss (km^3)
La. Nasser/Nubia	Nile	1963	107.0	10.0
Sennar	Blue Nile	1925	0.6	0.3
Roseires	Blue Nile	1966	2.7	0.4
Khasm el Girba	Atbara	1964	1.3	0.1
Jebel Aulia	White Nile	1937	3.5	2.5
Total			115.1	13.3

Notes:1. Storage given is the original storage without loss due to sedimentation.
2. Evaporation loss is that which is additional to loss from natural river.
Source: (Sir M. MacDonald & Partners Ltd., 1988)

Municipal and industrial use

Few data are available for municipal and industrial uses but published estimates for Egypt are given below in Table 5 (Egyptian Ministry of Public Works, 1981). The net consumptive use for municipal and industrial uses shown in Table 5 is the total water abstracted for these purposes less useful waste water return flows. In some cases (e.g Alexandria) the return flow is not available for re-use as it is discharged directly to the sea and lost from the system. Similar data for the Sudan is not at hand, but it is estimated that the total present water use for these purposes is less than 0.1 billion cubic metres per year.

Table 5 Estimated municipal and industrial water use in Egypt

	1980 (km^3 actual)	2000 (km^3 projected)
Potable Water	1.78	3.47
Industrial Water	0.32	1.37
Total	2.10	4.84

Note:1. Source: (Egyptian Ministry of Public Works, 1981)

Water use following the Nile Waters Agreement of 1959

The construction of the Aswan High Dam was made possible by the 1959 Nile Waters Agreement between Egypt and the Sudan which formed the framework for the necessary co-operation between those two countries.

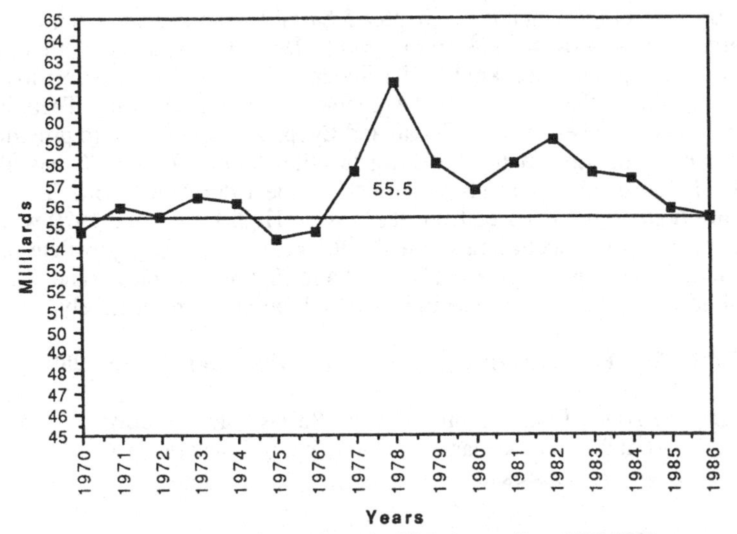

Figure 6: Releases from the High Aswan Dam 1970-1987

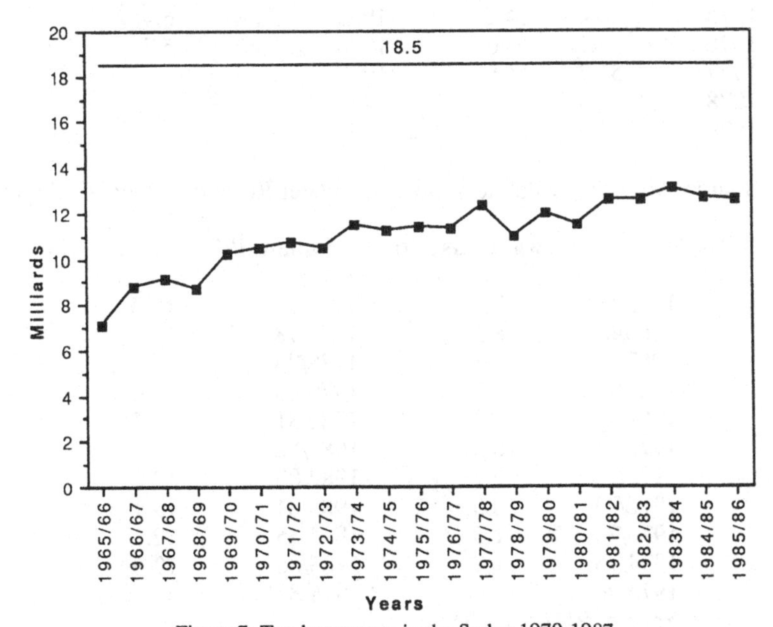

Figure 7: Total water use in the Sudan 1970-1987

As will be seen in more detail in Chapter 5 based on the average natural flow of the Nile at Aswan of 84 km^3 (from 1900 to 1959) the Agreement allocated 55.5 km^3 to Egypt and 18.5 km^3 to the Sudan, with assumed reservoir losses of 10 km^3. Under the terms of the Agreement a Permanent Joint Technical Commission (PJTC) between the Sudan and Egypt was established to provide a forum for co-ordinating matters relating to Nile Waters. The PJTC has also maintained records of water use, as defined by the 1959 Agreement. Table 6 and Figure 6 show the releases from the Aswan High Dam since 1970 which constitute Egypt's use according to the 1959 Agreement. The equivalent data for the Sudan water use is given in Table 7 and Figure 7. Tables 8 and 9 give the breakdown of recent water use in Egypt and the Sudan respectively.

Table 6 Releases from the Aswan High Dam 1970-1987

Year	Release (km^3)	Evaporation and seepage losses	Year	Release (km^3)	Evaporation and seepage losses
1970	54.7	9.3	1980	56.7	12.8
1971	55.9	10.7	1981	58.0	12.9
1972	55.5	12.4	1982	59.1	12.5
1973	56.4	8.0	1983	57.6	8.4
1974	56.1	10.8	1984	57.3	9.7
1975	54.4	14.2	1985	55.8	6.4
1976	54.7	15.0	1986	55.5	5.7
1977	57.7	14.6	1987		
1978	61.9	13.9			
1979	58.0	13.1			

Source: Ministry of Public Works and Water Resources, Gov. of Egypt

Table 7 Total water use in the Sudan 1970 to 1987 (km^3)

1965/66	7.1	1976/77	11.3
1966/67	8.8	1977/78	12.3
1967/68	9.1	1978/79	11.0
1968/69	8.7	1979/80	12.0
1969/70	10.2	1980/81	11.5
1970/71	10.5	1981/82	12.6
1971/72	10.7	1982/83	12.6
1972/73	10.5	1983/84	13.1
1973/74	11.5	1984/85	12.7
1974/75	11.2	1985/86	12.6
1975/76	11.4	1986/87	

Source: PJTC

Table 8 Estimated present water use in Egypt (km^3)

INFLOW	
Aswan Release	55.5
OUTFLOW	
Edfina to sea	3.5
Canal tails to sea	0.1
Drainage to sea	13.2
Drainage to Fayoum	0.7
Sub-total	17.5
WATER USE	
Municipal and Industrial	2.4
Evapotranspiration (irrigation)	33.6
Evaporation from water surfaces	2.0
Sub-Total	38.0

Conclusions

The Nile has provided the basis of development in Egypt and Sudan since the start of agriculture some 7,000 years ago. Use of Nile water on any significant scale started in the nineteenth century but availability of stored water in the low river season rather than the total availability of water has until very recently been the main constraint to development.

The Sudan has yet to use its full allocation of Nile water but this is only because of economic dislocation in recent years which held back the development project which would take up the remaining water available in a relatively short time.

Egypt has been releasing its full allocation of Nile waters from Aswan in recent years as shown in Figure 6. This does not, however, give the full picture of the availability of water to Egypt since the enormous storage available in Lake Nasser masks the effect of individual years releases. Figure 15 in Chapter 2 shows the variation in the live storage behind the Aswan High Dam since 1968. This shows dramatically the declining storage in recent years as a result of releases being significantly in excess of inflows to the reservoir, less losses. It was only the extraordinarily high flood of 1988 that saved Egypt from potentially serious water shortages.

There is no doubt that the availability of Nile Waters is the main constraint to the expansion of irrigated agriculture in both the Sudan and Egypt and special efforts to conserve water and to use water more efficiently are now urgently required.

Table 9 Estimated present water use in the Sudan

(Data is for 1985/86 in km^3 as at Aswan)

WHITE NILE	
Kenana	0.82
Hagar Asalaya	0.18
Pumping Schemes	0.61
Sub-Total	1.61
BLUE NILE	
Gezira and Managil	5.40
Guneid	0.24
Es Suki	0.23
Abu Naama	0.06
NW Sennar	0.27
Pumping Schemes	0.75
Rahad	0.84
Sub-Total	7.79
MAIN NILE	
Pumping Schemes	0.90
Basin Irrigation	0.17
Sub-total	1.07
ATBARA	
Khasm el Girba	1.27
RESERVOIR EVAPORATION	
Sennar	0.30
Khasm el Girba	0.19
Roseires	0.37
Sub-total	0.86
TOTAL	12.60

Source: PJTC

References

Butzer K. W. (1976). *Early Hydraulic Civilisation in Egypt*. The University of Chicago Press. Chicago, and London.

Collins R. O. (1990). *The Waters of the Nile*. Oxford, Clarendon Press.

Council for Scientific and Technological Research. (1982). *Water Resources*

in Sudan. Khartoum, National Council for Research.
Coyne et Bellier, Sir Alexander Gibb and Partners, Hunting Technical Services Limited.
Sir M. MacDonald and Partners Limited (1978). *Blue Nile Waters Study.* Khartoum, The Republic of the Sudan.
Sir M. MacDonald and Partners Limited (1979). *Nile Waters Study.* Khartoum, The Republic of the Sudan.
Sir M. MacDonald & Partners Limited (1983). *Jonglei Canal Project. Final Report.* The Republic of Sudan.
Sir M. MacDonald & Partners Limited (1988). *Rehabilitation and Improvement of Water Delivery Systems in Old Lands. Final Report.* May. Arab Republic of Egypt, Ministry of Public Works and Water Resources.
Egyptian Government (1920). *Report of the Nile Projects Commission.* Cairo, Government Press.
Egyptian Government (1921). *Nile Control, Volume 1.* Cairo, Government Press.
Egyptian Ministry of Public Works (1981). *Water Master Plan* (WMP), 17 Vols. Cairo. UNDP/EYG/73/024
Goldman, C. R., McEvoy, J. III, Richardson P. J. (1973). *Environmental Quality and Water Development.* San Francisco, W.H.Freeman and Company.
Hurst, H. E. (1952). *The Nile.* London. Constable.
Hussein Sirry Pasha (1937). *Irrigation in Egypt.* Ministry of Public Works, Government Press, Bulaq, Egypt.
International Bank for Reconstruction and Development (1959). *Report of the Technical Mission of Sudan Irrigation.*
International Bank for Reconstruction and Development (1987). *World Tables (1987), Fourth Edition.* The World Bank and International Economics Department, Socio-Economic Data Division.
International Bank (1980). *Impact Evaluation Report-Sudan.* Roseires Irrigation Project.
Republic of Sudan (1957). *Sudan Irrigation.* Khartoum. The Ministry of Irrigation and Hydro-Electric Power.
Republic of Sudan (1955). *The Nile Waters Question; The Case for the Sudan and the Case for Egypt and the Sudan's Reply.* Khartoum, Ministry of Irrigation and Hydro-Electric Power.
Russell, J. C. (1966). *The Population of Medieval Egypt.* J. am. Res. Cent. Egypt, **5**.
Shahin, M. (1985). *Developments in Water Science, 21, Hydrology of the Nile Basin.* Amsterdam-Oxford-New York-Tokyo, Elsevier.
Waterbury, J. (1979). *Hydropolitics of the Nile Valley.* New York, Syracuse University Press.
Willcocks, W., Craig J. I., (1913). *Egyptian Irrigation, Volume I and II,* London and New York, E. F. N. Spon Ltd and Spon & Chamberlain.

4

East Africa's water requirements: the Equatorial Nile Project and the Nile Waters Agreement of 1929. A brief historical review.

P. HOWELL

From 1898 until the late 1940s attention on the waters of the Nile focused almost entirely on the irrigation needs of Egypt and the Sudan. The possible future requirements of Ethiopia, occupying almost the entire catchment of the Baro/Sobat, Blue Nile and Atbara and providing over 80 per cent of the average Nile flow at Aswan, were ignored, their liberty to exploit their own resources having been limited by treaty. Equally, the countries at the headwaters of the White Nile, with substantial annual rainfall, were not thought of as areas where irrigation might be needed, regardless of the unpredictability and uneven distribution of rainfall in some parts and a high level of aridity in others. From the early 1950s, however, Egypt's proposals for storage in the Great Lakes and Uganda's plans for hydro-electric power at Owen Falls set in motion studies to determine East African water resources in general and irrigation needs in particular, as well as the level of constant discharge through the turbines required to meet growing electricity demand and the planning of future power stations. Professional staff, both as advisers and as members of the Water Development Departments in each of the three East African territories, were appointed with qualifications to match those of officials in the downstream states and a number of studies and surveys was set in motion to examine the technical and economic factors involved. As background to Uganda's water needs and by extension those of Kenya and Tanzania, something of the history of Nile waters negotiations between 1948 and, in each case, independence may be of interest in connection with future discussions on the subject.

Hydrological calculations were based on records of Nile flows from 1905 (in some cases from 1898) to the mid-1950s. The massive rise in Lake Victoria levels and increased discharges at Owen Falls which occurred from 1961 onwards had not, of course, been envisaged and invalidate most of the calculations of the period, though the principles remain the same. Plans for possible further storage in Lake Victoria or more particularly a dam at the outlet of Lake Albert were overtaken in 1959 by the complete downstream control of Nile waters afforded by the Aswan High Dam. In the course of time, however, the need for storage in the White Nile system and the means of augmenting its flow will be increasingly apparent owing to growing water demand to meet population growth or, perhaps, owing to decreasing Blue Nile flows caused by climatic change (see chapters 6 and 12). A summary account of technical investigation in Uganda, Kenya and Tanganyika of 40 years ago may not only add to the history of Nile Waters in the context of the upstream states but provide clues to a number of factors that may affect future negotiations over their water needs.[1]

The Owen Falls Dam Agreement

Negotiations over the Owen Falls Dam between Egypt and the British Government (primarily on behalf of Uganda but also Kenya and Tanganyika because lakeside interests would be affected all round) began in 1948 and were completed in 1953 , the text of the agreement being summarised in a press communiqué of 10th January of that year as follows[2]:

> " The Royal Egyptian Government:
>
> (i) will bear that part of the cost of the dam at Owen Falls which is necessitated by the raising of the level of Lake Victoria and by the use of Lake Victoria for the storage of water;
>
> (ii) will bear the cost of compensation in respect of interests affected by the implementation of the scheme, or in the alternative, the cost of creating conditions which shall afford equivalent facilities and amenities to those at present enjoyed by the organisations and persons affected, and the cost of such works of reinstatement as are necessary to ensure the continuance of the conditions obtaining before the scheme comes into operation, such costs to be calculated according to arrangements to be agreed between our two Governments;
>
> (iii) will pay to the Ugandan Electricity Board the sum of £980,000 as compensation for the consequential loss of hydro-electric power, such payment to be made on the date when power for commercial sale is first generated at Owen Falls;
>
> (iv) agrees that for the purpose of the calculation of the compensation under the provisions of sub-paragraph (ii), all new flooding round Lake Victoria within the agreed range of three metres shall be deemed to be due to the implementation of the scheme."

The 'Draft Heads of Agreement', which had been reached earlier at technical level in 1948, were incorporated in the Exchange of Notes in so far as they referred to storage in Lake Victoria and, *inter alia*, specified that storage would be allowed with the effect of (periodically) raising the level of the water up to 1.3 metres above the previously recorded maximum (on the Entebbe gauge) within a range of 3 metres; the amount of compensation for damage to lakeside interests would be assessed by a 'compensation survey to be undertaken by a British firm of consultant engineers appointed by Egypt'; storage in Lake Victoria would not begin until five years after compensation had been paid; and during the 'temporary' or 'initial' period (i.e before storage began) the discharge at Owen Falls would be on the 'natural run of the river'. It was also tacitly agreed though not entered in either agreement that after the initial period the constant volume to be passed would be 505 cumecs - a figure closely linked to the compensation for loss of power mentioned above, since it was well below the mean recorded discharge to that date.

It is sometimes believed that Egyptian plans for storage were confined to Lake Victoria, but these Draft Heads of Agreement also included a scheme agreed at technical level for storage in Lake Albert - 'after building a dam at Mutir, storage will be allowed to the extent of raising the level of the lake to 14 metres on the Butiaba gauge, i.e 1 metre above the recorded maximum, and - to safeguard the Sudan from excessive flooding - up to a maximum level of 18.5 metres'. (Hurst *et al.*, 1946 and Collins, R.O., 1990)

It is important to note, since the issue may well come to the fore again at some future date, that Egyptian plans had at one time envisaged storage in Lake Albert up to 35 metres on the Butiaba gauge and a dam at Nimule much farther downstream on the Sudan border, rather than Mutir. This was rejected by Uganda as wholly unacceptable, largely because flooding round the perimeter of the lake would have affected the interests of too many people and partly because much of the Murchison National Park would be inundated. Below Mutir the floodplain of the Albert Nile extends for considerable distances and flooding would be very extensive unless protected by embankment. Moreover the height of the Murchison Falls, a potentially very important hydro-electric site, would have been reduced by some 20 metres.

Overall Nile Control: 'The Future Conservation of Nile Waters'. The Equatorial Nile Project

The storage element in Lake Victoria was in effect part of the plans for overall control of the Nile, as advocated by H.E. Hurst (Director-General of the Physical Department of the Ministry of Public Works in Egypt) and elaborated in Volume VII of *The Nile Basin* (1946); so far as the White Nile was concerned, what came to be known as the Equatorial Nile Project, Hurst *et al.*,(1946).[4]. This plan involved not only annual storage, with releases to suit seasonal irrigation requirements in Egypt, but storage to obtain a guaranteed and predictable flow every year, and the yearly discharge of the long-term mean, with variations to balance changes of flow in the Blue Nile. It also embodied the concept of 'Century Storage,' in other words storage capacity sufficient to maintain a given flow equal to the average of 100 years. (Collins, R.O. 1990, pp. 199-201)

The plan proposed at that time included not only storage in Lakes Victoria and Albert, but regulation in Lake Kioga and some kind of 'balancing reservoir' in the Nimule-Juba reach of the White Nile to control the torrents which come in flash flood form, mainly from the east, and contribute a substantial but unpredictable amount of water (at that date averaging 4 km^3) for a relatively short season in the year.

Below Juba the river enters the Sudd where huge amounts of water are lost by spillage into the swamps, the annual inflow being reduced by an average of 50 per cent.(Howell *et al.*, 1988, p.25) It was therefore essential that the plan should also include the Jonglei Canal, a diversion channel with 55 million cubic metres per day capacity and designed to carry water past the swamps with much reduced losses. Plans in other parts of the Nile valley envisaged control of the Blue Nile, including a reservoir for over-year storage in Lake Tana in Ethiopia and a dam and reservoir at Roseires, and, below, a series of dams along the main Nile for annual storage and release.

Temporary discharge at Owen Falls

By 1955 a number of considerations had changed Uganda's objectives. In the first place, the discharge figure agreed during the construction period of 600 cubic metres per second no longer applied since the dam had been completed and according to the agreement must revert to the 'run of the river'. However, it quickly became apparent that with the very rapid growth of electricity demand and for the purpose of forward planning the Uganda Electricity Board during the 'temporary period' (5 years after the completion of the compensation survey) needed a more reliable constant figure than the run of the river which might, according to records to that date, vary between 1220 and 300 cubic metres per second (cumecs). The immediate objective in this context was to get an agreement on a higher and more certain figure; the ultimate objective was to obtain agreement to a constant discharge as near as possible to the long-term mean. To this end the Uganda Electricity Board aimed at a 'temporary' constant discharge of 630 cumecs, a proposal which was at technical level accepted as favourable by the Egyptians,[4] subject to the proviso that this did not give Uganda a vested right and that the discharge would eventually revert to the agreed figure of 505 cumecs.[5] The Sudanese government also agreed in principle, subject to reservations about the possible downstream effects of flooding in the Sudd region and the assurance that compensation would be payable by Uganda if the effects were damaging.

Supplementary irrigation needs

In the second place, by the early 1950s the value of supplementary irrigation in increasing the yield of crops, even in areas of relatively high but unevenly distributed rainfall, had become apparent. The East African Royal Commission of 1953/55 had concluded that increasing population and pressure on the land gave point to the need for supplementary irrigation. In Uganda the Madhvani and Mehta sugar estates near Jinja had already demonstrated by experiment the beneficial effects of spray irrigation on the yields of sugar crops and a number of other potential areas in Uganda had been identified for this kind of development. Uganda had reached the stage of needing water for irrigation and had engaged the services of Sir Alexander Gibb and Partners to carry out a country-wide water resources survey to identify her requirements in irrigation water and the scope for swamp reclamation.

The Nile Waters Agreement of 1929

Water from the Nile was not, however, available to Uganda or the other East African territories 'save with the previous agreement of Egypt'. This constraint stemmed from the Nile Waters Agreement of 1929[6] and the Egyptians had been careful to stress the existence of this agreement in the opening gambit of the Owen Falls Agreement.[7]

Negotiations had started in 1920 by the appointment by Egypt of the 'Nile Projects Commission' which - in a report entitled *Nile Control* (1926) -

produced a forecast of the ultimate water requirements of Egypt and the immediate requirements of the Sudan. Protracted exchanges continued between Lord Lloyd, the British High Commissioner in Cairo, and Ziwa Pasha, Egyptian Minister of Foreign Affairs, and in an Exchange of Notes dated 7th May, 1929, agreement was reached whereby Egypt would receive annually 48 billion cubic metres (km^3) of water and the Sudan 4 km^3. This, as with the subsequent 1959 Agreement, assumed a mean discharge as measured at Aswan of 84 km^3. The difference of 32 km^3 ran to waste in the sea.

The report of the Nile Commission - with minor amendments - was regarded as an integral part and it is significant that in its conclusions it states that:

> "The Commission has been impressed by the fact that future development in Egypt may require the construction of works in the Sudan and neighbouring territories, such as Kenya, Uganda and Tanganyika, and it feels that Egypt should be able to count on receiving all assistance from the administrative authorities in the Sudan in respect of schemes undertaken in the Sudan, as well as from the British Government in any questions concerning the neighbouring territories."

The dominant interests of Egypt are particularly noticeable in this document. Moreover, the agreement of 1929 [Clause 4 (ii)] goes on to say:

> "Save with the previous agreement of the Egyptian Government, no irrigation or power works are to be constructed or taken on the River Nile or its branches, or on the lakes from which it flows so far as these are in the Sudan or in countries under British administration, which would, in such a manner as to entail any prejudice to the interests of Egypt, either reduce the quantity of water arriving in Egypt, or modify the date of its arrival, or lower its level".

Clause 4 (vi) continues, somewhat optimistically, to state that:

> "It is recognised that in the course of the operations here contemplated uncertainty may still arise from time to time either as to the correct interpretation of a question of principle or as to technical or administrative details. Every question of this will kind be approached in a spirit of mutual good faith".[8]

For the East African territories this was a notorious agreement made long before their development potential or future needs had been considered at all, an agreement which has subsequently been repudiated by independent Tanzania, later followed by Kenya, under the so-called 'Nyerere Doctrine' of 1960 on the grounds that they could not be bound by treaties made by the former colonial power. The 1929 Agreement showed little regard for the interests of the East African countries and was to prove a constraint and embarrassment to those negotiating for water on their behalf during the 1950s and early 1960s.

It could, of course, be argued that the construction of hydro-electric power works does not fall within the terms of the agreement so long as only the natural run of the river is allowed to pass through the turbines or sluices, since this neither reduces the flow nor affects the date of arrival in Egypt. More effective and economic power production can, however, be achieved if the river flow is controlled and the discharge held as near to the mean as possible. Since power sites in Uganda are numerous and the potential very great [9], this constraint was of enormous significance then, as it could be in the future.

In the case of irrigation even a small abstraction of water can be held to 'reduce the quantity of water arriving in Egypt', though clearly this must take into account the massive losses of water in transit, particularly in the Sudd.

Restrictive though the 1929 agreement was to the East African countries, it must not be forgotten that it applied to the Sudan too, a country that by the 1950s was waiting to expand its irrigation output but restricted to the 4 milliards (km^3) allocated to it. The Sudan was, during the 1950s, struggling to reach a new agreement with Egypt for a larger quota, as well as acceptance of plans to build a further dam at Roseires on the Blue Nile for the extension of irrigation in the Gezira.

The East African Nile Waters Co-ordinating Committee

In 1955 the East African Nile Waters Co-ordinating Committee was established to represent the interests of Uganda, Kenya and Tanganyika and to determine a common policy. This consisted of the Ministers who included responsibility for water development in their portfolios in each country, though the conduct of the work involved usually fell to their Permanent Secretaries or their representatives[10].

The first act of this committee was to define the different categories of water likely to be the subject for negotiation. These were:

(i) 'Natural or basic water': the natural flow of the river unaffected by control works.
(ii) 'New Water': conserved and made available by artificial means through the implementation of the Equatorial Nile Project or any other control works carried out in the interests of Egypt and the Sudan.
(iii) 'Additional water': water made available by swamp reclamation and similar works round the perimeter of the lakes which would not otherwise reach the Nile system[11].

In this context the East African Governments had declared their inherent and indisputable right to a share in the first, to a share in the benefits of control works constructed within their territories outlined in the second; and an absolute right to the third. H.M.G. had already reserved the right of East African territories to negotiate an agreed share of the water of the Nile in a Note to the Egyptian and Sudanese Governments dated 22nd November, 1955.

In the meantime, Uganda - which at that moment in time had not only a vital interest in the question of the constant discharge for hydro-electric power production but the need for a rather larger immediate quota of irrigation water than the other countries - had come to consider what bargaining factors were available, factors which she was, incidentally, prepared to use for the benefit of East Africa as a whole.

Storage in Lake Albert

At that stage, when the Hurst plan, including the Equatorial Nile Project, was still the aim of the downstream states, the question of storage in Lake Albert was crucial. Meanwhile, in 1954, the Jonglei Investigation Team, set up to investigate the impact of the project on the Sudanese inhabitants of the Nile Valley, had pointed out that excess releases in time of heavy discharge would jeopardise remedial alternative livelihood schemes to be implemented to meet grazing and other losses which the project would cause. Flood protection was also essential to Sudanese interests and their report had stipulated that a top level of not less than 25 metres (rather than 18.5) in Lake Albert on the Butiaba gauge was required for the purpose, a figure which, of course, was based on maximum discharge figures to that date and before the dramatic increases of the 1960s had occurred.

The main feature of the new controlled river regime under the Egyptian plan was to be 'Timely' and 'Untimely' releases to suit seasonal irrigation requirements in Egypt, roughly speaking the reverse of the natural seasonal rises and falls in river levels which in the Sudan produced pasture during the dry months of the year - an essential feature of the local pastoral economy. Remedial schemes financed from compensation to be paid by Egypt would have been necessary and it was essential to protect them from excessive flooding. (Sudan Government, JIT 1954, p.lxviii and 676; Howell, P.,*et al.*, 1988; Collins, R.O. 1990)

For Uganda, therefore, the first step was to examine the effects of flooding to this level round the perimeter of the lake. A full-scale consultants survey was regarded as unnecessary and too expensive at that juncture, especially as such a survey at the expense of the promoters of the project would be written into any agreement in the future.

Instead, Dr J.V. Sutcliffe (see contributors) came out in 1956 to head a party from the Uganda Water Development Department to undertake a somewhat cursory but wholly adequate survey needed to indicate the approximate areas of flooding and the effects on the local population. The object of what was essentially a reconnaissance survey was to indicate the nature and size of the problem to a degree of accuracy on which decisions in principle could be reached and was:

> "to collate what information was available at the start, to check such contours as had been put on the map by interpolation and extrapolation from air photographs and the 20 m contour on the ground by the Egyptian Irrigation Service in 1931, and to obtain further information about the soils and land utilisation as was possible in the very limited time available".

The results were presented in a document [again secret since the contents related to delicate international negotiation] which broadly speaking showed that though the effects were substantial in areas flooded, that, with the exception of Jonam, the land was generally poor and there was plenty of good land for resettlement. The survey revealed lands likely to be subject to flooding at levels ranging from 14 up to 25 metres on the Butiaba gauge, the latter showing 1090 km^2 of land periodically inundated, of which 300 km^2 was graded as 'very good' or 'good', and 90 km^2 currently used for cultivation. This would directly affect about 20,000 people.

Despite opposition from local administrators, who thought that the proposals when understood would meet with strong local opposition, it was reckoned that the economic effects would not be all that damaging for Uganda, particularly when weighed against the benefits in Nile water sought[12].

Full compensation for land not only utilised but of potential value, the cost of resettlement, including water for irrigation if needed, would be stipulated as well as the costs of material assets such as port installations. Compensation for loss of head and hence power at Murchison Falls would also be claimed as well as the cost of clearing inundated forest land for fisheries. Compensation for losses could be used to develop the area, hitherto retarded for lack of funds, and communications might be improved because an extension of the railway from Soroti to the dam site might well be necessary for construction, an extension of which could subsequently be retained by the E.A.Railways and Harbours with consequent benefits to the West Nile District where lack of communications was a constraint on economic progress.[13]

This report was presented to the Executive Council of Uganda on 18th May, 1957. The Council agreed that taking into consideration all these factors and the relative bargaining position acceptance of a higher maximum level would afford, negotiators should be authorised, if and when need be, to discuss with downstream states storage in Lake Albert up to a maximum of 25 metres on the Butiaba gauge, subject to the stipulations outlined above[14]. In fact, as will be seen later, no such negotiations ever took place because the question of storage in Lake Albert was shelved when the Egyptians decided in 1959, with the agreement of the Sudan, to construct the Aswan High Dam.

Uganda's irrigation requirements

The water resources survey as applied to irrigation and swamp reclamation had been completed by Sir Alexander Gibb & Partners in September, 1955. The conclusions were that there were 110,230 acres suitable for irrigation requiring 0.535 km^3 of water per annum. There were two main crops suitable for irrigation - sugar cane and paddy rice, with the possibility of high grade cotton in selected parts of the country. Since there was then no large-scale systematic irrigation in Uganda, much field experimental work would be needed before development took place. Pilot schemes in all three crops were recommended. The report also identified 215,000 acres of predominantly clay swamps, many of which when reclaimed would release water, the 'additional water' referred to earlier, into the Nile system, though no attempt could be made to quantify this (Uganda Government, 1955). Similar surveys were completed in Kenya and Tanganyika by May 1957, when all figures were collated and printed in *East Africa's Case*, a secret document circulated among the technical departments concerned and to the Foreign and Colonial Offices in London[15].

Estimates (from 1957) of Water requirements for the East African Territories for development in the next twenty-five years

The same document declared the East African Governments' 'inherent and indisputable right to a share of the 'natural and basic' waters of the Nile, claimed a share in the benefits of 'new water' made available by control works constructed in their territories in the interests of Egypt and the Sudan, and their absolute right to 'additional water' made available by swamp reclamation.

The document continues: 'the East African Governments consider that the agreement of the downstream Governments to the following estimated quotas for the three East African territories should be obtained subject to the Notes attached.'

The document stresses: (a) that these estimates should be regarded as an absolute minimum acceptable for the next 25 years and (b) that the East African territories should have the right to further water should flooding displace persons as the result of storage in Lake Victoria and require resettlement and alternative livelihood involving irrigation.

Conditions and reservations to the claim covered any 'additional water' made available by works implemented at the expense of the East African Governments which was to be regarded as outside the Nile Waters Agreement; the right of periodical review at stated periods; the right to negotiate for an increased quota of water at any time; and the right of 'virement', i.e the right to use water for other schemes which might prove more profitable or desirable within the agreed quota. There were also stipulations, largely for political reasons, against the right of downstream users to place within East African territory permanent staff to monitor the administration of East Africa's quota[16].

Provisional Estimate of East Africa's minimum irrigation requirements for the next 25 years in milliards of cubic metres (km^3) per annum

Territory	Estimate by Sir A. Gibb & Pnrs	Other Irrigation Schemes	Contingencies 10%	Total
Kenya	0.297	0.080	0.038	0.415
Tanganyika	0.478	0.110*	0.059	0.647
Uganda	0.535	0.049+	0.058	0.642
Totals	1.310	0.239	0.155	1.704

Provisional Estimates of East Africa's Minimum Industrial and Mining Requirements For The Next 25 Years In Millards of Cubic Metres (km^3) Per Annum

	Industrial		Mining
Kenya		0.007	
Tanganyika	0.001		0.038
Uganda		0.020	
Total		0.066	

(*= 13,000 acres sugar -+ =10,000 acres tea)

Notes accompanying East Africa's Case:

" i) No account is taken of the use of water for domestic and stock consumption since it has always been assumed that the use of such water is outside the terms of the Nile Water Agreement.
ii) Estimates for water required for the production of hydro-electric power are not given because water so used is 100 per cent returnable and does not represent an extraction or cause a reduction in the amount of water available. Alterations in seasonal flow for power schemes on rivers feeding Lake Victoria would be balanced by other factors in the lake itself and the effect at Owen Falls would be insignificant. The question of the regulation of discharge at Owen Falls more directly concerns Uganda, the objective of the Uganda Government being to obtain a constant discharge at Owen Falls as near to the mean as possible, less the reduction caused by extractions to meet irrigation, industrial and mining requirements as given above.
(iii) Water for mining and industrial purposes has been given separately in the estimates. It may be held that water for these purposes does not come within the terms of the Nile Water Agreement.
(iv) It should be noted that all irrigation data used in the calculation of East Africa's requirements must be checked by experiment as recommended by Sir Alexander Gibb and Partners, and all estimated figures may be subject to revision at some future date. An additional element to cover contingencies has therefore been included.
(v) Of the total requirements (including industrial and mining) of 1.770 milliards very approximately 0.385 would be extracted below the Owen Falls leaving a total of 1.385 milliards to be extracted from Lake Victoria and its catchment. This represents a theoretical reduction of 44 cumecs in the discharge at the Owen Falls. The objective of the East African Governments is to obtain agreement to a constant discharge of the mean less any reduction in the water available caused by extractions for the irrigation, industrial and mining needs recorded above, but whatever the outcome in reaching agreement to a system of regulation giving a minimum or constant discharge of more than 505 cumecs, it is understood that any claim by the Egyptian or Sudanese Governments to alter the seasonal variations in such a way as to reduce the discharge below 505 would be resisted by the East African Governments" [17].

In the same document the East African Governments demanded full representation on any international negotiating body or conference at which the control of the Nile would be under discussion; that the British Colonial

Office should keep them informed of any meetings in which their interests came under discussion, and stipulated (a mutual arrangement) that the secretary of the Co-ordinating Committee should keep all three governments informed of any meetings, technical or political, within the the Nile Basin, at which the interests of any of the three territories would be discussed. To some extent this was an expression of suspicion of Uganda's dominant role in these matters and her prime interest in discharges through the turbines which could place limitations on the irrigation abstractions of the other two countries.

Meanwhile the political situation in East Africa was changing, and because legislation governing the compensation survey and the appropriation of lake-shore land would by then be subject to the concurrence of the unofficial members of the legislature who had raised the issue,[18] the document further stipulated that the compensation survey, agreed under the Owen Falls Agreement of 1953, should not be allowed to start unless Egypt and the Sudan recognised beforehand the right of the East African territories to a share in Nile waters.

The Aswan High Dam and Report on the Nile Valley Plan

The Aswan High Dam (the *Sadd al 'Aali* Project) was by then receiving the serious attention of Egyptian engineers. Moreover, it had also become something of a symbol of national aspirations under the Nasser regime. There seemed, even then, some prospect that it would supercede the Equatorial Nile Project since full seasonal control would be achieved at Aswan within Egyptian territory. If so, storage in Lake Albert and seasonal releases to meet irrigation requirements downstream would no longer be necessary at any rate in the foreseeable future, incidentally nullifying one of Uganda's most powerful bargaining factors.

The Sudanese, however, did not favour the High Dam and instead were by 1958 pressing for the 'Nile Valley Plan' put forward by H.A.W. Morrice, at that time Irrigation Adviser and W.N. Allan, Consultant to the Republic of the Sudan. This was in effect a refinement of schemes earlier put forward by Hurst and, with the use of computers, elaborated a very detailed plan for control throughout the Nile Valley. (Morrice and Allan, 1958)

East African Irrigation abstractions versus constant power discharges: a conflict of interests

It is worth mentioning in parenthesis that Kenya and Tanganyika had by 1959 become aware of the possible conflict of interest between themselves and Uganda over the constant discharge at Owen Falls. Storage to maximise the mean discharge would have an effect on future levels in Lake Victoria and therefore lakeside installations and in some areas cultivated land; irrigation abstractions would, at least in theory, reduce the discharge through the turbines at Owen Falls and any future power works below.

At a meeting of the East African Nile Waters Co-ordinating Committee in Nairobi in 1959 it is recorded that 'irrigation water is irreplaceable but there are additional hydro-power station sites between Lake Victoria and Lake Albert. Therefore if conflict arises between irrigation and power requirements, irrigation must take precedence'.

This rather naive and abrupt decision ignored the economics of power production which are, as we have seen, much affected by the amount of the discharge available. The issue is worth mentioning here because in the course of time it may have to be resolved between the now independent East African countries, probably including Rwanda and Burundi, although their primary interest is in hydro-electric power[19]. The Kagera River is after all the largest single source of inflow to Lake Victoria. Much depends on future levels in the lake.

A more tempered compromise was subsequently reached by the amendment of paragraph 47 of '*East Africa's Case*, which, *inter alia.,* included the words '... the East African Governments appreciate the importance attached by Uganda to a high rate of constant discharge for the generation of power. The Ugandan Government agrees to keep the governments of the other two territories currently informed of the rate of discharge on which the Ugandan Electricity Board intends to base its future development, and agrees that any proposed increase in discharge should be subject to consultation between them before the necessary negotiations are initiated with downstream users. It is likewise recognised that any constant discharge shall be open to negotiation between the East African territories should further water be required for irrigation or other purposes'[20].

Report on the 'Nile Valley Plan'

Meanwhile the Sudan, also handicapped by the Nile Waters Agreement of 1929, was continuing, at times in very acrimonious circumstances, to negotiate with Egypt for a more generous share. The history of the evolution of the Nile Valley Plan with the use of electronic computer techniques (which had become available to Nile engineers in 1953/54) has been recorded elsewhere (Morrice,and Allan, 1958; Collins, R.O. 1990, p.258 *et seq*). In the opinion of experts 'no more economical and effective approach to the problem of planning the Nile Valley as a whole could have been devised'.

It incorporated all previously proposed control works, including those which had only been tentatively in mind, e.g. a revised Jonglei diversion canal - two parallel canals each of 45 millions m^3 per day capacity[21], the Baro reservoir for the Sobat catchment, dams at Lake Tana and Roseires and an upper Blue Nile reservoir, a dam at Khashm el Girba on the Atbara, dams along the main Nile and a modified High Dam at Aswan. Like Hurst's 'Future Conservation of the Nile', it aimed, among other things, to store water in areas of low evaporation, unlike the Aswan High Dam where evaporation is highest.

Significantly for East Africa, controls and storage in the Great Lakes were included, with levels in Lake Albert at a maximum of 25 metres on the Butiaba gauge and East Africa's immediate irrigation requirements (given as 1.8 km^3) taken into account.

The hydrological assumptions, of course, were based on the flows of 1905 to 1952 and were, as we have seen, overtaken by the greatly increased rainfall and massive rises in level in Lake Victoria from 1961 onwards which had not been envisaged [21] (Kite, G.W. 1982), but Morrice's summary of the predicted effects in Uganda is worth quoting verbatim. As his note says 'the electronic computer can easily repeat the calculations with the new figures'.

"The Effect on Uganda of the Nile Valley Plan"

Introduction

1. The Nile Valley Plan has been drawn up by me working in collaboration with Mr. W.N. Allan. The two volumes describing it in detail have been printed in London, and copies are now in the hands of the Minister of Irrigation and Hydro-Electric Power at Khartoum. I expect that he will shortly authorise the distribution of this Report officially, but meanwhile I have handed a set of two volumes to Mr. P.P. Howell, Permanent Secretary of the Ministry of Corporations and Regional Communications in Uganda.

2. The object of this note is to explain briefly how the Nile Valley Plan will affect Uganda directly. Those who require further details can obtain them from the report. The effects can be conveniently grouped under the following headings:-

- (a) Levels in Lake Victoria.
- (b) The discharge in the Victoria Nile between Lakes Victoria and Kioga.
- (c) Levels in Lake Kioga.
- (d) The discharge in the Victoria Nile between Lakes Kioga and Albert.
- (e) Levels in Lake Albert.
- (f) The discharge in the Albert Nile between Lake Albert and Nimule.

These will now be discussed in succession.

Lake Victoria

3. The Nile Valley Plan is based on an analysis of the 48 years, 1905-1952. The net inflow to Lake Victoria has been found by adding the discharge from the lake to the increase in its contents. For the discharge at Ripon Falls the figures published in *The Nile Basin* have been reduced by 4, to allow for a possible error in the coefficient used to compute the official figures. The true error may be less than this and is certainly not greater; accurate information will not be available until the Water Development Department has carried out certain experiments. Meanwhile the figures used for the Nile Valley Plan are such that any error will be on the safe side because the volume of water available will be underestimated.

4. The Nile Valley Plan assumes that the contents of the Lake will be the same at the beginning of the period under analysis as at the end. It allows for irrigation water to be abstracted from the Lake at the constant rate of 0.12 milliards per month, which is equivalent to 46 cumecs. The average net inflow to the Lake calculated as in the previous paragraph was 641 cumecs. Therefore after irrigation has been fully developed, there remains an average discharge from the Lake of 595 cumecs.

5. As explained in the next section of this Report, the discharge from the Lake will not be allowed to fall below 579 cumecs. In these circumstances Lake levels will vary by 1.80 metres, which corresponds to a variation in Lake content of 120 milliards. This variation is independent of the starting level, and is very much less than the 3-metres for which provision has been made. If we make the reasonable assumption that the Lake oscillates symmetrically about the midpoint of the 3-metre range, the levels will vary between 10.60 and 12.20 metres on the Entebbe gauge.

From Lake Victoria to Lake Kioga

6. The Nile Valley Plan provides for a minimum discharge of 579 cumecs at the Owen Falls Dam. This is 15 per cent greater than the discharge of 505 cumecs which is the guaranteed minimum in the present agreement. It would not be physically possible to raise this figure to more than 595 cumecs, and it has been found that any attempt to do so would have undesirable effects on Lake Kioga.

7. Until irrigation starts upstream from Owen Falls, the guaranteed minimum discharge can be 625 cumecs. It will fall steadily to 579 cumecs as irrigation develops. The experiments mentioned in paragraph 3 of this Note may show that more water is available than the Nile Valley Plan at present allows for. If this happens, the electronic computer can easily repeat the calculations with the new figures, and the result will be a slight increase in the guaranteed minimum discharge. Therefore the Nile Valley plan guarantees 579 cumecs or better at the Owen Falls Dam.

8. For much of the time the discharge at Owen Falls will be steady at 579 cumecs; but it will be greater sometimes and very occasionally it will approach 1200 cumecs, which is the designed capacity of the Dam.

Lake Kioga

9. The content of Lake Kioga is to be kept as nearly as possible constant at 7.50 milliards (km^3), which corresponds to 11.77 on the Masindi gauge. For more than three-quarters of the time the gauge reading will lie between 11.00 and 12.50 metres. Its maximum value will be slightly less than the 13.50 which occurred under natural conditions in 1917. In order to maintain the required minimum discharge downstream (as described in the following paragraph) Lake Kioga will not be allowed to fall below 10.74 on the Masindi gauge.

From Lake Kioga to Lake Albert

10. Eventually this reach will be developed for hydro-electric power, and it is therefore desirable that the minimum discharge shall be as high as possible. Under the Nile Valley Plan provision has been made for a minimum of 563 cumecs. After allowance has been made for irrigation from Lake Victoria (see paragraph 4 above) the mean discharge is 599 cumecs. The planned minimum is thus 94 per cent of the mean, and it could not be raised further without causing difficulties in Lake Kioga.

11. If the Kafu Diversion Scheme was ever carried out, it would probably be desirable to lower the level of Lake Kioga. This would reduce evaporation losses, and so increase the available discharge downstream. In the absence of detailed information it is impossible to estimate even roughly what this increase would be. It might, however, be large and, for this reason an accurate survey of the Lake and the surrounding swamps is urgently required.

12. During the interim period, when irrigation is being developed from Lake Victoria, the discharge available below Lake Kioga would fall gradually from 609 to 563 cumecs. The maximum planned discharge from Lake Kioga is 1290 cumecs. This is slightly in excess of the designed capacity of the Owen Falls Dam.

Lake Albert

The Nile Valley Plan calls for a storage capacity of 100 milliards in Lake Albert. This corresponds to approximately 25 metres on the Butiaba gauge. The reason why such a large reservoir is required is that the very high guaranteed discharges upstream make it impossible to use effectively the full storage capacity of Lake

Victoria. It is reasonable that the downstream powers should derive some benefit from the storage capacity of the equatorial lakes, and it is also reasonable that Uganda should be able to make good use of the great hydro-electric potential of the Victoria Nile. These two aims can be combined by providing adequate storage capacity in Lake Albert.

14. Under the Nile Valley Plan the Lake Albert reservoir will be full only very seldom, but levels of 22 metres or so at Butiaba will not be uncommon. The reservoir will occasionally fall below 10 metres on the Butiaba gauge. Variations in level will not be seasonal. Allowance has been made for irrigation at the rate of 0.03 milliards per month.

From Lake Albert to Nimule

15. It is proposed to operate this reach in such a way that the southern Sudan is protected against flooding. For this reason the discharge at the Mutir Dam will never be allowed to exceed 1460 cumecs. If the torrents between Mutir and Mongalla are discharging heavily, it may be necessary to reduce the discharge at Mutir to zero, unless there is a balancing reservoir at Nimule. For this reason a balancing reservoir there is desirable, if only to maintain navigation at all times.

16. It would be possible to generate a certain amount of hydro-electric power at Mutir, but discharges and heads would be highly variable and the site is remote. Moreover, without a balancing reservoir at Nimule there would be no firm power, and even with such a reservoir the firm power would be negligible when the reservoir was low.

(H.A. Morrice, Hydrological Consultant, Entebbe, 18 August 1958)".[23]

The Morrice/Allan Nile Valley Plan was on the whole welcomed by the East African Nile Waters Co-ordinating Committee, particularly as it envisaged a lesser variation of level in Lake Victoria (1.8 metres instead of 3 metres) and allowed for a constant discharge of 579 cumecs at Owen Falls after deduction of their estimated irrigation requirements (46 cumecs) for the next 25 years.

Ultimate irrigation requirements in East Africa

Hitherto the East African territories had intended to submit the estimates prepared by Sir Alexander Gibb and Partners for a period of 25 years followed by review, with also scope for review at any time if circumstances revealed the need for more. This had been favoured by East Africa's former advisers, C.G.Hawes and Sir Douglas Harris, on the grounds that downstream users would only accept claims for irrigation water backed by economically and technically viable plans, and that these might have to pass the test of international arbitration if the need arose.

Morrice, who was appointed Consultant to the East African Governments in 1958, thought otherwise, arguing that Egypt and the Sudan would welcome estimates of longer term requirements. The text of Minute 12/59 - *Reassessment of Future Water Requirements* reads as follows: "Mr Morrice said, with regard to claims for water, that he was not convinced that the best way of safeguarding the future was by periodical review of the claim with a view to increased allocations. He also said that he felt it a mistake to support requirements for irrigation water on economic grounds and to reject other possible schemes as being uneconomic. In his view it was quite impossible to say that any particular irrigation scheme would be uneconomic at some future date because conditions constantly changed".

Morrice's recommendations were that the longer-term claim should be based upon the 'potentially irrigable land in the Drainage Area, irrespective of whether some of the projects might or might not in the future prove viable'. Alternatively, the demand might be based on the amount of water available for irrigation within the Drainage Area since water might be the limiting factor[24]. Although much information was available in the Alexander Gibb reports, Morrices's approach to ultimate requirements depended very much on further scientific data on the areas where land and water might be available for irrigation. One aspect was hydrology and was certainly a requirement which inspired the East African Committee subsequently to initiate, after appropriate consultations with Egypt and the Sudan, an approach to the United Nations for support for a hydro-meteorological survey of this region. This, under UNDP funding and World Meteorological Organisation management, took off in 1966 and was one of the most significant international collaborative projects between riparian states in the history of the Nile in its upper waters actually to take place (Collins, R.O., 1990, pp. 287-294).

Meanwhile the argument between the Sudan and Egypt continued unabated on what in effect was to be for them the revision of the Nile Waters Agreement of 1929. Negotiations between these countries are not part of this chapter, but it is worth noting that the Sudan had denied being party to the Nile Waters Agreement of 1929 which had been concluded long before her independence and asserted that she was not bound by it. This was to anticipate a similar declaration by Tanganyika on 4th July 1962 already noted; and it is also interesting to note that Egypt had herself at one stage denounced the 1929 Agreement as a 'wicked imperialist plot'[25]. The Sudan by 1958, by completing the enlargement of the Gezira Main Canal, had effectively ignored the 1929 Agreement, which the Egyptians had conveniently resuscitated to frustrate Sudanese plans to expand the Gezira Scheme and build the Roseires Dam.

As H.A.W.Morrice, having noted that neither Ethiopia nor the Belgian Congo (which then included Rwanda and Burundi) were party to the 1929 Agreement, commented in a letter dated 26th June 1959, 'the present position is therefore that the Nile Waters Agreement is being taken seriously by the East African Territories and no one else......my recommendation is that the East African Territories should urge HMG to give the Nile Waters Agreement the *coup de grace* on the grounds that it is a dead letter outside East Africa' [26].

The British Government and the East African territories had frequently considered simply ignoring the Agreement by helping themselves to the water required rather than formally challenging the Agreement's validity. [27] The trouble was that they were not yet ready to abstract water in any quantity; irrigation schemes awaited experimentation and trial. Moreover, while Uganda was prepared to offer storage concessions in the Great Lakes in return for recognition of East Africa's irrigation requirements and a satisfactory net discharge in the Victoria Nile, she was also prepared to say that unless these were recognised by the downstream states within a specified period of time, the offer would be withdrawn for all time.

The situation was, in fact, a curious political triangle since the British Foreign Office had been willing to consider abrogation of the Agreement so far as Egypt was concerned but reluctant to run the risk of offending the Sudanese; yet the Sudanese were themselves trying to abolish or evade its terms. It must also be remembered that much of the discussions between the East African territories described above took place under the shadow of the Suez crisis. Diplomatic relations between Britain and Egypt had been severed and were not renewed until 1961. Notes could be passed through a third power, but formal negotiations could not begin. Despite this, technical discussions between members of the East African Nile Waters Co-ordinating Committee and their Egyptian and Sudanese counterparts had continued on an amicable basis throughout the period in Entebbe, Nairobi, and Khartoum.

Prominent engineering staff from both the downstream states, including, for example, Dr Mohammed Amin Bey, a former Director-General of the Physical Department of the Ministry of Public Works in Cairo and later Technical Adviser to the Egyptian Government, visited East Africa on more than one occasion. Sayed Zaghirun al Zain al Zaghirun of the Sudan was also a frequent attendant at these meetings.

The Nile Waters Agreement of 1959

Neither Britain nor the East African Territories nor Ethiopia nor the Belgian Congo was consulted about the final agreement between Egypt and the Sudan, which was announced in Cairo and Khartoum on 4th November, 1959. As recorded in other chapters this allocated 55.5 km^3 to Egypt and 18.5 km^3 to the Sudan with 10 km^3 regarded as the average loss by evaporation in the reservoir behind the Aswan High Dam, which the Sudanese had conceded in return for their increased allocation and Egypt's agreement to the Roseires dam.

The reaction in East Africa was that the downstream states had simply divided the then recognised average flow of the Nile of 84 km^3 between themselves, leaving nothing for East Africa or other upstream countries.[28] The Equatorial Nile Project as modified by the Nile Valley Plan - which had always been resisted by Egypt - was dead. Mention is made, it is true, of 'projects for the utilisation of lost waters in the Nile Basin'. This refers in particular to projects for conservation in the Flood Region of the southern Sudan, (see Chapter 12) meaning the diversion canal or canals through the Sudd and adjacent swamplands. The Jonglei Canal stage II, however, would involve storage in Lake Albert, so the Agreement envisages an ultimate return to savings of water in the Great Lakes. But it also specifically refers to the sharing of the net gains, yet makes no mention of the rights of countries in which such works might be carried out. Recognition of the rights of upper riparians is only referred to obliquely by underlining the need to agree a 'unified view' on negotiations with them and if, agreement is reached on the 'acceptance of allotting an amount' to one or other of the upstream states, that 'amount shall be deducted from the shares of the two Republics in equal parts, as calculated at Aswan'. Given shortages due to natural causes and the needs of growing populations in Egypt and the Sudan, and given peace in the southern Sudan, a return to plans for storage in East Africa is almost bound to occur, but Egypt and the Sudan will be wise to negotiate in a more conciliatory way.

In the period preceding the 1959 Agreement, Nile Waters were very much on the international agenda. Vice-President Nixon had visited Uganda in 1958, had looked at the Owen Falls Dam and received a written exposition of the problems of the Nile from the Uganda Government.The World Bank had been interested in the Aswan High Dam until the Russians committed themselves to its implementation, but came forward with finance for the Roseires Dam in the Sudan. The President, Eugene Black, visited the area, including Uganda, and at one moment there had been hopes of an international conference at which all problems of the Nile Basin were to be resolved. This interest waned after the completion of the 1929 Agreement.

Diplomatic relations were resumed in 1961 and the British Mission returned to Cairo in April. For the first time for five years it was possible to raise the issue of East Africa's water requirements at political level direct. Accordingly the Hon. Bruce Mackenzie, Minister of Agriculture and Irrigation representing Kenya and the author representing Uganda flew to Cairo on 7th April, shortly after the return of the British ambassador, Sir Harold Beeley. Egyptian ministers, led by Zakariya Mohyeddin, Central Minister for the Interior and Chairman of the Committee on the High Dam, were cordial and appeared to recognise East African aspirations to a share in Nile waters for their own irrigation purposes, and had taken account of the Note outlining their case presented through the Swiss Embassy in the autumn of 1959.

The visit was something of a diplomatic success, but mainly because it provided during this very sensitive post-Suez period the opportunity to discuss matters unconnected with that unfortunate affair. A Press Release was issued in the Egyptian newspaper *Al Ahram* whose correspondent added a postscript that " the discussion was friendly and neither then - nor as far as I know during the entertainments that followed - did the UAR Ministers show any suspicion or raise irrelevant issues"![29]

No decisions were reached; instead the Egyptian ministers proposed a meeting between their very formidable Permanent Joint Technical Commission and an invitation to East African representatives to join them in Khartoum was issued and accepted.

Encouraging though the Cairo meetings had been, the East African delegates soon found they were dealing, to their disadvantage, with a different type of negotiator, accompanied by repeated protestations that the talks would be "technical and informal". They had nonetheless taken on a weightier image than earlier technical discussions which is evident from the nature of Egyptian and Sudanese representation. This included the Chairman of the Permanent Joint Technical Commission, Mohammed Khalil Ibrahim, an Egyptian, and Mahmoud Mohammed Gadein, the leader of the Sudanese delegation. Virtually all members of the Permanent Joint Technical Commission were present. The East African Governments fielded a strong delegation from all three territories. [30]

Two issues were under discussion: first and foremost the contents of the British Government Note of 11th August 1959, outlining East Africa's case for a share in Nile waters and their requirements for the next twenty-five years, subject to provisos about possible revision, as described earlier; and secondly East Africa's demand for interim recognition of very small amounts (0.16 km^3 for Kenya and 0.048 km^3 for Uganda) for irrigation projects of immediate urgency . At the request of Uganda the question of the temporary constant discharge at Owen Falls was also included.[31]

While the wording is slightly at variance with the original British Note, it is perhaps significant to quote verbatim the downstream users opening statement of their understanding of the East African case and their reactions to it:

"In a Note dated 11th August 1959, the British Government stated that the E.A governments consider that they have the following rights:-

1. Indisputable right to a share in the natural flow of the river unaffected by any control works.
2. The right to a share in the benefits of the new waters made available by using the E.A. Territories natural assets for storage purposes.
3 Absolute right in additional waters made available by carrying out works in E.A. at their own expense to make use of waters which owing to evaporation and transpiration would not otherwise reach the Nile.

Having, in vague terms appeared to admit that East Africa was at least theoretically entitled to a share in the first category, they immediately proceeded to refute it by saying that this water was subject to the established rights of other countries (themselves) which "had already made use of these waters" and that according to this principle the E.A. Territories had a right to a share only in the "excess waters, if any (sic), of the Upper Bahr el Gebel natural supply arriving from the Great Lakes"; and adding that "the Bahr el Gebel supply is almost (all?) being used by the Republic of the Sudan and the UAR and there is therefore no excess basic water available."

As for the second category, that, they said, related to the Equatorial Nile Project (by then incidentally abandoned in favour of the High Aswan Dam), followed by an exposition of the local uses of flood water in the Sudd by the inhabitants of that region (which they themselves had only recently intended substantially to reduce)[32], and to navigation requirements, adding that "these provisions constitute an established water right and that future control of the waters entering the Sudd region by means of storage in the Equatorial lakes will not in any way increase the water quantities entering the Sudan but will simply help to improve the means of using the present natural supply on which the two Republics have already based their programme of development." It is worth mentioning in parenthesis that Uganda had not even agreed to storage works in the Great Lakes other than limited storage in Lake Victoria under the Owen Falls agreement; this supposition appeared premature, and it was simply not true that there would be no benefit in water as the result of such works; why else was it referred to as *new* water?

They then quoted "Clauses III, IV and V" of the Nile Waters Agreement of 1929 which, they said, gave them the right to construct such works in British Government Territories, alleging that only the agreement beforehand of the local authorities on the measures to be taken for safeguarding their interests was required. The East African delegation refused to discuss these contentions on the grounds that this raised legal issues they were not authorised to consider.

So far as "Additional" water was concerned the Egyptian stance becomes less clear, but despite stressing that the U.A.R. also had rights under the 1929 Agreement to carry out works locally of a similar nature in their own interests, they appeared to recognise that water saved by minor swamp reclamation works could be claimed by East Africa, the only source available to them, always assuming that if the water was used for irrigation and could in any way affect the flow of the Nile, they would be consulted and, by inference, would have powers of veto. The fact that East Africa only intended small and scattered swamp reclamation schemes for agricultural purposes and that savings in water were not only doubtful but coincidental appears to have escaped them. All these points were also set out in a document circulated at the meeting.

The East African delegation was aghast. It was apparently admitted that they were in theory entitled to a share in the waters which originated and passed through their territories, but only to water in excess of the requirements of the downstream states and there was none. Moreover, there would be no excess water available to them from storage works constructed in their own territory. Accordingly they enquired about the status of the downstream users' statement and the document which accompanied it and whether these represented a formal reply to the official Note on East Africa's requirements conveyed through the British Government in 1959. The downstream delegation replied that they were not official, but "rather a basis for discussion" and the comments of the East Africans would be welcome.

The East African delegation replied that it had made reservations concerning its water requirements as early as 1956 and had indicated their needs again in 1959. "The document presented to it and the views expressed in the Egyptian and Sudanese opening statement, if accepted as they stood, closed the door to further discussion and were unacceptable."

The downstream delegation then adopted a marginally more co-operative attitude and requested that discussion on each point should proceed. In the context of 'basic' water (the natural flow) the East African representatives reiterated that they disputed "the right of the two downstream states to the total flow of the Nile either by current usage or planned commitment. East Africa's right to a share in this water had been reserved before the 1959 Nile Waters Agreement, on which in any case they had not been consulted and to which they were not a party. These reservations would continue until firm agreement as to their immediate and ultimate quota had been reached."

Both for 'basic' water (the natural flow) and 'new' water made available by storage in the equatorial lakes, East Africa could not accept that there existed established rights by downstream users to the total water entering the Sudd region in the Sudan or to the total outflow from East Africa augmented by storage. Nor could they accept the contention that Clauses III, IV and V [actually clauses 4 (iii), (iv) and (v)] gave Egypt "a right to construct such works (or any type of irrigation works) in British Government Territories". Clause 4(ii) gave Egypt the right to veto irrigation works by upstream states which in any way might affect downstream interests; no clause gave them the positive right to construct anything without prior agreement. The most that the 1929 Agreement inferred was that they would be granted facilities for hydrological study and Britain would "use its good offices" to facilitate such works initiated by Egypt and agreed by all parties in regions under British influence.[33] They added - significantly as things turned out - that the conditions of the Agreement were unlikely to be acceptable in the case of countries which had achieved independence or were about to do so.[34]

Given this situation, the most that could be hoped for was agreement to hold further meetings and discussions at which a healthy compromise might conceivably be reached. This was, in fact, agreed and was possibly the only achievement of this meeting.

The East Africa delegation went on to say that in the meantime there were two issues on which early agreement would be welcome - agreement to the use of the very modest amounts required for urgent works in Kenya and Uganda already stated. Acceptance of these small amounts by the downstream states would do much to influence public opinion in East Africa in their favour. The PJTC, apparently blind to the benefits of good public relations acceptance of the abstraction of even such limited quotas would generate, again employed delaying tactics by saying that they were prepared "to approach their Governments in this connection as soon as the quantities of these requirements and the periods of their full consumption are available".

The meeting ended with a set of agreed conclusions formulated by the PJTC:

"1. Technical Informal Talks and Discussions can be left open and initiated any time in future, and should take place at frequent intervals.
2. The urgent requirements of the East African Territories can be looked into as they arise and be put forward to downstream states.
3. Ultimate requirements and periods of utilisation will be left for further discussion and scrutiny.
4. Subject to further technical scrutiny it is agreed that the question of temporary regulation at Owen Falls will be mutually discussed with a view to reaching a satisfactory conclusion at the next meeting.
5. It is understood that the above agreed conclusions are informal and not binding."

The Egyptians and Sudanese, dominated by their Permanent Joint Technical Commission had, after three full days of discussion, succeeded in a masterpiece of prevarication, the only success so far as H.M.G. and the East African Territories were concerned being the agreement to further discussion. It is worth adding, however, that outside the meeting informal agreement was reached in principle on Uganda's proposal that application be made to the United Nations Development Programme to fund a hydro-meteorological survey of Lake Victoria (later extended to all the lakes), a particularly important project which began in 1966.

The next technical meeting, held in Nairobi in July 1962, was preceded by a meeting of the Nile Waters Co-ordinating Committee held at ministerial level in Dar es Salaam and somewhat optimistically set the following objectives:

"(a) To keep the informal talks open so that during the transitional political phase in East Africa we continue to have direct contact with the Permanent Joint Technical Commission on Nile Waters

(b) To continue the arrangement reached during the Khartoum talks in October 1961 when it was agreed at technical level that all three East African Territories could take water from the Nile Drainage Area for those projects in which development would be retarded if permits for the use of water could not be issued by the Governments concerned [35]

(c) To put forward as a basis for discussion, and without prejudice, a figure representing the long-term requirements of East Africa (50 to 75 years). This was considered to be of importance because failure to have a figure recorded was to invite the risk that the downstream states - which are embarking on enormous and costly conservation projects - might later claim that they had a prescriptive right to all water, except, perhaps, the quantity claimed through diplomatic channels in August 1959 and recorded in the blue book, *East Africa's Case*. This quantity of 1.704 milliards (km^3) is related to a 25-year period of which at least three years have already elapsed.

(d) To try to get direct discussions started on a revised figure of regulated discharge at the Owen Falls Dam during the so-called temporary period before storage in Lake Victoria is authorised.[36]

The meeting in Nairobi was attended again by strong representation from the PJTC. The first objective of the East African Governments, (a) above, was agreed; continuity appeared to be assured. The PJTC would be ready to discuss "immediate requirements", meaning urgent needs, and not the figures put forward in the Note of 1959, [37] in the context of a five-year programme which could be corrected whenever informal technical discussions were held, adding - a rather unexpected request - "that in order that immediate requirements could be related to downstream planning they would also like to reach an agreement on ultimate requirements".

While it was therefore agreed that water urgently needed (defined as "that which was required for actual use during the next five years), ultimate requirements - given as a tentative 5 milliards (km^3) by the East African delegation - were not acceptable on the grounds that the case was not supported, as was the earlier statement of requirements in August 1959, by technical data - meaning, as before, technical feasibility and economic viability, a rebuff predicted by Hawes and Harris long before.

The PJTC also said that a reply to the Note of 1959 would be sent, but they objected not so much to the quantity but to the way the claim was framed - the reservations made and "the lack of finality."

It was reckoned that working relations had been a big improvement on the

Khartoum meeting, which had been decidedly acid at times. Nevertheless, very little progress was made in reaching agreement on this occasion either. Tanganyika had already formally declared the Nile Waters Agreement of 1929 invalid as described above - a situation the PJTC chose to ignore - and Kenya followed shortly after. There are no records available of any subsequent meetings.

The Egyptians and Sudanese had failed to recognise East Africa's so-called immediate requirements put forward formally in 1959, and, as might have been expected, had turned aside the less carefully prepared case for ultimate needs, even though they had asked for this figure.

These episodes revealed the inflexible attitude of Egypt in particular which was to be repeated whenever the subject of any significant upstream development was raised in subsequent decades. Egypt's decision to set such store on the criteria of water use, technical feasibility and economic viability, as central to negotiating issues was to have a major impact on the attitudes and behaviour of both upstream and downstream states. Egypt has always given high priority to recording, analysing and evaluating the Nile Resource. Both Egypt and the Sudan have a long history of competent hydraulic engineering, and knowledge of the upper White Nile gathered by their engineers as well as expatriate staff from the beginning of the century is probably unparalleled in any other river in the world.

Upstream states in East Africa, on the other hand, despite the local knowledge made available by the UN sponsored Hydro-Meteorological Survey of the lacustrine area, have often felt themselves disadvantaged and in the absence of fully completed water master plans for their countries are disinclined to reveal their hand before it is ready. The East African Nile Waters Co-ordinating Committee served a useful purpose in determining the water needs of the region, at least in the short term, and provided a well informed body which was able to negotiate at professional level with the Permanent Joint Technical Commission of Egypt and the Sudan of the time. The East African countries, who may have conflicting as well as common interests in the Nile waters, require an equivalent body. The Lake Basin Development Authority (LBDA), formed in Kenya in 1982, deserves special technical assistance and financial support from external aid sources with this in mind.

General references

Collins, R.O. (1990). *The waters of the Nile: Hydropolitics and the Jonglei Canal, 1898-1988*. Oxford, Clarendon Press.

Howell, P., Lock, M. and Cobb, S. (1988). *The Jonglei Canal: Impact and Opportunity*. Cambridge University Press.

Hurst, H.E., Black, R.M. and Simaika, Y.P. (1946). The Nile Basin **Vol. VII**. *The Future Conservation of the Nile*. Cairo, Ministry of Public Works; Physical Department (with subsequent modifications by Dr Mohammed Amin and H.G. Bambridge).

Kite, G.W.(1982). Analysis of Lake Victoria Levels. *Hydrological Sciences*

Journal, **27**, pp. 99-110.

Morrice, H.A.W. and Allan, W.N. (1958). *Report on the Nile Valley Plan.* 2 Vols. and (1959). Planning of the Ultimate Hydraulic Development of the Nile Valley. Proceedings. *Institute of Civil Engineers.* No. 6372. **Vol XIV.**

Sudan Government. Jonglei investigation Team (1954) - JIT. *The Equatorial Nile Project and its Effects in the Anglo-Egyptian Sudan.* Khartoum.

Uganda Government (1955). *Water Resources Survey of Uganda Applied to Irrigation and Swamp Reclamation 1954/55.* Report by Sir Alexander Gibb and Partners.

Notes and archival references

1. Most of the correspondence between the East African Territories and the Colonial and Foreign Offices in London from 1948, when negotiations for storage in Lake Victoria and hydro-electric power production by Uganda began, until their independence in the first years of the 1960s, was classified as secret and, under the official British 'Thirty Year Rule', not available to historians who have studied the history of negotiations over the waters of the Nile. What follows is taken from official documents in the possession of the author, correspondence between him and the late H.A.W. Morrice, CMG, MA, MICE referred to as Howell Papers up to 1962 and held at Rhodes House, Oxford, and FCO Records Office archives. The author was Chairman of the East African Co-ordinating Committee (see Collins, 1990, p. 260) from 1956 to 1961 and Consultant to the East African Territories until 1962, while Morrice, formerly Irrigation Adviser to the Sudanese Government, was Irrigation Adviser to the British Secretary for State for the Colonies and Hydrological Consultant to the three East African territories from 1958 to the end of 1959 when he died, a tragedy for East Africa and all the Nile valley countries. Previous advisers to the Uganda Government in the 1950s were Brigadier C.G.Hawes, CIE, MC, MICE and Sir Douglas Harris, KBE, CSI, MICE, who had been a senior irrigation officer in the Government of India.
2. Treaty Series No 85 (1955), Cmd. 964.
3. See also (Secret) *The Equatorial Nile Project and the Nile Waters Agreement of 1929: East Africa's Case.* (1957). Government Printer Entebbe, Uganda. Howell Papers and FCO Records Office archives.
4. In fact by then the Egyptians were themselves interested in a discharge higher than 505 cumecs, but evidently wished to reserve their position. Technical Adviser (Dr. Mohammed Amin), Ministry of Public Works (21.10.1956) to Minister of Natural Resources, Uganda; (Enclosure to Khartoum Despatch to Entebbe No 2) Ministry of Foreign Affairs, Sudan (3.11.56) to Governor, Uganda; and Howell, Permanent Secretary Corporations and Regional Communications, Uganda (5.3.58,) to Irrigation Adviser, Khartoum. Howell Papers.
5. The discharge has necessarily to fluctuate with the variation in power load. These fluctuations are frequently sudden and appreciable. They in

no way affect the average discharge, as they are adjusted every day and a final adjustment is made within every month so as to keep the total monthly discharge within the prescribed limits.

6. Treaty Series No. 17 (1929). Cmd.9132.
7. 'The Royal Egyptian Government and His Brittanic Majesty's Government, in accordance with the spirit of the Nile Waters Agreement of 1929, have agreed to the construction of a dam at Owen Falls in Uganda for the production of hydro-electric power and the control of the Waters of the Nile'. (Cairo, 30th May, 1949). Cmd.9642.
8. Exchange of Notes Between His Majesty's Government in the United Kingdom and the Egyptian Government in regard to the use of the Waters of the Nile for Irrigation purposes. HMSO (1929). Cmd 3348.
9. This is of current significance. The demand for electricity in Uganda and Kenya already exceeds the capacity of the Owen Falls station and a decision on the development of the next site is urgently required. Much depends on lake levels since high flows favour Bujagali, 7 km below Owen Falls; low levels favour the Murchison Falls Site. Plans, as before, assume an average flow from Lake Victoria of 630 cumecs. Other sites are at Kamdini 2 km above Karuma on the Victoria Nile; Ayago, roughly half way between Chobi bridge and Murchison Falls where there are two potential sites; and Murchison itself. Total potential installed capacity is currently estimated at 1501 MW.
10. In the case of Uganda the responsibility followed the author (who was also chairman) in three different ministries, ending in 1958 in the Ministry of Commerce and Industry - which covered the Uganda Electricity Board as well as East African Regional Communications, including East African Railways and Harbours whose installations would be affected by levels in Lakes Victoria, Kioga and Albert.
11. *East Africa's Case.* (1957), p.19.
12. Secret Document. *The Equatorial Nile Project and the Nile Waters Agreement of 1929: Uganda's Case. Vol. II. Upstream Effects of the proposed Mutir Dam on the Albert Nile.* (1.3.1957) Uganda Government, Entebbe. Howell Papers; FCO Records Office Archives.
13. The mode of regulation of the proposed dam would also be of great importance in considering both effects and possible remedies on which Uganda would demand detailed negotiations and safeguards.
14. Ex.Co (57) 120. Memorandum by Minister of Natural Resources. Howell, Natural Resources (22.4.57). Howell, Natural Resources, to Mathieson, Colonial Office (1.5.57); and Governor, Uganda to Secretary of State for the Colonies, S/188/J of 18th May, 1957. Howell Papers and FCO Records Office archives.
15. *East Africa's Case.* (1957).
16. This did not include Egyptian staff stationed at Jinja to monitor the terms of the Owen Falls Agreement.
17. *East Aftica's Case* (1957), p. 20 *et seq.*
18. By then public opinion was for the first time beginning to realise the limitations placed upon the East African Governments by the Nile Waters Agreement of 1929. The issue was raised by Legislative

Councils, Chambers of Commerce and in the Press.

19. Minutes of the 5th Meeting of the E.A Nile Waters Co-ordinating Committee. Nairobi 31.3 - 1.4.1959 Minute 13/59, (ii) a. Howell Papers. Kenya and Tanganyika do not seem to have appreciated the bargaining advantage of Lake Albert storage which Uganda was prepared to bring into play on their behalf.
20. *East Africa's Case.* (1957).
21. Curiously, because Morrice had been the first chairman of the Jonglei Investigation Team and was well aware (as indeed was Allan) of the local adverse effects of reducing spill into the Sudd swamps, no mention is made of the environmental impact in the Sudan.
22. For example, the annual discharge at Owen Falls increased from an average of 20.6 km^3 (1905-52) to 50.5 km^3 in 1964.
23. Howell Papers.
24. 5th Meeting of the Nile Waters Co-ordinating Committee, Minute 12/59. Howell papers. It is worth noting that at this date (1.4.59), Morrice was still confident that the 'Nile Valley Plan' rather than the Aswan High Dam would be adopted, though it would take much time to put his plan into effect.
25. Howell Papers.
26. H.A.W.Morrice to R.Crauford-Benson, Ministerial Secretary, Ministry of Lands and Mines, Tanganyika. 26.6.59. Howell Papers.
27. Howell to Morrice, 8.7.58. Howell Papers.
28. In 1959, J.F.Glennie, Chief Planning Officer of the Department of Water Development and Irrigation in Tanganyika (and formerly senior irrigation engineer to the Jonglei investigation) argued that 'it would be impossible to prove that any diversion or irrigation works upstream of Lake Victoria would alter the timing of arrival in Egypt'. Moreover, if all the water leaving Lake Victoria and Lake Albert were held back, the flow at Aswan in any one year would be reduced by less than 1/3 x 1/7 or 1/21, less than 5 per cent. Morrice regarded this as something of an exaggeration, and pointed out that in any case Glennie's calculations assumed no diversion canal through the Sudd, but agreed that the net effects downstream of moderate abstractions would be small. The Egyptians and Sudanese must themselves have been well aware of the relatively small net effect downstream of abstracting 1.7 km^3 from the East African catchment.
29. Sir Harold Beeley and accompanying documents (Cairo) to Sir Roger Stevens (FO) 11.4.61. FCO Archives and Howell Papers.
30. Led by the author who returned from the Middle East for the occasion as consultant.
31. Record of Meeting held in Khartoum 16/17.10.61 between representatives of Egypt the Sudan and East Africa. Howell Papers.
32. Sudan Government. J.I.T. (1954) and Howell, P.P.,(Ed). The Equatorial Nile Project and its Effects in the Sudan. *Geographical Journal.* **Vol. CXIX**. Part 1, Mar. 1953.
33. Both the verbal statement made by the P.J.T.C. at the meeting and the accompanying document appear to be a misinterpretation - possibly

deliberately - of the terms of the Exchange of Notes Between H.M.G. and the Egyptian Government in regard to the use of waters of the River Nile for Irrigation Purposes. Cairo, May 7 1929 [HMSO Cmd 3348]. Clause 4 (ii) is quoted in full on page of this chapter. It is worth quoting the other three relevant clauses here:

"Clause 4 (iii). The Egyptian Government, in carrying out all necessary measures required for the complete study and record of the hydrology of the river in the Sudan, will have all the necessary facilities for so doing.

Clause 4 (iv). In case the Egyptian Government decide to construct in the Sudan any works on the river or its branches, or to take any measures with a view to increasing the water supply for the benefit of Egypt, they will agree beforehand with the local authorities on the measures taken for safeguarding local interests. The construction, maintenance and administration of the above mentioned works shall be under the direct control of the Egyptian Government.

Clause 4 (v). His Majesty's Government in the United Kingdom of Great Britain and Northern Ireland shall use their good offices so that the carrying out of surveys, measurements and works of the nature mentioned in the two preceding paragraphs is facilitated by the Governments of those regions under British influence."

As drafted Clauses 4 (iii) and (iv) apply only to the Sudan.

34. Tanganyika was already independent; Uganda to follow in 1963, Kenya shortly after.
35. This is not in the 1961 record but may, as was so often the case, have been agreed informally outside the meeting.
36. Major General J.K. Edwards (Kenya) to F.D. Webber (Colonial Office) 15.9.62. Howell Papers.
37. There was some confusion among the Egyptians and Sudanese over the difference between "Immediate Requirements" meaning the 25-year quota listed in the Note of 1959 (cf. *East Africa's Case*) and much smaller amounts needed for urgent projects awaiting implementation.

5

History, hydropolitics, and the Nile: Nile control: myth or reality?

R. O. COLLINS

Introduction

The dependence upon the waters of the Nile for Egypt and the Sudan led naturally to the question whence do the waters flow and thence how best to utilize them - Nile Control. In the nineteenth century Muhammad Ali began to build regulators to promote land reclamation in Egypt, but it was not until the British occupation of Egypt in 1882 that the construction of barrages and dams for flood control and irrigation to feed Egypt's rapidly expanding population was systematically undertaken. During the political domination of Lord Cromer at Cairo and the sweeping conception of the hydrological unity of the Nile Basin by his hydrological adviser, Sir William Garstin, Nile Control from its sources to the Mediterranean became a reality seriously contemplated by the corps of British and later Egyptian hydrologists and engineers who embarked upon the research required to develop the Nile Basin for the full utilization of its waters, particularly after the inclusion of the Sudan and East Africa within the British imperium, combined with British diplomatic domination at Addis Ababa.

Garstin was the first to envisage the hydrological development of the Nile Basin in 1904, but his majesterial report was followed by the refinements of Sir Murdoch MacDonald in *Nile Control* (1921). After a lacuna during the interwar years, the famous proposal for the full utilization of the Nile waters by H.E. Hurst in *The Future Conservation of the Nile* (1946), and the later brilliant mathematical model by H.A.W. Morrice and W.N. Allan in the *Report on the Nile Valley Plan* (1958), was the most efficient design for Nile Control conceived. All of these proposals, however, were frustrated by the parochial politics of the riverain states once the British imperium had been removed from the Nile Valley. To be sure, 'The Nile Waters Agreement of 1959' cleared the obstacles to the construction of the High Dam, the Roseires Dam, and the Jonglei Canal, but despite the creation of the Permanent Joint Technical Commission (PJTC) between Egypt and the Sudan to propose further regulators for Nile Control, the instability within the upstream states of the Nile Basin, particularly the civil war in the Southern Sudan, and the declarations by the East African countries and Ethiopia of the right to a share in the water have frustrated the search for the full and efficient utilization of the Nile waters for the benefit of all the peoples within its great basin.

The Nile is one of the great natural wonders of the world. It is the longest river flowing from south to north 6,825 kilometres over thirty-five degrees of latitude. Its basin embraces some 3,100,000 square kilometres of equatorial and northeast Africa. It flows through every natural formation from towering mountains and well-watered highlands to the most barren of deserts. Within this vast area man in Egypt and the Sudan has watched the rich, brown waters of the Nile flow northward to the Mediterranean Sea anxiously waiting for the life which came with them, for without the river there would be only sand and rock and wind. Man cannot survive where there is no water and cannot multiply where there is no fertile soil. The Nile provides both these necessities of human existence.

This is what distinguishes the Nile from the other great rivers of the world. Throughout the northern reaches of the Nile man and his civilizations have been dependent upon the river for their very survival. This has been so for many millennia. At times the river has given too much, resulting in disastrous floods which swept away his habitations huddled by the Nile. In other years the waters did not come, and there was drought and famine in the land. And so from earliest times man has pondered the question whence came the waters and what forces controlled the flow of life? No one who has ever drunk from the waters of the Nile or who has lived by its fertility, no matter in what age, could not have asked these two fundamental questions.

Dependence upon the Nile led naturally to the question whence do the waters come? Men speculated. Pharaohs sent out expeditions to seek the source. Nero ordered his centurions up the river. All were turned back by the Sudd, the great forbidding swamps of the Nile, and the speculations remained. Like the timelessness of the Nile itself, the centuries slipped by without man knowing the source of the life-giving waters. To be sure the Portuguese Jesuit priest, Pedro Paez, gazed upon the origins of the Blue Nile at the holy spring of Sakala in 1613 as did James Bruce in 1770, but where was the source of that broad and majestic White Nile flowing out of Africa past Khartoum? In the words of the Victorian Africanist, Sir Harry Johnston, the quest for this source had become 'the greatest single geographical secret after the discovery of America' (Johnston 1905 p.vii), and between 1858 and 1877 the luminaries of African exploration - Burton, Speke, Grant, the Bakers, Livingstone, and Stanley - plunged into Africa to chart the equatorial lakes whence flowed the beginnings of the White Nile.

Although the Victorians may have discovered the Nile, they did not understand it. Indeed, by the last quarter of the nineteenth century, they really knew little more than the Pharaohs except, of course, its sources. The British occupation of Egypt in 1882 changed all that, and by 1890 knowledge of the natural flow of the Nile had become crucial to the engineers in Cairo who were responsible for making the desert bloom. The problems facing Britain in Egypt were twofold.

First, the introduction of perennial irrigation in the early nineteenth century by Muhammad Ali had transformed the productivity of the irrigated fields of Egypt by a continuous flow of water via regulators to produce two or even three crops in any given year. By the end of the century, however, these regulators and barrages of the great viceroy and his successors could sustain only three million feddans under perennial cultivation at a time when the population of Egypt was increasing beyond the means to feed it.

Second, by 1890 Lord Cromer, Britain's imperious proconsul in Egypt, had convinced the prime minister, Lord Salisbury, that Britain's paramountcy in Egypt was not going to be temporary, as was supposed at the time of the conquest in 1882. Thereafter, the primary consideration of Britain's imperium on the Nile and at Suez was the security of its waters. If Britain were to remain unchallenged in Egypt, British engineers must not only increase the supply of water available to irrigate the land to feed the expanding Egyptian population but to defend any threat to that precious water throughout the length and breadth of the Nile Basin to secure Britain's strategic position at Cairo and Suez.

To carry out these objectives Sir Colin Scott-Moncrieff, the under-secretary of state for public works in Egypt, recruited a remarkable group of hydrologists and engineers largely from the Indian Irrigation Service to begin the regeneration of the irrigation system of Egypt. Meanwhile, Salisbury and Cromer set about using all their diplomatic skill and, if necessary, British military might to preserve all the Nile waters from their source to the Mediterranean for the sole security of Britain in Egypt. This quest, like the Victorian explorers before them, inexorably dragged Britain far up the Nile until the basin of the White Nile from Lake Victoria to Khartoum was firmly within the womb of the British Empire.

Economic or strategic imperialism: Nile control

Those who argue that Britain's conquest of Egypt was motivated by economic concerns, rather than the security of the Nile and Suez, should examine the perceptions and actions of the principal British officials in the Nile Valley at the turn of the century if further evidence is required to convince them that the safety of the Nile waters was their highest priority to which all other British interests in Egypt were deferred. If there are still doubts, they should recall that the two great liberal powers of Europe, Britain and France, were swept to the brink of war until the French challenge for control of the Nile at Fashoda was defused in 1898 by British naval and military strength under General H.H. Kitchener commanding the Anglo-Egyptian army which had defeated the Sudanese forces of the Khalifa Abdullahi at the battle of Omdurman. Following his confrontation with Captain Marchand, Kitchener immediately sent Sir William Garstin, friend and adviser to Cromer, and in 1898 the under-secretary of state for public works in Egypt, up the White Nile to report on the navigability of the Bahr al-Jabal through the infamous Sudd. Garstin reported that British paramountcy in the Nile Basin was contingent on clearing the Bahr al-Jabal of sudd. Men and money were no object; Cromer made it abundantly clear in his despatches that the Nile and its control were essential to his governance of Egypt. Subsequently, Sir William Garstin made three reconnaissance expeditions to the upper Nile Basin, which he called the Lake Plateau, between 1899 and 1903.

His first report was submitted in 1901 in which he made several proposals for the control and efficient use of the Nile waters. He perceptively observed without the benefit of adequate data that some 60 per cent of the water entering the Sudd at Mongalla was lost and could be conserved by banking the Bahr al-Jabal to prevent the Nile waters dribbling off into the remote swamps. After further expeditions he produced his majesterial *Report on the Basin of the Upper Nile, 1904*. It is a remarkable document from which all further studies of the upper Nile take their beginning and return ultimately to Garstin's conclusions. By 1904 Garstin had abandoned his idea to bank the Bahr al-Jabal to conserve water, and while advocating proper surveys and further studies, he adopted the suggestion of J.S. Beresford, the former inspector-general of irrigation in India, to dig a direct canal east of the Sudd from the Sobat mouth to Bor capable of carrying 55 Mm^3/day, the same scheme proposed by Dr. H.E. Hurst forty years later as part of the overall design of water development for the Nile Valley.

The total area of the great swamps of the upper Nile can never be measured with precision since its size varies with the day, the month, the season, and the year, all governed by the fluctuations of the outflow from the equatorial lakes, the annual increment from the torrents between Lake Albert and Mongalla, and the amount of rainfall. Nevertheless, the annual mean size of the Sudd floodplain, including its two principal zones, the permanent and the seasonal swamps, has been estimated during this century, 1905-80, at 16,931 km^2 (Sutcliffe and Parks 1982 Fig 1). After the great rains over the equatorial lakes between 1961 and 1964, the extent of the Sudd floodplain nearly doubled to 30,600 km^2 (Sutcliffe and Parks 1987, p. 148).

Despite their profusion and the transmission of water vapour by transpiration, the aquatic plants of the Sudd are not primarily responsible for the massive loss of water. Evaporation, not transpiration, is the principal cause of the disappearance of billions of cubic metres of water from the great swamps of the Nile. Since the rate of evaporation over the past fifty years has been virtually constant and does not vary significantly between open water and that in the permanent or seasonal swamps covered by aquatic vegetation, the volume of water lost by evaporation is determined by the area of flooding which can vary dramatically on a plain of such insignificant slope that a rise of flood waters by a few centimetres can inundate hundreds of square kilometres, resulting in a loss by evaporation from a low of 40 mm^3/day or 22 per cent of discharge into the Sudd to a high of 120 Mm^3/d or 61.2 per cent of discharge as measured at Mongalla (JEO 1975 pp. 48-9). Thus, between 1905 and 1980 the total mean annual inflow into the Sudd at Mongalla has measured 33 km^3 while the total mean annual outflow remains 16 billion m^3 or approximately half (50 per cent) the discharge of the Bahr al-Jabal at Mongalla. This enormous loss of water in the Sudd is the equivalent of 20 per cent of the total mean annual flow for the same period (84 km^3) of the river Nile at Aswan (Sutcliffe and Parks 1987).

Not only was Garstin's intuitive observation of a loss of 60 per cent in the swamps in 1904 not far wrong, but it made certain that his 'cut' to by-pass the Sudd, later known as the Jonglei Canal, became the first priority for Nile control of the Lake Plateau. It was to be the lynchpin in a chain of regulators at Lakes Victoria, Kioga, and Albert all of which could, of course, be irrelevant if there were no means to pass the waters of the lakes through the

Sudd into the White Nile and thence to Egypt. Garstin set immediately to work with the enthusiastic support of Lord Cromer. C.E. Dupuis from the Ministry of Public Works was sent off forthwith to the United States to study dredging practices and in 1906 funds were appropriated to purchase dredgers, steamers, and barges to begin construction. The Cromer-Garstin era of Nile development came to an end and with it Garstin's cut.

And it is perhaps a chilling but historical axiom, too often forgotten in the twentieth century's obsession with technology, that decision making still remains dependent upon the human factor. Lord Cromer retired in 1907 and Garstin the following year. The intimate relationship between these two remarkable men combined with their vast experience and vision deprived the young, bright, and eager British hydrologists then being recruited by the Egyptian Ministry of Public Works for Nile Control of knowledgeable leadership. The position of Adviser to the Ministry of Public Works, which Garstin held until his retirement, was the single individual ultimately responsible for the overall planning of Nile Control. Following Garstin it was held for brief and unproductive periods by Sir Arthur Webb and C.E. Dupuis until the appointment in 1912 of Lord Kitchener as British High Commissioner and Sir Murdoch MacDonald as Adviser to the Ministry respectively, the combination inaugurating a new era in Nile planning. Neither, however, had the experience or knowledge of Cromer and Garstin, and although they pushed through the approval of the Blue and White Nile Dams at Sennar and Jabal al-Auliya in 1914, the outbreak of the war interrupted the consideration of any additional projects.

The end of Nile control

The return of Lord Kitchener to Great Britain and the secondment of MacDonald as a Colonel in the Royal Engineers to plan the defence of the Suez Canal and water supplies in the Sinai dealt a devastating blow to the rational planning of water use in the Nile Basin. Moreover, MacDonald was singularly ill-suited to play the role of Adviser to the Ministry, particularly without strong political support from the Residency. After all Kitchener was an engineer. MacDonald himself was a 'construction' man with drive, charm, and a gift for facile conversation, but his appointment deeply divided the British officials of the Egyptian Irrigation Service and particularly infuriated Sir William Willcocks, the architect of the dam at Aswan completed in 1902, who regarded himself as the patriarch of Nile studies and was personally determined to destroy MacDonald. Taking charge of the anti-MacDonald faction Sir William launched a personal vendetta against Sir Murdoch MacDonald which at best can be described as scurrilous, vindictive, and vicious and at worst criminal slander and libel for which Sir William was duly found guilty in the British Consular Court in 1921. Although he was remanded over on his good behaviour because of his age, the publicity accompanying the feud and subsequent trial in the Egyptian Press did much to compromise Britain's administration of Egypt and was seized upon at every opportunity by the nationalists to discredit the British in Egypt.

Although vindicated by the Court and a special Nile Commission appointed to review the fifteen charges brought by Sir William Willcocks against him, including falsifying data and deliberately destroying hydrological evidence, MacDonald's position had become untenable, and he resigned in 1921 to run successfully as the Liberal candidate for Inverness, a seat he held until 1950, and established a successful engineering consulting company. Behind he left his defence, *Nile Control*, in two volumes published in Cairo in 1920 with a second edition in 1921, and the future of Nile planning in total disarray. His position as Adviser to the Ministry of Public Works was never filled, and upon the independence of Egypt in 1922 the British High Commissioner no longer could play the political and financial role of Cromer, Gorst or Kitchener, while their successors became bogged down in the swamp of disputatious and frustrating negotiations between Britain and Egypt over their own relationships and the role each was to play in the Sudan. Leaderless, the younger hydrologists - Dr. H.E. Hurst, J.I. Craig, Dr. P. Phillips, E.M. Dowson, R.P. Black, W.A.C. Perry, A.D. Butcher, H.G. Bambridge and others thus continued to amass enormous quantities of data and almost mindlessly devised numerous schemes to bring down the water from the equatorial lakes without any coherent plan for control of the waters of the Nile Basin. Indeed the obsession with gathering statistics in distinction to planning led R.M. MacGregor, Director of the Sudan Irrigation Department, to remark sardonically in 1945: 'The very nature of the Sudd Region appears to have some oppressive effect on those who for too long have been too close to its detailed problems, and to make it all the more necessary that direction should be kept by some superior authority less liable to lose perspective.' (MacGregor 1945, p.7)

Dreams and schemes

Of all the permutations of the Jonglei Canal which emerged from the analysis by the Physical Department of the Ministry of Public Works and the Egyptian Irrigation Service supported by statistics obtained from the Sudan Irrigation Department, which had been created as a separate entity in 1925 after the forced removal of the Egyptian Army and officials from the Sudan, two schemes dominated discussions in Cairo - the Sudd Canal Project and the Veveno-Pibor Scheme during the 1920's and the Bahr al-Zaraf Scheme in the 1930's. The two former projects were largely the work of W.A.C. Perry, the Inspector of Irrigation, Upper White Nile Division of the Egyptian Irrigation Service. He first sought to carry out the Garstin Cut from the Sobat to Bor and an initial E£350,000 was appropriated for him and his officials to conduct a survey. By 1925 Perry had completed his surveys, abandoned the Garstin Cut as too expensive and introduced a variant by clearing the Bahr al-Zaraf from the White Nile to Ajwong and from there constructing a canal to the Bahr al Jabal at Jonglei. An initial credit of E£1,000,000 was approved for its construction. The Sudan Government and particularly its officials in the Upper Nile Province would have none of it. Having gotten rid of the Egyptian Army and Egyptian officials in 1924 the Sudan Government was not about to let either Egyptian labourers or Egyptian authorities into a province they did not control and hardly administered. Moreover, the Nile Commission of J. Canter Cremers, R.M. MacGregor, and Abd al-Hamid Pasha Sulayman had only been appointed six months before to work out a Nile waters agreement between the Anglo-Egyptian Sudan and Egypt which ultimately led to the Anglo-Egyptian Nile Waters Agreement of 1929 by which Egypt received a disproportionate share of the Nile waters, its 'historic rights' of 48 km^3 leaving for the Sudan a paltry 4 km^3. The Sudan Government immediately commissioned W.D. Roberts, the former Inspector-General of Irrigation in the Sudan for the Egyptian Ministry of Irrigation, to report on the environmental impact on the people of the Upper Nile of the proposal. By the time he submitted his report in 1928 that indeed the canal would have a deleterious effect on the Nilotic inhabitants, the Sudd Canal Project was dead. Perry and his team had an alternative - the Veveno-Pibor Scheme. In retrospect this Scheme was not only bizarre in conception but also costly in construction which was its ultimate undoing. Its merits were that it did not require a dam at Lake Albert nor disturb the truculent Nilotes. The Bahr al-Jabal would simply be diverted into a canal to the Veveno River which in turn reached the Pibor, Sobat, and ultimately the White Nile below Malakal. Prohibitively expensive, the Scheme was officially dropped in 1932, and the White Nile Division, now under the leadership of A.D. Butcher, revived the Sudd Canal Project, now known as the Jonglei Canal Scheme.

A.D. Butcher spent eight years in the Sudd at the end of which he produced *The Sadd Hydraulics*, then regarded as a *tour de force* by hydrologists, but in fact it provides no satisfactory explanation of water losses and only diverted his attention, as well as that of others, from a thorough examination of the Garstin Cut. The Jonglei Canal Scheme was approved in 1938 but varied from Perry's Sudd Canal Project by proposing a canal to run north from Jonglei to meet the Bahr al-Zaraf and then continue parallel to it. By having the Bahr al-Zaraf carry much of the water the canal itself could be small and consequently the impact upon the Nilotic peoples was relegated to two insignificant paragraphs.

The reaction to Butcher's Jonglei Canal was instantaneous. In 1939 John Winder, a British Commissioner in the Upper Nile Province, raised many of the same criticisms of Perry's Sudd Canal Project made ten years before by W.D. Roberts, causing quite a stir in Malakal and Khartoum at a time when the senior officials of the Sudan Government were embarrassingly aware of and not a little guilty about their past indifference to the Southern Sudan. Within a few weeks Winder was trekking through his district with A.E. Griffin, the Adviser to the Sudan Government Department of Irrigation, who reported that indeed the Jonglei Canal would inflict many hardships upon the peoples of the Upper Nile. This was followed by a more comprehensive and persuasive report by Winder himself in the spring of 1940 which convinced the Sudan Government that the time had arrived when it required, first, its own independent study to counter Egyptian proposals which were regarded with deep distrust and which did not deal with the impact upon the Nilotic inhabitants of the Canal Zone, and second, data by which to assess claims for damages from Egypt. And then the Second World War broke over the Sudan like a *habub*.

Despite the war the Egyptian and British hydrologists had not been idle.

Since the retirement of Sir Murdoch MacDonald in 1921 there had been a void in leadership and planning of the full utilization of the waters of the Nile Basin. A hodge-podge of projects had been produced in an *ad hoc* fashion - proposals for dams at Lake Tana, Lake Albert, and Lake Victoria, the Jonglei Canal. Masses of data had been collected, volumes of statistics published, and endless discussions concerning the hydrology of the Nile Valley consumed countless hours. Yet there was no overall direction and only the reports of Garstin and MacDonald to fall back upon. Into this vacuum moved Dr. H.E. Hurst who had joined the Survey of Egypt in 1906 and largely because of his mathematical abilities and contributions to measuring Nile discharges had been made the Director-General of the Physical Department of the Ministry of Public Works in 1919. His department was primarily engaged in analyzing hydrological readings from which gradually emerged during the interwar years the concept of long-term storage. Although he did not possess the political acumen of Garstin or MacDonald nor could he assume the powers of decision-making enjoyed by British hydrologists under the protectorate, these handicaps to leadership were compensated by the senior Egyptian officials themselves who were demanding that a rational and comprehensive scheme for Nile water development be drafted.

The revival of Jonglei and the Equatorial Nile Project

In 1945 the Egyptian Government informed Khartoum that it was prepared to enter into negotiations to consummate the Jonglei Canal Scheme. Although the Sudan Government had been considering the establishment of a team to gather independent information about the Sudd and its peoples, first suggested by Griffin and Winder back in 1939 and 1940, the war and shortages of qualified manpower had prevented this suggestion being brought to fruition. The best that could be done at short notice was to send H.A.W. Morrice, Divisional Engineer, Projects Division of the Sudan Irrigation Department, south to Malakal in May 1945 to begin the technical discussions with the Egyptian Irrigation Department concerning Jonglei, the idea being that Morrice would formulate the questions the Sudan survey team should tackle concerning Butcher's Jonglei Canal Scheme. Morrice and his superiors in Khartoum received a rude shock at Malakal. Morrice met with Butcher's successor, H.C. Bambridge, Inspector for Irrigation for the White Nile Inspectorate of the Egyptian Irrigation Department. To Morrice's astonishment, Bambridge expressed no interest in Butcher's Jonglei Canal Scheme and boldly proposed a canal running in a straight line from the Sobat to Jonglei - in concept similar to the Garstin Cut of 1904. The canal would be excavated by draglines developed during the war thus cutting the time for construction and therefore the expense. Little regard was expressed by Bambridge concerning the people of the Canal Zone; it was a straight-forward hydrological engineering scheme.

British officials in Khartoum were incensed. Throughout the whole history of the Sudd Canal the Egyptians, represented mostly by British hydrologists, had ignored the Sudan, carrying on their investigations with little consultation, creating in Khartoum only distrust for their proposals, not to mention their motives. General Huddleston, the Governor-General, announced that the time had come for the Sudan to appoint its own team to prepare for battle with the Egyptians. That same month the Jonglei Committee was established in Khartoum consisting of senior officials to review the reports of the Jonglei Investigation Team which was to operate in the field providing reliable information about the effects of the Jonglei Scheme on the interests of the Sudan and the inhabitants of the Upper Nile. The Jonglei Investigation Team lasted in theory ten years, but in practice, as a productive research group of scientists and technicians, it worked only four years from 1949 to 1953 under the leadership of Dr. P.P. Howell. By 1954 working under great pressure and by then virtually alone Howell published the final report of the Team in four volumes: *The Equatorial Nile Project and Its Effects in the Anglo-Egyptian Sudan: Being the report of the Jonglei Investigation Team.* It was an impressive and, given the time constraints placed upon the Team by the Jonglei Committee, who in turn were aware of the changing political position of Britain in the Sudan, a remarkable account of the hydrology, ecology, the peoples of the Upper Nile and their livelihoods, the effects of the canal and possible remedies and assessments for damages, as well as numerous specialized studies, surveys and maps based on the assumption by the Sudan Government that the Egyptian Government was presenting the Garstin Cut as a *fait accompli*, and that its responsibility was to acquire information to ameliorate as best as possible the impact of the canal on the 700,000 Nilotic people living in the Canal Zone and to translate the damage to them and the Sudan in to pounds and piastres, and hence the monetary compensation to demand from the Egyptians. In this sense the Report of the Jonglei Investigating Team was unlike the more massive studies undertaken in the 1970s and 1980s which projected an entirely different canal to coincide with a large-scale socio-economic development of the Canal Zone.

All of the fears aroused on the Upper Nile and Khartoum from the Morrice-Bambridge conversations at Malakal in May 1945 were soon confirmed by the publication in 1946 of *The Future Conservation of the Nile* by H.E. Hurst, R.P. Black, and Y.M. Simaika of the Physical Department of the Ministry of Public Works. During the aimless interwar years Hurst in co-operation with Dr. P. Phillips and later Black and Simaika had published under the auspices of the Ministry of Public Works six volumes and several supplements under the general title of *The Nile Basin.* These volumes contained largely statistical material with little commentary and no profound proposals. They did, however, form the basis of a total plan for the efficient hydrological development of the Nile Valley for which the Egyptian Ministers had been pressing. The result was volume VII in *The Nile Basin* series by which Hurst and his associates laid out a rational, detailed, long-range plan for Nile development along the lines of Garstin's report forty years before but in greater detail and refinement by introducing the concept of 'Century Storage'. In the preface to *The Future Conservation of the Nile* the Minister of Public Works, Abdel Kawi Ahmed Pasha, wrote with relief and pride: "This is the first time that the full development of Egypt has been considered in detail and a new idea, that of 'Century Storage' is introduced. The book makes it clear that we can no longer proceed by small stages leaving the ultimate development for future consideration. The new ideas show that on some important points a decision must be made now. The main projects are seen to be closely connected parts of one whole, and their connection is a complicated one'" (Hurst, *et al.* 1946).

The technical details of the interaction of the various proposed projects are complex, but the concept upon which they were founded is relatively simple and remains so to this day. The rapid increase of the population of Egypt had generated a constant increase in the area of cultivated land. In Egypt and the Sudan this required the most efficent use of the Nile water which in turn required the total control and regulation of all the waters in the drainage basin of the Nile. The goals of 'the most efficient' use were to supply not only additional quantities of water but a constant supply of water while at the same time preventing floods. To achieve these objectives Hurst utilized the reliable records accumulated by the Egyptian Irrigation Service during the past fifty years and published in the first six volumes of *The Nile Basin,* while demonstrating the interrelation between the individual projects which had emerged in the form of a variety of separate proposals during the preceding decades. The unique feature of Hurst's proposal for Nile Control was 'Century Storage', a refinement of MacDonald's over-year storage, by which water would be stored in the great lakes of equatorial Africa, where evaporation is balanced by rainfall and the geological configuration is such that an increase in volume can be accomplished without a substantial increase in the surface area exposed to evaporation.

Hurst proposed a great reservoir at Lake Albert supplemented by a regulator at the outlet from Lake Victoria and at Lake Kyoga. In order to make efficient use of the lake reservoirs under the concept of Century Storage, however, a large canal (the Garstin Cut) would be required to carry 55 Mm^3/day past the Sudd from Jonglei to the mouth of the Sobat. The whole general scheme became known as the Equatorial Nile Project and precipitated angry officials in Khartoum to establish the Jonglei Committee which in turn approved the work of the Jonglei Investigation Team to assess the environmental impact and damages to the peoples of the Upper Nile from the project. In February 1947 the Egyptian Government officially adopted Hurst's scheme for Nile Control of which the Equatorial Nile Project was its principal component. Presented with what they perceived as a *fait accompli* the Jonglei Committee accepted the approval of the Equatorial Nile Project and its big canal with apprehension and not a little bitterness. It did not matter. By the time Paul Howell produced the four volume comprehensive report of the Jonglei Investigation Team, British officials in the Sudan were more concerned about packing their bags and seeking new careers. Sudanese independence, civil strife in the Southern Sudan, and the spectre of the High Dam at Aswan distracted everyone's attention from the Equatorial Nile Project.

The High Dam, The 1959 Nile Waters Agreement and the Nile Valley Plan

As the Southern Sudan was torn by the ravages of war and the shadows of famine, refugees, disease, and death, the population of Egypt and the Sudan inexorably increased, and the demand for additional cultivation for domestic and external consumption became more vital and vocal. The pressure to resolve the hydropolitics of the Nile Valley fell heavily upon the revolutionary government of Gamal Abd al-Nasser and the independent government in the Sudan. The result was two decisions which satisfied immediate concerns but did not provide long-term solutions to population growth or the additional water to sustain it - the construction of the High Dam at Aswan and the Nile Waters Agreement of 1959. So long as Britain had dominated the Nile Basin its very presence had made possible the unified planning of the Nile Valley, while at the same time Egyptian fears of intervention with the Nile waters were allayed by the imperial shield of Great Britain. Once, however, independence swept through the Nile Basin, precipitated by the independence of the Sudan in 1956, Egyptian fears were aroused when seven other independent upstream states also laid claim to the Nile waters. These fears soon turned into an obsession. Upon seizing power in Egypt on 23 July 1952, Nasser and his eleven fellow officers were neither ideologues nor doctrinaire revolutionaries. They held no constituency outside the army, but they were not without fundamental goals - the independence and prosperity of Egypt. To translate these ambitious principles the Free Officers needed a spectacular and visible symbol to demonstrate their intentions to the Egyptian people and the world. The High Dam at Aswan became *That Symbol*, and within two months it was under active consideration by the Revolutionary Command Council from which there was no turning back. Politically it was a gigantic and daring scheme, a monument to the vision of the revolutionaries. Economically it provided water and power. Most important, and the clinching argument before which all the finely-honed mathematical proposals of British hydrologists for the rational development of the Nile Basin crumbled, the High Dam would free Egypt from being the historic hostage of upstream riparian states by providing over-year storage within the boundaries of Egypt. Whatever the demerits of the High Dam, they were rendered insignificant by that fact.

It is doubtful that any member of the Revolutionary Command Council had heard of the High Dam before coming to power. Although General F.H. Rundall had first proposed a high dam at Aswan as early as 1876, it was really the creation of Adrian Daninos, a Greek-Egyptian engineer, who in 1912 advocated the electrification of the Aswan Dam to promote industry in Egypt. He fell into obscurity until after World War II when in 1948 he, in cooperation with an Italian, Luigi Gallioli, conceived of one grand structure at Aswan to guarantee over-year storage in a vast reservoir capable of generating enormous quantities of hydroelectric power. He sought to sell his idea without success. Hurst cited the high evaporation rate; other problems of siltation, seepage, and scouring were discouraging. After the revolution Daninos went directly to two army engineers he knew, Samir Hilmy and Mahmud Yunis, who were members of the Free Officers and head of the Revolutionary Command Council's technical office. There the High Dam was born and duly delivered on 15 January 1971 after intense international controversy, bitter environmental and technical criticism, and war. 'Century Storage' as conceived by Hurst appeared dead.

The determination to build the High Dam was a unilateral decision by Egypt, but it could not wholly ignore the interests of other riparian states to the south. Relations with Uganda had been normalized by the exchange of notes in May 1949 by which Egypt had agreed to provide financial assistance for the construction of the dam at Owen Falls to supply East Africa with hydroelectric power so long as the discharges to be passed through the dam would be on the authority of the Egyptian resident engineer to meet Egyptian requirements downstream. In 1946 Britain and Egypt had even revised their agreement of 1935 to finance a dam at Lake Tana if the Ethiopians would agree. The Sudan, however, was the most significant of all the riparian states and, by virtue of her geographical position, the crucial partner with whom Egypt must come to terms if development of the Nile Valley was to continue south of Aswan.

Not only would the High Dam reservoir flood the fertile land in Sudanese Nubia and inundate the town of Wadi Halfa, but by 1952 it had become abundantly clear that the Sudan would require a substantial increase in the amount of water for irrigation beyond the 4 km^3 allocated to it by the 1929 Nile Waters Agreement. Contentious and competitive negotiations followed which transcended the years from the Anglo-Egyptian Condominium to the independent Sudan, characterized by informal and frequently productive discussions among Egyptian, British, and Sudanese engineers and acrimonious and rhetorical exchanges from their leaders. Since the Sudan needed additional water the Egyptians sought to bargain approval for a dam at Roseires, upon which the expansion of the Gezira cotton scheme was dependent, for Sudanese approval of the High Dam. No Sudanese minister in the full flush of independence could agree to such onerous terms and consequently the Sudanese ministry of irrigation and hydroelectric power unveiled its own comprehensive scheme for Nile Control, the Nile Valley Plan.

Through the initiative of H.A.W.Morrice, Irrigation Adviser, he and W.N.Allan, principal irrigation consultant to the Sudan Government, produced by computer analysis the *Report on the Nile Valley Plan* (2 volumes) published by the Sudan Ministry of Irrigation and Hydro-Electric Power in June 1958. It was Morrice who had first suggested the use of computers in 1955 and by the following year had drafted a computer program which was subsequently modified but remained essentially the program run by IBM computers in Britain and presented to a conference of hydrologists, engineers, and mathematicians in January 1957. On the one hand the *Report on the Nile Valley Plan* was vintage Hurst, confirming the principles of 'Century Storage' with the equatorial lakes as reservoirs and a large Jonglei Canal, or rather two parallel canals designed to carry as little as 19 Mm^3/d or as much as 89 Mm^3/d, to bring down the water. On the other the Nile Valley Plan was much more comprehensive than Hurst's and included with the aid of IBM's computer many more variables with greater emphasis on the generation of hydroelectric power. In the abstract the Nile Valley Plan was an elegant and efficient use of the Nile waters and provided the Sudanese with their own water development plan but in the world of political reality it has been ignored, overwhelmed by the destructive power of nationalism. The Nile

Valley Plan was vehemently rejected by the Egyptians and therefore ultimately abandoned. Today it is all but forgotten, to the loss of the peoples of the Nile Valley.

During the summer of 1958 relations between Egypt and the Sudan reached their nadir, accompanied by a continuous barrage of propaganda from Cairo about Nile waters, the High Dam, and the unity of the Nile, sprinkled with virulent personal attacks on Sudanese politicians. Finally, on 17 November 1958 Major General Ibrahim Abboud, a gentle pragmatist, was invited to take over the Sudan Government from the impotent politicians three weeks before the Soviet Union formally offered the financial assistance necessary for the construction of the High Dam. If Egypt wanted the High Dam, the Sudan wanted the Roseires Dam on the Blue Nile for a massive expansion of the Gezira and adequate compensation and remedial measures to which Egypt had a moral obligation to provide for those Sudanese who were to be displaced by the High Dam reservoir. Discussions were renewed early in 1959, and stimulated by the suspicions aroused by Britain's call for an international conference to establish an International Nile Waters Authority in which Great Britain would be a member, the Agreement for the Full Utilization of the Nile Waters was concluded between Egypt and the Sudan on 8 November 1959. By the terms of the agreement Egypt would receive 55.5 km^3, a gain of 7.5 km^3, but the Sudan's share would increase to 18.5 km^3 for a substantial increase of 14.5 km^3 from its allotment granted in 1929. Another important advantage to the Sudan was that any additional water which could be conserved or discovered would be divided on a 50-50 basis between Egypt and the Sudan. Since no additional water was likely to come from Ethiopia, the Equatorial Nile Project was revived and Jonglei was the key. Ironically, 'Century Storage' may have been made irrelevant by the High Dam, but if the equatorial lakes were no longer to be reservoirs for storage, they were soon to become in the great rains of the 1960s the repositories of large amounts of additional water which could not have been foreseen by Hurst.

Of equal importance as apportionment of the water by the agreement was the establishment of the Permanent Joint Technical Commission (PJTC) which would henceforth be responsible for the planning and implementation of all hydrological works on the upper Nile. For the first time since the resignation of Sir Murdoch MacDonald in 1921 there was now a body given the responsibility for directing Nile Valley water development - Nile Control. The membership of the PJTC was jointly held by Egyptian and Sudanese engineers and hydrologists with a rotating chairman every three years and including such internationally known hydrologists as Yahia Abdel Mageed, Abdullahi Mohammed Ibrahim, Mohammed A. Mohammedein, and Bakheit Makki Hamad. Now that the question of 'Century Storage' was resolved by the completion of the High Dam, Egyptian and Sudanese hydrologists and engineers had seventeen years to plan the means to tap the additional water falling upon the equatorial lakes while civil war raged in the Southern Sudan. Moreover, the Sudan itself was too preoccupied with the construction of the dams at Roseires and Khashm al-Girba and the Nubian resettlement scheme to undertake any immediate project in the Sudd. Besides, in 1958 there was still sufficient water to meet the needs of these development schemes. A generation later this is no longer the case.

The politics of water: The Nyerere Doctrine and the Blue Nile Plan

When Hurst's 'Century Storage' (the Equatorial Nile Project) was conceived the principal riparian territories in the upper Nile Basin were under British administration. Political unity made possible the comprehensive and rational development of the region, and this assumption was implicit in the Nile Valley Plan proposed by an independent Sudan. In 1961, 1962, and 1963 the British colonies of Tanganyika, Uganda, and Kenya respectively became independent sovereign states, enormously complicating the development of conservation projects in the Lake Plateau. The government of Tanganyika was the first to invoke what became known as the Nyerere Doctrine, by which neither Tanganyika nor subsequently independent Uganda and Kenya recognized the 1929 Nile Waters Agreement. The Nyerere Doctrine legally destroyed the Equatorial Nile Project, for henceforth any conservancy schemes in the equatorial lakes would have to be negotiated by Egypt and the Sudan with the riparian states. The primary goal of the planners, saving the waters of the lakes, has become, however, ever more tenuous, not through legal constraints nor the construction of the High Dam, but by the political and frequently tumultuous political realities in the East African states, particularly Uganda.

After the signing of the 1959 Nile Waters Agreement, the tall shadow of the Ethiopian highlands spread across the plains of the Nile Valley. Egypt, and less so the Sudan, had always been wary of their somnolent riparian overlord whose geographical position could not be denied. On 6 February 1956, only one month after the independence of the Sudan, the imperial Ethiopian Government announced in the official newspaper, the *Ethiopian Herald*, that Ethiopia would reserve for her own use those Nile waters in her territory, 86 per cent of the total Nile flow. This public declaration was followed several months later by official notes to the diplomatic missions in Cairo by which Ethiopia 'reserved its rights to utilize the water resources of the Nile for the benefit of its people, whatever might be the measure of utilization of such waters sought by riparian States.' (Whiteman 1964 pp. 1011-12).

Suddenly in 1958, in the midst of the Nile waters negotiations in Cairo and undoubtedly encouraged by them, the Ethiopians launched a major study of the water resources of the Blue Nile for irrigation and hydroelectric power, carried out by their ally the USA and specifically by the Bureau of Reclamation of the Department of the Interior.

The Blue Nile Plan required five years of intensive research; it is in striking contrast to the more methodical, but plodding, British inspired studies of the Egyptian Irrigation Service. With characteristic conviction the US Bureau of Reclamation included not just the river but the whole of the Blue Nile basin in its investigations, embracing 'its hydrology, water quality, geology, physiography, mineral resources, sedimentation, land use, ground water and local economy' (Guariso *et al.* 1987, p. 108). Although a brief encounter compared with the efforts of the Ministry of Public Works, in terms of total tonnage the seventeen volumes and appendices of *Land and Water Resources of the Blue Nile Basin: Ethiopia* outweigh the tomes of *The Nile Basin.*

The Bureau of Reclamation proposed four major dams on the Blue Nile with a combined storage of 51 km^3 equal to the mean annual flow of the Blue Nile with a hydroelectric capacity three times that of the Aswan High Dam. Of more immediate interest was the effect of the four dams on the natural flow of the Blue Nile and, of course, on irrigation in Egypt and the Sudan. The annual flood of the Blue Nile would be virtually eliminated, the flow into the Sudan becoming constant, and the total quantity of Blue Nile water reduced by 8.5 per cent. If all the projects were made possible, the amount of land put into cultivation in Ethiopia would be equal to 17 per cent of the current land under irrigation in Egypt and would require six km^3 of Nile water.

Such a prospect, no matter how remote, could not but rekindle ancient Egyptian fears of Ethiopian control of the life-giving waters - an historic fable, to be sure, but one that is still believed and which concrete could make a reality. Here were the ghosts of Friar Jordanus, Prester John, Jean de Lastic, Ariosto, James Bruce, and even Sir Samuel Baker emerging from the mists of the past in the bizarre form of the US Bureau of Reclamation to bestow upon the Ethiopians the plans by which to control or at least to intervene in Egypt's historic rights to the Nile waters. If the Egyptians were going to construct the wasteful dam at Aswan to secure sufficient water in an Egyptian reservoir, they could hardly be expected to regard with equanimity projects which would seemingly deprive them of their national security. This historic paranoia now came into conflict with the vigorous nationalism of the riparian states. First the demands of the Sudanese for additional water, then the Nyerere Doctrine, and finally the Blue Nile Basin Development Plan all loomed as a threat to Egypt's precious water. And the Ethiopians would not let them forget. At the United Nations Water Conference at Mar Del Plata in 1977, Ethiopia reasserted its rights to the waters of the Blue Nile, and in June 1980, at the meeting of the Organization of African Unity (OAU) in Lagos, the Ethiopian representative peremptorily charged Egypt with planning to divert Nile water to Sinai illegally since there was no international agreement as to its distribution. These fears, stoked by nationalism, have destroyed, and are likely to continue to destroy, any holistic development of the Nile Valley for the benefit of all its peoples, not just Egyptians, the Sudanese or the Ethiopians.

Ironically, the Blue Nile Plan, if properly managed would not substantially affect the water available to Egypt and the Sudan. Under appropriate working arrangements the amount of water for irrigation throughout the Nile Basin could actually be increased; 'even if Ethiopia were simply to pursue its own objective of managing the reservoir to maximize hydropower production, without considering the interest of Egypt and the Sudan, the amount of water available to the downstream riparians would not be substantially affected' (Guarsio, *et al.* 1987, p. 111). Even if Ethiopia could implement the Blue Nile Plan, drawing off six km^3, Egypt and the Sudan would still benefit from the construction of the reservoirs in the Blue Nile, if properly managed, with a maximum loss of only 2.5 km^3, a trivial amount compared to the total Nile flow and sources of additional water available in the equatorial lakes.

If the projects recommended by the Bureau of Reclamation were completed, Ethiopia would capture the Nile flood but in return would release 46.9 km^3 - or substantially more than the current mean annual discharge at Roseires - because of a loss of only 3 percent by evaporation against a loss of over 12 per cent in the Aswan reservoir. This is in essence the Blue Nile plan as it was in the century storage of Hurst and the Nile Valley Plan of Morrice and Allan - store water in regions of low evaporation at the Nile sources and thereby gain the additional water to quench the thirst of expanding populations downstream. According to the Blue Nile Plan the Sudan, for instance, would receive 2.7 km^3 more than its present allocation under the 1959 Nile Waters Agreement. Water stored in the four Blue Nile reservoirs and managed in conjunction with a Roseires reservoir freed of debris and heavy siltation could then be released in May to reach Egypt when its water requirement is the highest without sustaining the great loss by evaporation now experienced at Aswan. That both the Sudanese and Egyptian allocations could still be higher than their share under the Nile Waters Agreement is simply due to the Aswan reservoir being operated at relatively low levels, thus reducing evaporation losses below the estimates of the treaty' (ibid p. 112). Egypt, however, would no longer be the beneficiary of additional water in years of high flood, which would then be stored and regulated in the Blue Nile reservoirs, not at Aswan. Moreover, lowering the level of Lake Nasser in order to limit the evaporable loss would concomitantly reduce the hydroelectric power, but in return Egypt would receive additional water for irrigation. Ethiopia could, of course, malevolently withhold water it did not need in a year of low rainfall to threaten disaster in the Nile Valley. The Egyptians have historically deeply feared this threat to their survival, and such an action would be tantamount to an act of war. It was just such a fear, in the jungle of predatory nation states, which determined the construction of the High Dam at Aswan.

The Great Rains and the Egyptian Master Water Plan

While the scientists and engineers of the Bureau of Reclamation were toiling through the Blue Nile gorge, nature was capturing the attention of the hydrologists at the equatorial lakes. The dramatic but unpredictable rainfall between 1961 and 1964 increased the level of the equatorial lakes by approximately 2.5 m producing extensive damage around their shores and disastrous flooding of the Sudd floodplain, doubling its size (13,100 to 29,800 km^3) and destroying an estimated 120,000 head of livestock and tens of thousands of Nilotic lives to add to the miseries of civil war.

The East African countries did nothing to alleviate the flooding of the Southern Sudan, obviously more concerned to ameliorate their own damage and when their representatives met with the PJTC in October 1961, they studiously ignored the predicament of the rising waters in the upper Nile. The PJTC offered 0.75 km^3 to the East African states for irrigation between 1961 and 1965 to relieve the flood-water which these states could not use and did not need. The immediate decision taken in October was to provide relief for the areas surrounding Lake Victoria by an increase in the outflow through the Owen Falls dam at the expense of the hapless Nilotes downstream in the Southern Sudan. Between 1961 and 1962 the discharge at Owen Falls increased from 20.6 to 38.6 km^3 and then rose again to 44.8 km^3 in 1963 to reach the astronomical flow of 50.5 km^3 in 1964 (*Hydromet* 1974 i, p. 593). Allan and Morrice had argued that adequate flood protection for the Sudan was as important as additional water for Egypt. Their successors appear to have forgotten this historic law of the river Nile, as the great floods of 1988 have returned to remind them.

The unprecedented rise in the lake levels was not, however, completely ignored. Under the initiative of the East African Nile waters coordinating committee a survey of Lake Victoria had first been proposed by Sir Andrew Cohen, Governor of Uganda, and the creation of Dr. P.P. Howell (See Chapter 4). This committee was soon overwhelmed by the independence of the East African territories, but the idea did not die and upon the recommendation of the PJTC and the financial support of the UN, the hydrometeorological survey of the equatorial lakes was established in 1968 and extended to the Kagera basin in 1971 in which the East African countries, Egypt and the Sudan actively participated, while Ethiopia and Zaire joined in 1971 and 1977 respectively but only as observers. The hydrometeorological survey of the Lake Plateau was soon accompanied by a proposal on the part of Egypt and the Sudan for a commission charged with the total planning of the water resources of the Nile Basin. Nothing came of this proposal, but despite the suspicions of Ethiopia and the East African states both Egypt and the Sudan have continued to press for such an agency which they would obviously dominate by virtue of their historic and technical experience and their political and military power.

While failing to persuade their riparian neighbours to accept a Nile Basin commission or even to come to the negotiating table, the PJTC had not been idle. They drafted plans for a much reduced Jonglei Canal from that envisaged by the Equatorial Nile Project to be included in a more ambitious Egyptian Master Water Plan (EMWP) of 1981 to conserve additional water losses in the upper Nile Basin besides those in the Sudd. All of these plans are as far into the future as those of the Blue Nile and the Kagera basins and consequently remain shrouded in an aura of economic and political fantasy. Nevertheless, they are taken very seriously by the PJTC relentlessly seeking ways to squeeze additional water from the upper Nile Basin despite the fact that civil strife and political violence in the region have prevented the detailed studies required for their technical execution, let alone the financial, international, social, and environmental implications of the proposals. Taken together there are four projects collectively referred to by the PJTC as Jonglei Phase II, or more accurately the Nile Master Water Plan dams at Lakes Victoria, Kioga, and Albert, the expansion of the Jonglei Canal (Phase I), the Machar Marshes scheme, and the Bahr al-Ghazal dams and diversion canals (See Chapter 12). The Sudd may be the single greatest waster of water in the upper Nile Basin, evaporating a mean annual loss from 1905 to 1980 of 16.9 km^3, but the implementation of Jonglei Phase II could yield as much as an additional 14-20 km^3, albeit at great but undetermined cost.

There was imagination and grandeur in all of these plans and dams but an equal unreality. The land in which all of these conservancy projects would be built has been racked by civil war for many years. In all the calculations by the hydrologists of the PJTC there is little consideration of, and less consultation with, the people who will be most affected by the concrete and construction. Economically, the proposals can only be completed at great cost, supported by dubious financial equations. Unfortunately, these massive projects, on a scale to dwarf those of the Pharaohs, are to be built in one of the most remote regions of the world, devoid of functioning infrastructure and crisscrossed by an incalculable number of natural obstacles. Perhaps these hurdles could be overcome by an international basin agency, mobilizing the combined resources of the riparian states and with the authority to coordinate and implement the hydrological works in a manner commensurate with social programmes and sensitive to local concerns, but the British imperial shield has been replaced by narrow nationalism, personal hubris, and ignorance. These are not the building blocks for international cooperation nor the instruments to overcome historic, cultural, and religious animosity. Perhaps the weight of history lies too heavy in the silt of the Nile Valley, but man will always need water; and in the end this may drive him to the river to drink with his traditional enemies.

In 1969 General Gaafar al-Numayri came to power in Khartoum by a coup d'etat with the avowed purpose of ending the debilitating war in the Southern Sudan. He was immediately challenged on the Right by the Ansar, the followers of the Mahdi, and then on the Left by the Communist Party, each challenge was crushed with uncharacteristic bloodshed. By 1972 he was ready to deal with what had become known as 'the Southern Problem' at a time when the Southern insurgents, the Anya-Nya led by Colonal Joseph Lagu, were gaining ever greater strength. Numayri granted his plenipotentiaries virtual *carte blanche* so long as they ended the civil war, and with the timely intervention of Emperor Haile Selassie the Addis Ababa Agreement was concluded in February 1972 which ended the fighting. The ink was hardly dry on the Addis Ababa accords when the PJTC came forward in 1973 with a new proposal for the Jonglei Canal, still the lynchpin in the hydrological development of the upper Nile. Three events had occurred since the work of the Jonglei Investigation Team which conditioned the revision of Jonglei - the great rains of the 1960s in the equatorial lakes, the construction of the High Dam at Aswan, and the technological breakthrough in canal construction by the invention of 'the Bucketwheel'.

The Bucketwheel

The heavy rainfalls, particularly between 1961-1964, on the equatorial lakes created in a few short years an amount of water which Hurst had calculated, using normal rainfall records, would take decades. The second event, of course, was the completion of the High Dam just a year before the signing of the Addis Ababa accords which in one stroke, rather wasteful to be sure, resolved the quest for 'Century Storage'. The third event was the creation of the 'Bucketwheel'. One of the reasons why the Lake Albert Dam had achieved primacy in British planning for the hydrological development of the upper Nile was the fact that the Jonglei Canal, at the time estimated to be 280 kilometres in length, would require twenty years to excavate with the dredgers available. The draglines developed during the Second World War would shorten the time required to construct the canal, but even so Jonglei was still regarded as a long term project. The invention of the 'Bucketwheel' dramatically changed all these assumptions. Built by the German firm of Orenstein and Koppel for the French conglomerate, Compagnie de Constructions Internationales (CCI), it first proved its capabilities in 1964 in constructing the Chasma Jhelum Link Canal from the Indus to the Jhelum River in Pakistan. Digging the Jonglei Canal, the 'Bucketwheel' has excavated up to 2500 to 3500 cubic metres an hour, 300,000 to 500,000 cubic metres a week, and extended the canal at a rate of between one and two kilometres every six days.

The proposal presented by the Permanent Joint Technical Commission to their two governments conceived of an entirely new Jonglei Canal from that devised by Hurst as part of the Equatorial Nile Project. It did not envisage any modification of the flows from the equatorial lakes. It would simply divert 25 million cubic metres per day, a draw-off in an amount that would not dramatically change the regimen of the Sudd. It was consummated by an agreement signed by the Egyptian and Sudanese ministers for irrigation in June 1974. The project presented by the PJTC was strictly a hydrological design for a canal dug in a straight line from the Sobat 280 kilometres to the insignificant village of Jonglei, capable of delivering at Malakal 4.7 km^3 of additional water annually which, after losses in transmission, would increase the yield as measured at Aswan by 3.8 km^3.

The canal itself would vary in width from 28 to 50 metres with sections of greater width for steamers to pass and a depth of 4 to 7 metres with regulators at either end with a designed flow of $3\frac{1}{2}$ kilometres per hour (two miles per hour) to prevent weed growth as well as special equipment at either end hopefully to prevent the infestation of the ominous water hyacinth. A raised all-weather road would be built parallel to the canal, completing the long dream of a Cape-to-Cairo route and shorten the distance by water from Malakal to Juba by 300 kilometres. Egypt and the Sudan would share the cost equally as well as the benefits of the extra yield. There was only scant reference to the impact of the canal on the Nilotic inhabitants, a serious flaw, but a matter outside the responsibility of the PJTC. Tenders were put out for bid, and on 28 July 1976 the final contract was signed between the Egyptian and Sudanese ministers of irrigation and CCI. The cost of transporting the Bucketwheel from Pakistan to the Sobat was absorbed by CCI, leaving the cost of excavation and contingencies to be borne jointly by Egypt and the Sudan at the modest price of LS14,966,310 ($42,953,304; LS = $2.87) made possible by the efficienty of the Bucketwheel. (On 13 March 1980 the PJTC revised the costs of construction precipitated by the realignment of the canal to extend it to Bor or 360 km at a total cost of LS 43,502,776 or $124,852,967, nearly three times the estimated construction costs in 1976 in order to adjust for the dramatic rise in the price of fuel in the late 1970s).

The 'Bucketwheel' itself can only be described as awesome. Five stories high, weighing 2300 tons and propelled on eight caterpillar treads, this behemoth had to be fully automated since it was absolutely essential to keep the great wheel, to which are attached twelve buckets each with a capacity of three cubic metres simultaneously revolving and rotating 180 degrees, precisely on grade in a plain where the slope is only five centimetres per kilometres. Thus the margin of error above or below grade could be only three centimetres, a task quite impossible by manual operation and was thus accomplished by a laser beam fixed on grade a mile ahead to which the machine automatically adjusted within the fixed limits. The wheel itself completed a revolution every minute while rotating in five, the buckets depositing the soil onto a conveyer belt which straddled the canal to dump the material on the east bank creating an elevated road. The 'Bucketwheel' was powered by eight caterpillar engines consuming 40,000 litres of fuel every twenty-four hours, 365 days a year, being shut down only on Saturdays and Sundays for ten hours of maintenance or by mechanical or more frequently by electrical failure (some 3000 in 1981) in a system containing 33,000 separate electrical circuits exposed to the heat and humidity of the Sudd. Transported from Pakistan to Port Sudan, the Bucketwheel finally reached Kosti by train, camel, and truck where a supply center was established and thence up the White Nile by two company steamers to the Sobat where a large base was erected. It took two and a half years to bring the pieces of the Bucketwheel from Pakistan to the Sobat, but in June 1978 it sputtered to life and the excavation of the Jonglei Canal commenced.

The Jonglei controversy and the future of the Nile waters

Even before the Bucketwheel had taken its first bite into the clay of the Upper Nile a barrage of criticism concerning the impact on the population and ecosystem of the Sudd was loosened against the project by Southerners within the Sudan and the environmentalists from without. The Southern Sudanese opposition to the canal was expressed principally in the Southern Regional Assembly in Juba led by Joshua Dan Diu from Fanjak and the late Benjamin Bol Akok the representative from Aweil and the Deputy Speaker. In the Canal Zone itself there was discussion among the chiefs but little organized opposition since the interests of the people were centered upon the excess water which had destroyed so much of their pastureland since the 1960s. There was, however, much vociferous opposition to the canal among the secondary school students who were always quick to seize on issues that were not perceived to be in the best interests of the South. The principal argument of the opposition was that the agreement to construct the canal had been signed before any feasibility studies had been undertaken to identify what the impact of the canal would be on the inhabitants. Certainly, the intellectuals were well aware of the Equatorial Nile Project and the criticisms of the Jonglei Investigation Team, but this was a telling argument, the rationale of which was soon obscured, however, by bizarre rumors that 6000 Egyptian *fallahin* would be settled in the Canal Zone and arguments that the canal would drain off enormous quantities of water from the Sudd. It would deplete the supply of fish, disturb weather patterns and rainfall, and even lead to an increase in desertification.

Those who opposed the Southern Regional Government, particularly Benjamin Bol and Joseph Oduho, enlisted other dissidents to use Jonglei to incite the students in the schools to boycott classes and demonstrate against the government in October 1974. When rumours were circulated that Egyptian troops were already in Khartoum preparing to escort Egyptian *fallahin* to the Upper Nile, the students of Juba Commercial Secondary School staged a protest, marching through Juba to the offices of the High Executive Council. They were joined by many citizens of Juba who had no interest in the canal but saw it as a means to bring down the government led by Abel Alier and dominated by the Nilotes who were regarded by the Equatorians of Juba almost as much the enemy as the Northern Sudanese or Egyptians. Blocked by the police a melee ensued which quickly turned into a riot after three people were killed by the security police. It took the army two days to restore order but not after considerable damage had been done to public and private property.

The government responded with uncommon dispatch. Three committees were hastily formed from members of the High Executive Council and the Regional Assembly and packed off to the provinces to dismiss the rumours and emphasize the benefits of the canal. Promises were made. The inhabitants of the Canal Zone were to receive schools, medical and veterinary care, clean drinking water, agricultural extension services, and bridges and ferries for livestock and people to cross the canal to reach the *toic* pastures. All of these activities were to be directed by the National Council for the Development of the Jonglei Canal Area appointed instantly by Presidential decree. Abel Alier, the respected President of the High Executive Council, issued a calming statement to the Regional Assembly which was widely distributed disclaiming the rumours and emphasizing that his government was not simply going to accept meekly continued underdevelopment, backwardness and poverty. 'I wish to say that although this (Jonglei) is a Central Government project, the Regional Government supports it and stands for it. If we have to drive our people to paradise with sticks, we will do so for their own good and the good of those who come after us' (Abel Alier 1974).

Overnight what had been a large-scale engineering scheme had blossomed into an ambitious programme of social and economic development. The National Council for the Development of the Jonglei Area was charged by President Numayri with 'formulating socio-economic development plans for the Jonglei area and the promotion of studies of the effects of the construction of the canal on the lives and livelihood of the local inhabitants' (Presidential Order No. 284, 1974). The Council was responsible for implementing development projects, the procurement of funds, and the establishment of programmes of agriculture, industry, and public works. The council itself was an omnibus body with a broad spectrum of politicians and civil servants which met irregularly, the actual work being delegated to the Jonglei Executive Organ (JEO) led by a succession of commissioners and directed by a permanent staff. The appointment of the Council for the Jonglei Area, or rather its executive, the Jonglei Executive Organ, however, suddenly created a rival to the Permanent Joint Technical Commission, for the plans being generated for the socio-economic development of the Canal Zone by the JEO were not always compatible with the designs of the PJTC. The engineers and hydrologists of the PJTC were principally concerned with digging a canal. They had many years of experience and money and suddenly found their project challenged by social, cultural, and economic considerations emanating from a rival for control, the JEO. There were bound to be tensions; that there were not more is a tribute to the common sense and compromise by the influential members of these two potentially antagonistic agencies.

Criticism of Jonglei was not confined, however, within the Sudan. Emanating from the United Nations Environmental Programme with its headquaters at Nairobi the hue and cry of the environmentalists was picked up in the European press, particularly in Germany and France. Sometimes thoughtful, often strident, frequently ignorant, a steady stream of copy poured forth from the media led by a coalition of environmental groups in Europe and the United States known as the Environmental Liaison Centre with headquarters in Nairobi demanding an immediate moratorium on any construction work in the Canal Zone. The academics held seminars; the United Nations conferences. In 1977 the United Nations Conference on Desertification held in Nairobi provided a global forum for those who denounced the canal as advancing the march of sand southward while drastically altering the climate of Sudan's neighborus. The discussions culminated at a packed meeting sponsored by the Royal Geographical Society in London on 5 October, 1982 entitled 'Impact of the Jonglei Canal in the Sudan'. (Howell,1983)

The criticisms of the environmentalists are many but can be segregated into charges that Jonglei will drastically affect climate, underground water supplies, silt and water quality, the destruction of fish, and changes in the life style of the Nilotic people. The spectre of thousands of Egyptian *fallahin* descending upon the Upper Nile Province lingers on even after that issue had been put to rest following the Juba riots. With the exception of the change in the lives of the Nilotes, there is little substance to these criticisms. The charge by Professor Richard Odingo of Nairobi University that 'Here is a canal being built in an area that could easily be Africa's next desert' (Environmental Liaison Centre 1977) is nonsense. After thorough research the Jonglei Investigation team had demonstrated twenty-five years before that the canal would have no significant effect on precipitation. Current research has only confirmed this conclusion. Others have feared that ground water in the Sahel region to the north will dry up by the off-take of water from the Sudd by the canal. Again the critics have not done their homework. The Sudd, in fact, is underlain by a strata of impervious rock known as the Umm Ruwaba upon which rests a layer of impervious clay. The wells of the Sahel and Sahara are recharged by the floods of occasional rains, not from the swamps of the Nile. Nor will water quality be affected. The ecological imbalance resulting in a massive deterioration in plant cover produced by the reduction of nutrients contained in the water drawn-off by the canal is hardly likely to occur when the mineral content of the Bahr al-Jabal upstream of the Sudd and on the White Nile downstream are virtually the same. Moreover, increased flooding accelerates the rate of evaporation therefore increasing the salt content of swamp water. And will the fish, which play a crucial role in the annual life cycle of the Nilotic people, disappear? The Jonglei Investigation Team did predict a reduction in potential fish production by nearly half, but they were taking into account a much larger canal and a permanent swamp with its extended floodplain half the size of the Sudd today. Indeed, the multi-volume study of the Canal Zone undertaken by Mefit-Babtie Srl. from 1979-1983 concluded that 'an operational canal therefore should have little impact on these fishing grounds and fish stocks. Shrinkages of the swamps and floodplains will have a minimal effect upon them except to make land access easier in places' (Mefit Babtie Oct. 1983 *Final Report*, p. 70).

Of greater importance to the environment of the canal zone would be the effect of Jonglei on the flooding from the Bahr al-Jabal of the permanent swamp and seasonal floodplains. Since the canal will reduce the amount of water flowing into the Sudd, not only will the outflow be diminished but the area of permanent swamp and seasonal floodplain will decrease, changing the ecology of the Sudd, with a concomitant effect on the human and animal populations. The PJTC has calculated a host of scenarios, based on measured flows from 1905 and controlled by the regulator at Bor, whereby the amount of water being diverted through the canal would vary from 15 to 25 Mm^3/day, depending upon the natural flow of the Bahr al-Jabal and the seasons, wet (May-October) or dry (November-April). First, the effect of the canal is understandably greater on the permanent swamp than on the seasonal floodplain. Second, the reductions in the area of both the permanent and the seasonal swamp are naturally greater when 25 rather than 20 Mm^3/d are passed down Jonglei. Thus, at a canal flow of 20 Mm^3/d measured against a river flow of seventy-five years (1905-80) the permanent swamp will shrink from 9500 to 6200 km^2 (35 per cent) the seasonal swamp only from 7400 to 5800 km^2 (22 per cent). At a canal flow of 25 Mm^3/d, the reduction in the permanent swamp will be greater, from 9500 to 5500 km^2 (43 per cent), the seasonal swamp from 7400 to 5400 (27 per cent)-(Sutcliffe and Parks 1982 p. 46). But there are also a quarter of a million people scattered throughout the canal zone. Before the disruptions of the current civil war the majority of the southern Dinka lived in permanent settlements west of the canal. North of Kongor, the Gaawar Nuer villages are scattered on either side, while the Lou Nuer live to the east; but not all migrate west to the *toic*, preferring instead to go north to the Sobat. The slight rise in the land along the Duk Ridge makes these habitations possible; this area will expand west of the canal as the size of the permanent and seasonal swamp is reduced and creeping flow from the east is blocked by the canal itself. Moreover, the possibility of improved access and communication afforded by the canal, as well as services, trade, and administration, will undoubtedly concentrate villages along the line of the canal, with concomitant advantages and disadvantages. The canal will shorten the navigation route from Khartoum to Juba by some 300 km and will hopefully provide a wide range of new economic alternatives to a people dependent upon herding, cultivation, and fishing. At the same time the Nilotes living west of the canal on the Zaraf Island and in the Bahr al-Ghazal will remain as isolated as they are today in time of war; for even in peace the steamer traffic along the Bahr al-Jabal will become less frequent, if it does not cease altogether, and the river channel will be subject to the accumulation of sudd obstructions (See also Chapter 12,). Oil exploration may serve to keep the Bahr al-Jabal open, just as the canal would allow easier access to the heart of the Upper Nile lying to the east; but the exploration and exploitation of oil reserves in the Sudd, like the completion of the canal itself, all depend upon an elusive peace in that war-torn land.

Much of the optimism about the development of the canal zone during the decade of peace in the Southern Sudan has been perhaps misplaced, based as it was on the prospect of the numerous benefits to be derived from the raised roadway. The road is only a panacea even if it is properly constructed. Roads throughout the Upper Nile have always been notoriously difficult to build. The cotton soil, which turns to cement in the dry season and viscous mud in the wet, is totally unsuited for road construction. Not a stone nor rock nor pebble can be found in the Upper Nile, the nearest road-making material being the ironstone plateau many long and tortuous kilometres to the south in Equatoria. The terrain of the Upper Nile is broken by cracks, hummocks, and holes. The heavy rains relentlessly destroy the construction of the previous dry season, and what remains is soon turned to deep and impassable ruts by the few vehicles which venture along the tracks of the province. Even the expectations generated by the raised roadway are premature. Not only has the PJTC insisted that the road remain on the east bank of the canal, thus rendering it of little use to the populous southern Dinka whose villages are situated on the west bank, but the surface will not be sealed by gravel, asphalt, or cement, so that the clay debris from the canal will be serrated and gutted during the rains by even a few passing cars and trucks, and by migrating cattle and wildlife crossing the canal. Despite slope and camber to facilitate the run-off of rain-water, erosion will be heavy; and without intensive labour, expensive grading equipment, and constant maintenance, as CCI has learned during canal construction, this road will not become the all-weather, northern link to the south, but just another impassable quagmire in the Upper Nile.

A more important and immediate advantage of the canal will be a permanent water supply from the canal for the inhabitants. Although the water quality of the Bahr al-Jabal does not meet the standards of the World Health Organization, people and animals will drink it if they can reach it. The astonishing failure to install off-take pipes between kilometres 40 and 309, as recommended by CCI and the JEO, will result in cattle concentrating along the banks of the canal in order to drink, pulverizing the road and the canal banks, and devouring the limited pasture near the canal instead of watering from pools and *hafirs* fed by pipes and located a sufficient distance away from the line of the road and canal, where grazing is more plentiful. This irresponsible dereliction and the determination to continue the road along the east bank were decisions which have bitterly alienated the inhabitants - decisions known to all, not just the southern élite, and regarded as showing a hostile lack of sensitivity to the needs of the Nilotes, confirming Southern Sudanese deeper feelings of neglect and discrimination, set against meagre financial savings.

Flooding along the east bank from creeping flow has already affected the life of the peoples and animals living east of the canal. Although the impounding of water along the 267 km of excavated canal has not reached the extent that many feared, the accumulation of local precipitation and creeping flow is an obstacle to migrating herds of cattle and wildlife. But, as with so many other aspects of the canal, what is perceived as harmful has in fact presented opportunities. The impounded water can be directed and dissipated, as CCI has demonstrated in its pilot scheme, to produce nutritious and palatable grasses in the dry season. The construction of a network of dikes 3-10 km east of the canal would not be very costly, and would open additional grazing and provide drinking water during some months of the dry season. 'Uncontrolled, it (the flooding) may also cause damaging inundation of permanent habitations and their adjacent cultivations; indeed, it had already caused considerable hardship in this way before November 1983 when the matter was raised at the JEO's Annual Coordinating Conference at Bor. The failure of the PJTC to acknowledge the serious nature of this problem caused much public concern.' (Howell *et al.* 1988, p. 422).

While the Nilotes grumbled about the failure of the government to fulfill the promises made in 1974 of socio-economic improvements in the form of schools, dispensaries, and veterinary services, they had experienced only a few of these services in the past and consequently regarded their appearance in the present with some scepticism. As the Bucketwheel chewed through the cotton soil of the Upper Nile, however, the canal and its impact upon the region suddenly became a reality. The schools, dispensaries, and cattle vaccines may appear in the promised future, but the excavated canal was now, the present. The failure to provide off-take structures and to control creeping flow produced instant resentment; but the presence of the canal was already disrupting the traditional migratory patterns essential for the very survival of Nilotic cattle and wildlife as early as 1982. Cattle had to move through the intermediate lands, at the end and the beginning of the rains, to the *toic* pastures. During this migration most of the people and their animals in the canal zone had to cross the canal.

Not all the Dinka and Nuer need to cross the canal. The Gok Dinka, south and east of Bor, do not have to cross the canal to reach the Bahr al Jabal and the Athoic Dinka north of Bor have their permanent settlements west of the canal, as do the Gaawar Nuer farther north. Most Dinka and Nuer, however, must cross the canal twice, some four times, each passage entailing considerable risk for their cattle and demanding much tiresome work for the herdsmen. The number of annual cattle crossings has been estimated at 700,000; sheep and goats 100,000; people 250,000. Clearly, the proposed ferries and bridges are a totally inadequate solution to these daunting statistics; but the failure to find funding for their implementation produced as much resentment as their inadequacy. Cattle can swim, as they do across the Bahr al-Jabal at Bor and across the upper Sobat, but before entering the water hundreds of thousands of hooves will wreak havoc upon the surface of the canal embankment. Goats and sheep are poor swimmers, as are the young, the old, and the infirm human beings who do not remain in the permanent settlements. Indeed, the resourcefulness of the Nilotic cattlemen will probably prove in the end more effective in getting across the canal than any palliatives the government may offer. The Dinka and the Nuer are adroit herdsmen, and they will undoubtedly exploit new options presented by the canal while seeking means to overcome it's obstacles In 1988, five years after the resumption of civil war in the Southern Sudan, the peoples of the canal zone experienced the worst of all possible worlds - a land ravaged by war and an empty canal providing neither water nor benefits. The challenge to their determination and ingenuity has resulted in their digging out the canal embankment and constructing ramps to drive their cattle across to the intermediate lands and the *toic* pastures.

The renewal of civil war in the Sudan in 1983 has brought the return of famine, disease, and death to the peoples of the upper Nile, with effects even more debilitating than the great floods of the 1960s, the seventeen years of civil war from 1955 to 1972, or an incomplete (or, for that matter, a finished) canal. To be sure, the grievances created by the excavation of Jonglei were substantial ingredients in the overflowing stew of discontent among the Southern Sudanese, but they were neither cause nor catalyst for the resumption of war against the Sudan Government by Colonel John Garang and his followers in the Sudan Peoples Liberation Army. Even if the canal were to produce no benefits, only costs, human and material, the present conflict has been far more destructive to the livelihood of the Nilotic peoples than would be any changes in their way of life to accommodate themselves to an environment altered by the presence of the canal. Few water projects in the world have undergone such scrutiny as Jonglei; and although further research is always needed in the hope of finding more definitive answers, what has already been accomplished can be the basis of realistic conclusions if and when the Jonglei Canal is finally taking water to the Northern Sudan and to Egypt. If properly managed, with sensitivity to the concerns of the inhabitants and a reasonable effort to meet their needs, the potential benefits from the Jonglei for the Nilotic peoples far outweigh the potential losses occasioned by the changes in their way of life. No longer can the denizens of the Sudd remain in splendid isolation; independence, floods, and civil war have shown that. It appears on balance that in a changing world it is more prudent to have new options (which Jonglei would provide) than to rely on a tradition challenged from without and from within.

Textual Sources

Alier, Abel (1974). 'Statement to the People's Regional Assembly on the Proposed Jonglei Canal'. Khartoum.

Butcher, A.D. (1938). *The Sadd Hydraulics,* Cairo.

Collins, Robert 0. (1983). *Shadows in the Grass: Britain in the Southern Sudan, 1918-1956,* New Haven.

____ (1985). 'The Big Ditch; The Jonglei Canal Scheme', in Daly, M.W.(ed.), (1985). *Modernization in the Sudan: Essays in Honor of Richard Hill.* New York.

____ (1987). 'The Jonglei Canal: The Past and Present of a Future'. 6th Trevelyan Lecture. University of Durham. Durham.

____ (1990). *The Waters of the Nile: Hydropolitics and the Jonglei Canal, 1900-1988.* Oxford.

Egyptian Ministry of Public Works (1981). *Water Master Plan.* 17 vols.

Cairo.
Garstin, Sir W. (1901). *Report as to Irrigation Projects on the Upper Nile.* London.
_____ (1904). *Report upon the Basin of the Upper Nile*, London, Guariso, Giorgio, *et al* (1987). 'Implications of Ethiopian Water Development for Egypt and Sudan', *Water Resources Development* 3.
Howell, Dr. Paul P. *et al.* (1954). *The Equatorial Nile Project and its Effects in the Anglo-Egyptian Sudan: Being the Report of the Jonglei Investigation Team.* 4 vols. Khartoum.
_____ (1983). The impact of the Jonglei Canal in the Sudan. *Geographical Journal*, **Vol.149,** Part 3.
_____ Lock, M. and Cobb, S. (1988). *The Jonglei Canal: Impact and Opportunity.* Cambridge.
Hurst, H.E., Black, R.P., and Simaika. Y.1. (1946). *The Future Conservation of the Nile.* vol. VII. *The Nile Basin.* Cairo.
Hydromet (1974) *Hydrometeorological Survev of the Catchments of Lakes Victoria, Kyoga and Albert.* 4 vols. Geneva.
JEO (Exccutive Organ for Development Projects in the Jonglei Area) (1975). 'The Jonglei Project: Phase One'. Khartoum.
Johnston, Sir H.H. (1906). *The Nile Quest.* London.
MacDonald, Sir Murdoch (1920, 2nd edn., 1921). *Nile Control* 2 vols. Cairo.
McGregor, R.M. (1945). 'The Upper Nile Irrigation Projects". W. N. Allan Papers. Sudan Archive. Durham University.
Morrice, H.A.W. and Allan,William Nimmo (1958). *Report on the Nile Valley Plan.* 2 vols. Khartoum.
Mefit-Babtie Srl. (Oct. 1983). *Development Studies in the Jonglei Canal Area: Final Report.* vol. I. Glasgow, Khartoum, Rome.
Numayri, Gaafar al (1974). 'Presidential Order No. 284). Khartoum.
Odingo, Richard (1977). 'Environmental Liaison Centre Press Release'. Nairobi.
Sutcliffe, J.V. and Parks, Y.P. (1982). 'A Hydrological Estimate of the Effects of the Jonglei Canal on Areas of Flooding'. Wallingford.
_____ (1987). 'Hydrological Modelling of the Sudd and Jonglei Canal'. Hydrological Sciences Journal. 32/2.
U. S. Department of the Interior (1964). *Land and Water Resources of the Blue Nile Basin:Ethiopia.* 17 vols. Washington, D.C.
Whiteman, Majorie D. (1964). *Digest of International Law.* vo1.3. Washington, D.C.

Part II

Nile management and factors affecting future management

6

Global climate change and the Nile basin

M. HULME

Introduction

Precipitation over the Nile Basin, and consequently Nile discharge, has fluctuated both historically and pre-historically. Precipitation zones have migrated latitudinally over the middle Nile by as much as 600 km over the last 20,000 years. Within the twentieth century, decadal precipitation changes have been up to 20 per cent of these Holocene changes. Our current understanding of precipitation over the Nile Basin suggests its sensitivity to a variety of factors operating on different time and space scales. These factors include changes in the Earth's orbit, global ocean temperature anomalies, migrations of the Inter-Tropical Convergence Zone, and land-cover changes within the African continent. Future global climate, however, may be subject to a new anthropogenic forcing, the impact of which on precipitation over northeast Africa, and hence Nile discharge, is still poorly understood, namely, progressive increases in the global atmospheric concentration of greenhouse gases.

This chapter presents a range of climate scenarios specifying the likely magnitudes and rates of change in Nile Basin precipitation over the next few decades. These scenarios are constructed from historical analogues, recent instrumental data, and General Circulation Model experiments. It is suggested that twentieth century precipitation characteristics are unlikely to be a reliable guide to twenty-first century precipitation. Relative contributions to Nile discharge from the Blue and While Niles may change, and increased evaporative water loss may be as important for assessing future Nile discharge as changes in precipitation. There will need to be greater flexibility in Nile Basin water management.

The Nile Basin encompasses nine countries in northeast Africa and has a surface area of about 2.9 million km^2. The river itself is 6700 km from source to mouth, the second longest in the world (Figure 1). The population of the Nile Basin is about 140 million of whom at least 50 per cent are heavily dependent on the Nile waters for their economic and domestic existence. The reliability of Nile discharge is fundamental for the well-being of northeast Africa, and especially of Egypt and the Sudan, the two major downstream nations in the Nile Basin.

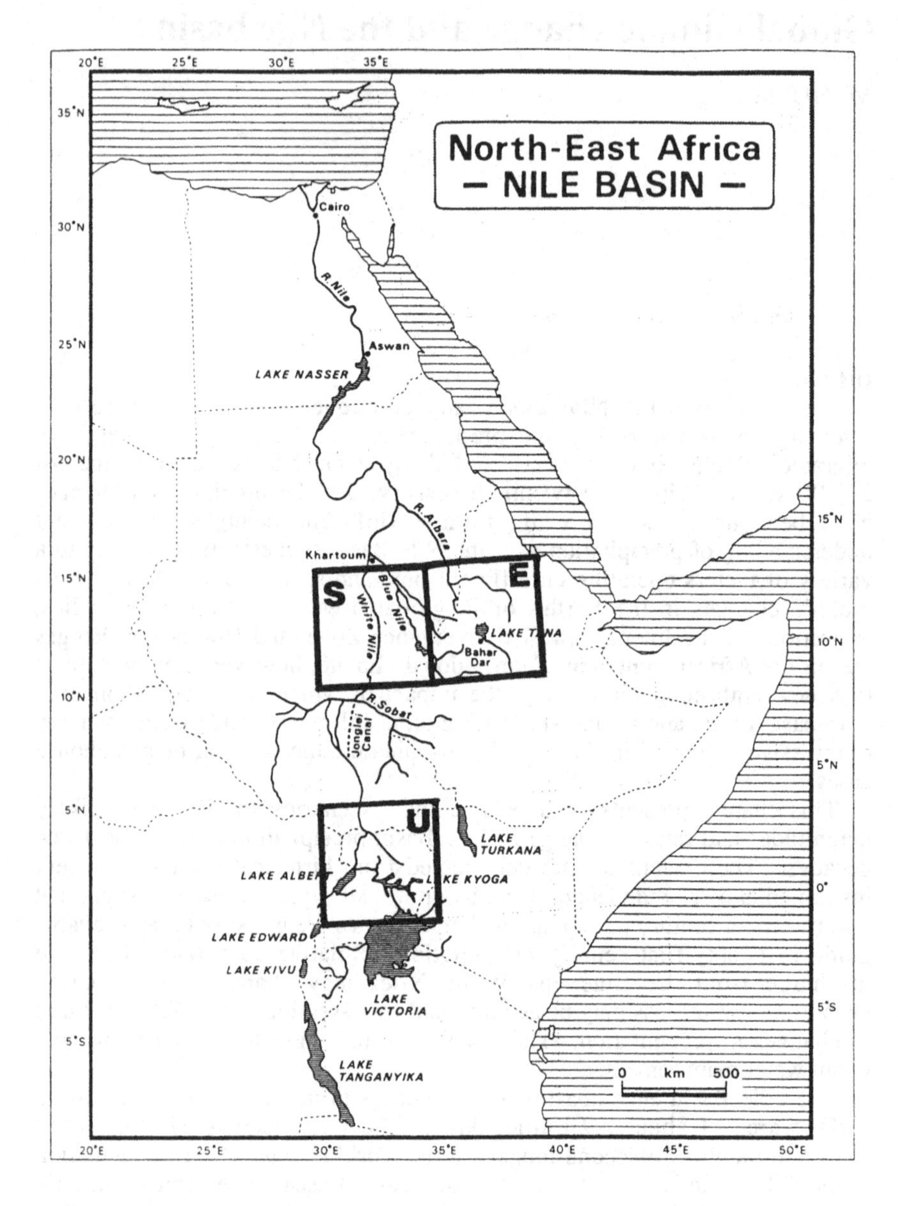

Figure 1: The Nile Basin. The three regions used in the instrumental data are shown (U= Uganda, E= Ethiopia, S= central Sudan)

Nile discharge has fluctuated extensively over time. Well-established periods of substantially lower flows than present occurred during the last glaciation around 20kyr BP (Williams and Faure, 1980), and within the Holocene smaller fluctuations occurred resulting both in increased (e.g. 11 to 7kyr BP, Wickens, 1975), and decreased (e.g. 1180 to 1350 AD, Hassan, 1981) discharge. Within the twentieth century substantial interannual variability in Nile discharge has occurred with a maximum annual yield of 120 km^3 in 1916 and a minimum of only 42 km^3 in 1984. Interdecadal variations within the present century have also been substantial, with mean annual discharge between 1900 and 1959 reaching 84 km^3 compared to a mean of only 72 km^3 between 1977 and 1987, a 15 per cent decline.

Such fluctuations in discharge reflect a variety of both natural and anthropogenic processes operating on different time scales. Natural processes include changes in precipitation regimes over the headwaters of the Blue and/or White Niles, changes in evaporation, and changes in catchment vegetation which affect runoff. These natural processes which determine discharge variations are also affected by anthropogenic changes within the Basin environment. Thus precipitation may be affected by, for example, human-induced land-cover change (Nicholson, 1988), evaporation by the creation of artificial water bodies, and vegetation by the extraction of woodfuel (Ahlcrona 1988). An additional anthropogenic control on total Nile discharge is the extraction of water for domestic, agricultural, industrial, or power generation purposes.

This chapter addresses one of the above controls on Nile discharge, namely variations in the precipitation input over the Nile Basin. The precipitation of the Nile Basin is sensitive to both natural and anthropogenic forcing factors which operate on a variety of time and space-scales. Any future management plan of the Nile waters requires an assessment of the likely precipitation inputs over the Nile Basin. Such an assessment has several important components, requiring to distinguish between 1) changes in the precipitation inputs over the Blue and White Nile catchments; 2) changes in the interannual variability of precipitation and changes in mean seasonal precipitation; 3) changes in the frequency-distribution of short-term (e.g. one to five day) and longer-term (e.g. seasonal to decadal) precipitation characteristics.

In order to assist in distinguishing between these different sensitivities of Nile Basin precipitation to climate change, we first present a hierarchy of Nile Basin precipitation forcing factors in terms of their spatial and temporal scales. These factors have been proposed as operating in either the recent or more distant past, or are likely to operate in the near future. A number of these factors is discussed briefly, giving examples of how Nile Basin precipitation has been affected by each.

The chapter then describes precipitation changes over the Nile Basin during the present century, and the projected changes for next century. Precipitation changes during the instrumental period since the nineteenth century have been substantial, but the high temporal resolution of the data available over this relatively short period emphasises the large interannual variability of Nile Basin precipitation. Future precipitation may be assessed by a number of methods, each of which generates estimates on distinctive time and space scales and which are the result of particular forcing factors. The results from one such method are presented, namely a climate scenario for the Nile Basin derived from General Circulation Model (GCM) experiments exploring the effect of rising concentrations of greenhouse gases on global climate. We conclude by emphasising the need to estimate future Nile Basin precipitation using a model incorporating all of the forcing factors identified at the beginning of this chapter. Such a capability has not yet been established.

A hierarchy of factors affecting Nile Basin climate

When seeking explanations of both weather and climate anomalies the question of scale is all important. Although the precipitation climatology of the Nile Basin is not uniform (we distinguish later, for example, between the Blue and White Nile catchments), there is a number of forcing factors which determines the precipitation input over large parts of the Nile Basin, and hence contributes to the resulting annual discharge of the Nile. These factors operate on distinctive scales which are defined diagrammatically in Figure 2.

Not many of these separately identified factors are truly independent. Thus, for example, El Ñino/Southern Oscillation (ENSO) events partly determine the characteristics of the Tropical Easterly Jet, and the location of the ITCZ partly determines the moisture characteristics of the local atmosphere. Furthermore, the space and time scales that each factor is identified with are also approximate. The diagram (Figure 2) helps, however, in demonstrating the complexity and hierarchy of factors which determine the precipitation input over the Nile Basin in any given year. Five of these factors will briefly be elaborated: orbital cycles, the greenhouse effect, ENSO events, regional land-cover changes, and migrations in the ITCZ.

Orbital cycles

Cyclical variations in the parameters of the Earth's orbit around the sun have been known about since the nineteenth century, and have been reasonably well quantified since the work of Milankovitch in the 1930s. The arguments over the extent of their influence on global climate have been only more recently resolved, largely through the use of GCMs in modelling the impact of the consequent changes in the distribution of solar insolation on the global circulation (Berger, 1988).

The shortest of these cycles (the precession of the equinoxes, which operates over 19,000 and 23,000 years) is the one of greatest relevance for our current purposes. The extensive work undertaken by the COHMAP (Cooperative Holocene Mapping Project) has combined observational and modelling evidence to demonstrate convincingly the sensitivity of tropical precipitation to this particular orbital forcing (COHMAP, 1988).

At the end of the last high-latitude glaciation at 18kyr BP, the highly reflective ice-sheets, the generally cold oceans, and the equatorward-extended sea-ice borders resulted in an equatorward displacement of the polar front and the mid-latitude westerlies. The African monsoon was less extended and precipitation over Africa, particularly sub-tropical Africa, was somewhat lower than today. Over the next 9000 years the boreal summer (JJA)

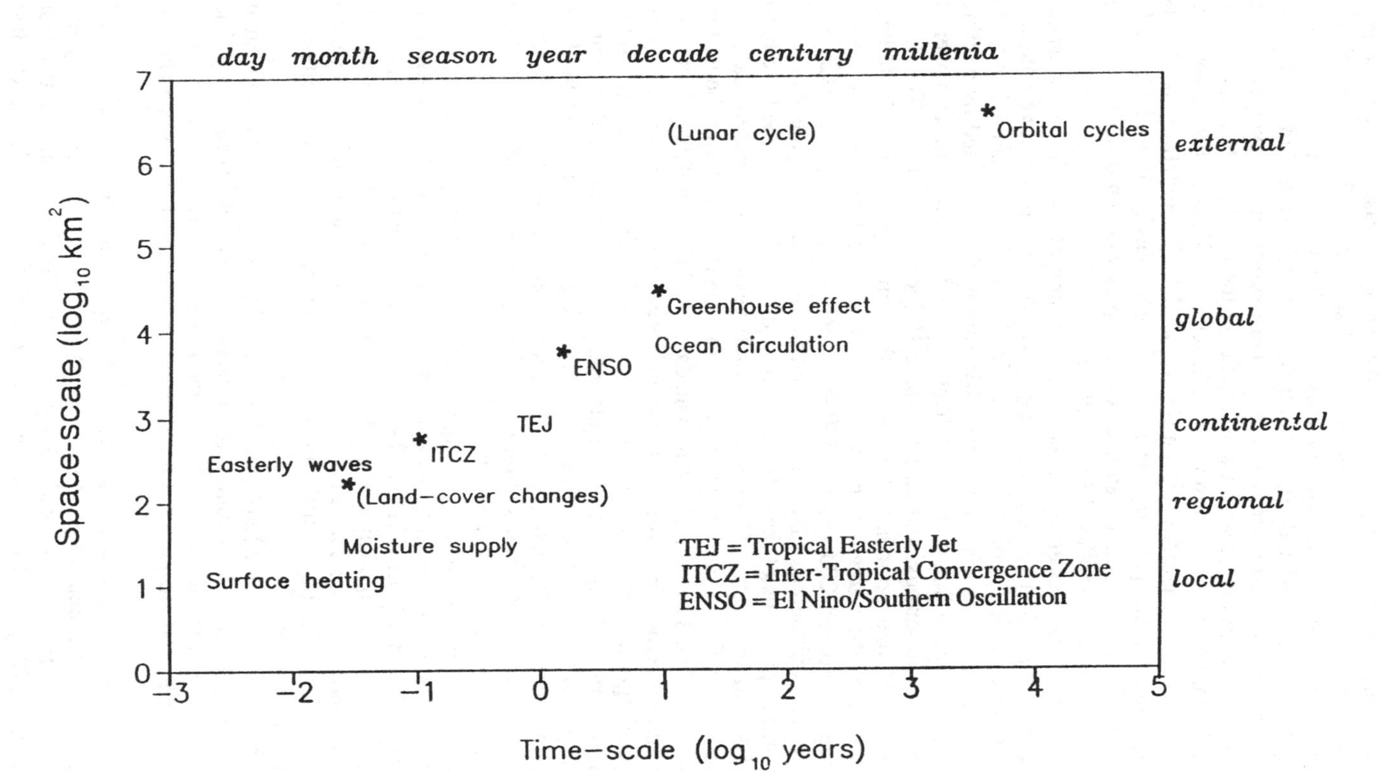

Figure 2: A heirachy of forcing factors on the Nile Basin precipitation arranged by time and space scales.

insolation received over the Northern Hemisphere increased by about 8 per-cent, with a corresponding decrease in austral summer (DJF) insolation over the Southern Hemisphere. The increased Northern Hemisphere summer heating led to increased thermal contrasts between land and ocean and enhanced monsoons in the tropics and northern sub-tropics. Lakes expanded on the Saharan margins, precipitation zones over the middle Nile in the Sudan were shifted up to 500 km northwards (Wickens, 1982), and Nile flows were more substantial than today (Adamson *et al.* 1980). Modelling work has suggested that at the peak of the mid-Holocene pluvial around 9kyr BP, precipitation levels in the African and Asian tropics were up to 25 per cent to 30 per cent higher than today (COHMAP, 1988).

The central Sudan lies at the current northern limit of the African monsoon, and hence precipitation yield in this region is a particularly sensitive indicator of the extent and vigour of the monsoon circulation. The latitudinal shifts in the precipitation zones over northeast Africa are well represented by plotting the position of the 400 mm annual isohyet. Figure 3 does this for a number of recent years and periods, and also for three Holocene centuries. The driest and wettest 30-year periods in the twentieth century produced a latitudinal shift in precipitation zones of between 50 km and 75 km. This movement increased to between 200 km and 300 km when individual years (1929 and 1984) were considered. By contrast, latitudinal shifts in the estimated century-mean positions of the 400 mm isohyet during the Holocene have been up to 500 km in extent, Over the last 9000 years to the present-day, the seasonality of solar insolation has returned to a pattern similar to the end of the last glaciation. The very different boundary conditions (notably global ice-cover) have meant, however, that tropical precipitation has not fallen to correspondingly low levels. Nevertheless, over the last few thousand years, the forcing of orbital changes on global circulation has been in the direction of reduced precipitation over the Nile Basin. This is especially true for the Blue Nile catchment which, being under a monsoonal regime, has been more sensitive to these global circulation shifts. Precipitation over the White Nile catchment has been less subject to orbital forcing.

Since the Earth is currently nearing the end of one of the precession cycles, after about 5kyr AP one might expect a strengthening of the monsoonal circulation over northern Africa as boreal summer insolation again begins to increase. Such a strengthening should reach a maximum at about 15kyr AP (Berger, 1988). For the immediate future, however, (less than 500 years) orbital forcing will be of little importance for Nile Basin precipitation.

Greenhouse forcing

Changes in the orbit of the Earth occur externally to the global climate system yet are one of the most important forcing factors on millenial time scales. The next example of forcing of Nile Basin precipitation operates internally to the climate system, and currently on the decadal to centennial time scale (Figure 2).

Global atmospheric concentrations of carbon dioxide (CO_2) have varied substantially throughout the history of the Earth. Within the Quaternary era they have varied from a maximum of about 300 ppmv during the last interglacial (120kyr BP) to a minimum of about 190 ppmv during the last

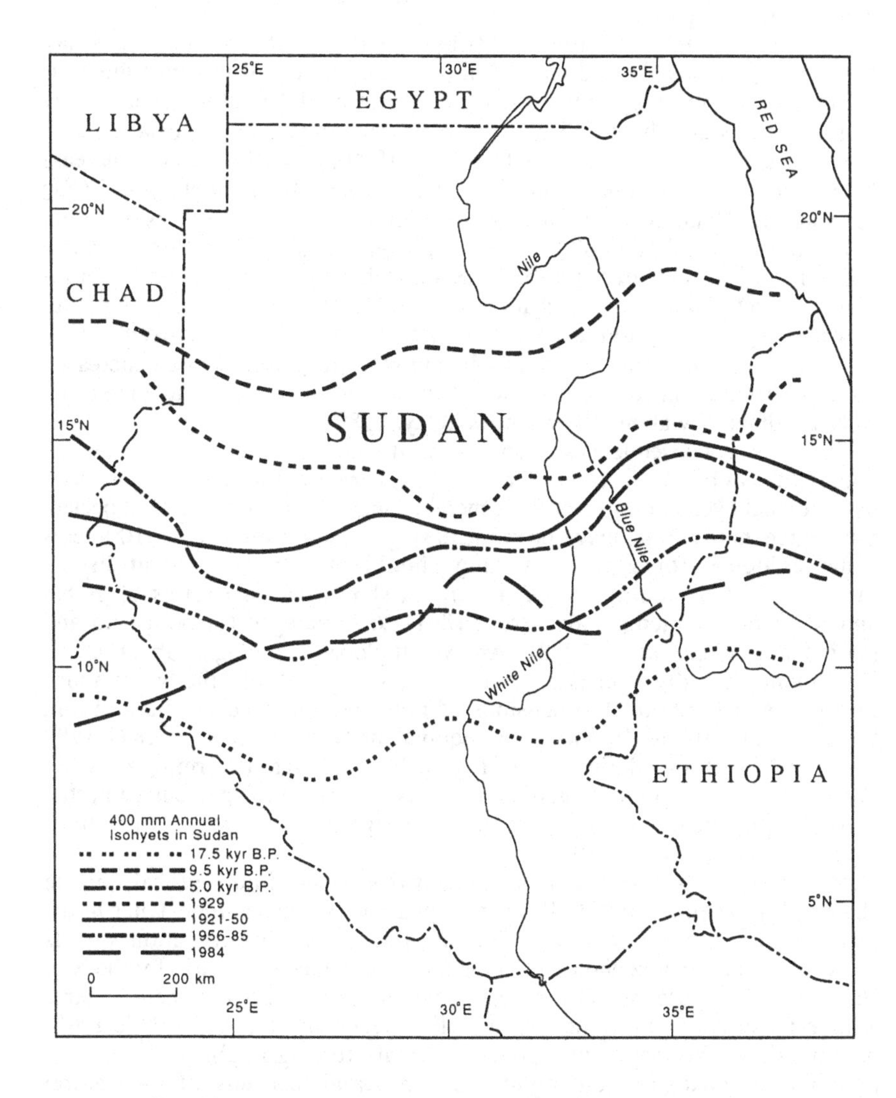

Figure 3. Position of the 400mm annual isohyet in central Sudan for various twentieth century years and periods, and for three Holocene centuries. Holocene estimates are from Wickens (1982)

glaciation (25kyr BP). For much of recent human history over the second half of the Holocene, global levels of atmospheric CO_2 have been relatively stable at between 240 ppmv and 280 ppmv.

The accelerating use of fossil fuels by society over the last 150 years has, however, again destabilised the global carbon cycle, the disturbance this time being clearly anthropogenic in origin. Current global atmospheric CO_2 concentrations are about 353 ppmv, which represents a 25 per cent increase over pre-Industrial levels (ca. 280 ppmv in 1800), and a 12 per cent increase over 1958 levels (315 ppmv), the first year of reasonably accurate global CO_2 estimation. When these increases are combined with increases in other 'greenhouse gases', notably methane, nitrous oxide, chlorofluorocarbons (CFCs) and low-level ozone, it becomes clear that the global climate system is being subject to a major perturbation through a substantial change in the long-wave energy absorption capacity of the atmosphere. The current CO_2-equivalent concentration of about 410 ppmv corresponds with an increased radiative forcing since 1800 of more than 2W m^2, equivalent to increasing solar irradiance by about 1 per cent (Wigley, 1989).

The climatic consequences of an enhanced global greenhouse effect remain uncertain. Global surface warming, stratospheric cooling and an enhanced hydrological cycle are three of the general changes which are widely expected. Since greenhouse gases have been increasing steadily over the last 100 years some evidence for global warming should already be evident in the instrumental temperature record. Figure 4a shows the time series of global annual surface air temperature from 1861-1989 averaged for both land and marine areas (Jones *et al.*, 1988). An overall global warming of about 0.5°C is evident, broadly consistent with, but not proof of, the greenhouse hypothesis. The regional dimensions of this warming need not be uniform, however, and Figure 4b shows the equivalent time series from 1871-1988 over the Nile Basin (defined as in Figure 1). An overall warming of again about 0.5°C is evident, although the majority of this warming occurred in the early decades of this century with little change in Nile Basin temperatures having occurred since the 1950s.

Determining the specific regional dimensions of any future global climate change, for example for Nile Basin precipitation, is much harder. Later in the paper we discuss methods of estimating the regional precipitation signals associated with the greenhouse effect and present a model scenario for the Nile Basin. None of these methods, however, is fully satisfactory and there remains a large uncertainty about the nature of the greenhouse forcing of Nile Basin precipitation. This new anthropogenic climate forcing might lead either to precipitation changes more rapid and substantial than any of the natural forcing mechanisms we are discussing, or to smaller changes which will be indistinguishable from the large natural variability of Nile Basin precipitation.

ENSO events

On time scales of less than a decade, the most significant forcing of global climate anomalies occurs through El Ñino/Southern Oscillation (ENSO) events in the eastern Pacific (Ropelewski and Halpert, 1987). Some of the pioneering

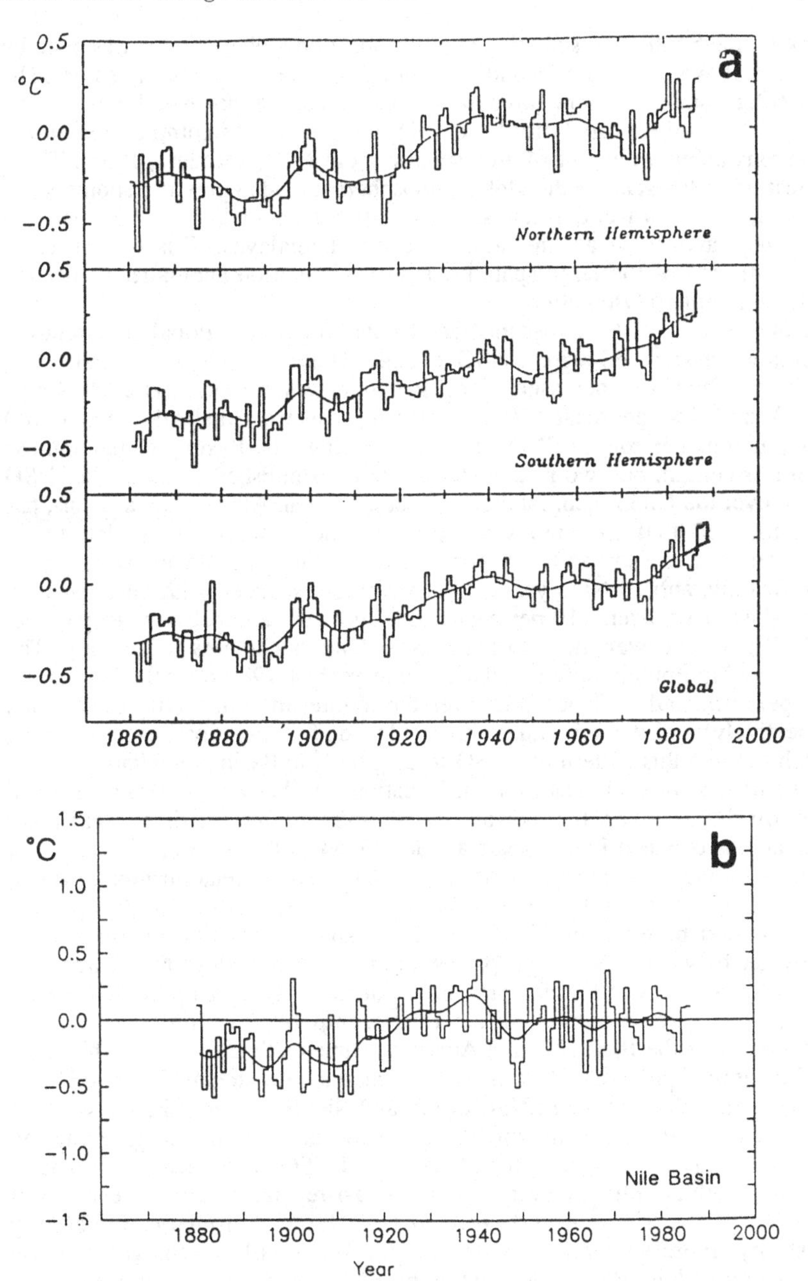

Figure 4: a+b, Time series of annual surface air temperature based on land and marine data (expressed as a °C anomaly from the 1951-80 mean).
a: Global and hemispheric series. (Source: Jones *et al*, 1988)
b: Nile Basin. (as defined in Figure 1)

work on the Southern Oscillation and its impact on global climate was done by Sir Gilbert Walker in the second and third decades of the present century. His particular concern was to forecast seasonal precipitation over India, and he achieved a limited capability by developing a series of complex regression equations using a number of atmospheric indicators (Normand, 1953). These indicators of the state of the global atmosphere included pressure anomalies at Darwin, Capetown and Buenos Aires, rainfall anomalies at Zanzibar and Salisbury, and snow accumulation over the Himalayas. This was an early demonstration of the large spatial extent of circulation anomalies associated with the Southern Oscillation.

More recent work has quantified the nature of the global precipitation anomalies associated with ENSO events. These include well established tendencies both for increased (Peru, Ecuador, western USA, and the Kenya coast) and for decreased (Mozambique, eastern Australia and India) precipitation. The role of ENSO events in forcing Nile Basin precipitation has been less certain, but two recent analyses have established a detectable ENSO signal over the Nile Basin. Janowiak (1988), from an Africa-wide analysis, has demonstrated that El Ñino years produce mean precipitation deficits of between -5 per cent and -15 per cent over the Nile Basin, while more significantly, anti-El Ñino years (El Ninas) are associated with mean positive anomalies of between +10 per cent and 25 per cent and between +5 per cent and 10 per cent over the White and Blue Nile catchments repectively. The relatively high precipitation of the El Ñino year of 1988 over the Nile Basin (+8 per cent and +13 per cent over the White and Blue Nile catchments respectively) and the subsequent high Nile flood, is the most recent example which fits into this pattern of ENSO forcing on Nile Basin precipitation.

Another recent study has examined relationships between ENSO events and Nile discharge over the period 1914-1988 and established significant associations between ENSO years and low flows in the Atbara, Blue Nile and Main Nile rivers (Attia and Abulhoda, 1990). No such relationship, however, was detected for the White Nile at Malakal. The explanation for the apparent contradiction between this latter exception and the positive relationship of Janowiak between ENSO and precipitation over the White Nile catchment most probably lies in the damping effect of the Sudd swamp on White Nile discharge. There is a more direct relationship between precipitation and discharge over the Blue Nile and Atbara catchments than between White Nile catchment precipitation and White Nile discharge downstream of the Sudd.

The relationship between ENSO events and Nile Basin precipitation suggests that between 10 per cent and 40 per cent of interannual precipitation variability may be accounted for in this way. ENSO phenomena are a natural part of the global climate system with a quasi-regular recurrence every four to seven years. There would seem to be a limited discharge forecasting capability (more so for Blue Nile than for White Nile discharge), and the modulation of Nile Basin precipitation by ENSO events certainly needs to be considered in any comprehensive assessment of future Nile discharge and its variability.

Regional land-cover changes

The possible forcing of Nile Basin precipitation by changes in regional land-cover characteristics occurs through two processes: the role of vegetation in modifying surface albedo and hence the energy flux and vertical instability of the local atmosphere; and the role of vegetation in facilitating the recycling of moisture through evapotranspiration. The importance of extensive vegetation change over continental Africa and South America in modifying regional precipitation has been widely debated (e.g. Salati *et al.*, 1979; Dickinson and Henderson-Sellers, 1988; Nicholson, 1988). The heart of the difficulty remains a paucity of observational evidence of both the magnitude of surface albedo changes in sub-Saharan Africa over time, and the exact contribution of recycled moisture (as opposed to oceanic moisture) to regional precipitation.

It has been suggested for some time that changes in the extent of the Sudd swamp in the southern Sudan, whether by natural high floods (e.g. the early 1960s), or by deliberate modification (e.g. the Jonglei Canal), would alter precipitation yield over the Sudd, and perhaps downwind of the Sudd over the eastern Sudan and western Ethiopia (see most recently El Tahir, 1987). Changes in swamp area would alter local evaporation rates, atmospheric humidities, and hence precipitation. There is little observational evidence, however, in support of such a relationship; indeed the Jonglei Investigation Team concluded that drainage caused by the canal would have no significant effect on precipitation (JIT, 1954). Moreover the large expansion of the Sudd in the late 1960s (for example, from an estimated 8300 km^2 in the 1930s to 22,100 km^2 in 1973; Sutcliffe and Parks (1987) did not lead to increased precipitation (Howell *et al.*, 1988; also see Figure 5 of this chapter). Equally, it seems unlikely that any substantial reduction in swamp area following the completion of the Jonglei Canal would lead to reduced precipitation yield. Increased areas of dry land would increase sensible heat flux and hence convective activity, which would counteract any tendency for reduced atmospheric moisture and hence reduced precipitation yield.

The few modelling experiments which have simulated atmospheric circulation and moisture sources for Sahel (and Nile Basin) precipitation, have suggested a relatively small contribution to such precipitation from the evaporative recycling of moisture (e.g. Cadet and Nnoli, 1987; Druyan and Koster, 1989). The largest source of moisture which precipitates over the Sahel and Blue Nile Basin is the tropical Atlantic, transported directly into the region in the strong low level southwest monsoon. Over the White Nile catchment, the moisture recycling role of Lake Victoria is more substantial, although again direct transport of moisture from the Indian Ocean contributes a larger proportion of precipitation.

If there is any role for land-cover change in altering the moisture supply to the Nile Basin it is most likely to be through the enhancement of pre-existing differences between wet and dry years. Druyan and Koster (1989) showed that following a wet start to the rainy season, the contribution of local evaporate to late-season precipitation was up to five times that in a year with a dry start to the rainy season, i.e. surface vegetation has a within-season feedback role in strengthening any given early-season precipitation anomaly. Consequently one might reason that if the regional vegetation resource is depleted, then this internal feedback role is weakened, reducing the amplitude of anomalously wet years. The frequency and magnitude of dry years would remain largely unaffected. There is some support for this contention from the empirical precipitation record for the Sahel and Blue Nile catchment over the last few decades.

At the present time we must conclude that the forcing of Nile Basin precipitation by regional land-cover change remains largely unsubstantiated. This is not to deny that large scale vegetation change has occurred over large parts of the Nile Basin, nor that such changes may yet be shown to have detectable effects on regional precipitation yield. Current empirical evidence, in contrast to some recent theoretical modelling, simply remains inconclusive. The lack of any association between the extent of the Sudd swamp and local precipitation in recent decades would seem to deny any substantial relationship operating on this particular space-scale (Howell *et al.*, 1988).

ITCZ migrations

On short time scales (less than one month), the main controls on precipitation over the Nile Basin are localised convective heating, the passage of easterly waves westward over the Blue Nile and middle Nile catchments, and the extent and timing of the seasonal migration of the ITCZ (Figure 2). These local controls may be forced by larger scale circulation anomalies as discussed previously. A recent example of where these local factors combined to generate an exceptional short-term precipitation event over part of the Nile Basin occurred in August 1988.

During a 14-hour period of the night of August 4/5 1988, an area approximately 500 km by 300 km in the north central Sudan experienced intense precipitation. At Khartoum, a total precipitation of 210 mm was recorded between 0800 hrs on 4 August and 0800 on 5 August, which is over twice the previous highest recorded 24-hour total (Hulme and Trilsbach, 1989). Severe flooding resulted around Khartoum and the main Nile valley to the north, partly as a direct result of this storm, and partly because the highest Blue Nile flood level at Khartoum since 1946 was reached a few days later (Sutcliffe *et al.*, 1989). The immediate causes of this exceptional precipitation event were threefold: a northward excursion of the ITCZ; a strong and deep southwesterly surface airflow with a high moisture content resulting from substantial rains in the southern Sudan over the previous few days; and the passage of an upper easterly wave over north central Sudan triggered by a strong surface low pressure trough (Ali, 1989). The unusual combination of these factors caused a precipitation event with an estimated return period of over 400 years (Hulme and Trilsbach, 1989).

Whether such events are likely to become more common in the future is obviously an important question. Such intense precipitation has a high runoff ratio and thus contributes more substantially to river discharge than several lesser events of similar total magnitude. The answer to the question is not straightforward. It requires an assessment of the specific local synoptic factors causing such events, and whether the combination of these factors is sensitive to any of the larger-scale forcing factors discussed above. It is not sufficient, for example, to suggest that more northerly migrations of the ITCZ will lead necessarily to increased heavy storm frequencies unless there is also an increased likelihood of vigorous easterly waves and deep, moist surface airflows.

Such complex atmospheric interactions argue strongly in support of integrated atmosphere-ocean-land General Circulation Models (GCMs) which, potentially, incorporate all of the forcing factors of Figure 2 over the complete space-time scale continuum. Such modelling is still at an early stage and is particularly hampered by the coarse spatial resolution of the grid into which the land-atmosphere system is divided. We will present some results for the future climate of the Nile Basin from the current generation of GCMs, but first the precipitation characteristics of the twentieth century are described.

Recent precipitation changes

The monthly precipitation data used in this paper are from the Global Precipitation Dataset (GPD) held in the Climatic Research Unit at the University of East Anglia. This GPD has been extended and updated from that used by Bradley *et al.*, (1987) and Diaz *et al.*, (1989). Data are held for about 400 stations in the Nile Basin, with records commencing in the 1880s and continuing through 1989.

Monthly precipitation series for three selected regions are constructed using a procedure outlined in Hulme (1989). Precipitation data are transformed to probability values using a root-normal distribution. Within each respective region the data are then averaged to produce a single probability value. This is converted back to mms using calculated region-average distribution parameters. This procedure produces precipitation series which are relatively insensitive both to missing data and to changes in station networks over time. The regions selected are 5° latitude by 5° longitude boxes and represent the upper White Nile catchment (referred to henceforth as 'Uganda'), the upper Blue Nile catchment ('Ethiopia'), and the middle Nile Basin including the confluence of the White and Blue Niles at Khartoum ('central Sudan') (see Figure 1). The Blue Nile contributes between 70 per cent and 80 per cent to overall Nile discharge, and the White Nile between 20 per cent and 30 per cent. The contribution of precipitation over the central Sudan to Nile discharge is small.

The twentieth century precipitation time series for these three regions is shown in Figure 5. Owing to differences in seasonality between the regional precipitation regimes, annual precipitation is shown for Uganda, whereas seasonal precipitation from June to August (JJA) is shown for the other two regions. JJA seasonal precipitation represents 54 per cent and 70 per cent of annual precipitation over Ethiopia and the central Sudan respectively.

Table 1 shows the direction of precipitation change over the twentieth century calculated by a simple linear regression. All three regions show a decline in precipitation totals, although this decline is minimal for Uganda. In relative terms the decline is greatest in the central Sudan where a 9 per cent reduction in seasonal precipitation has occurred over the twentieth century. In all three regions, however, most of the reduction has occurred within the last two decades, with substantial negative anomalies being recorded since 1965,

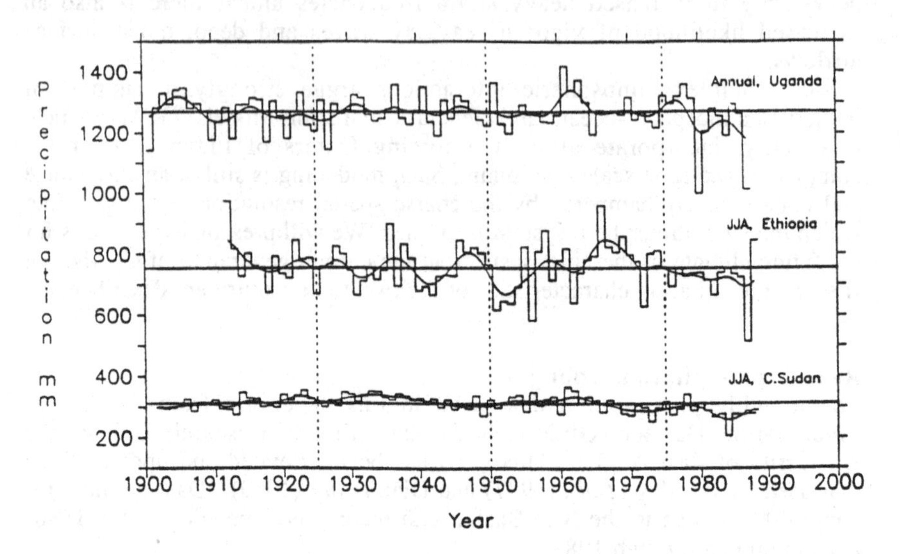

Figure 5: Precipitation time series for three regions within the Nile Basin. Horizontal lines indicated median precipitation for the full record. The smooth line isa 9-pt. padded Gaussian filter.

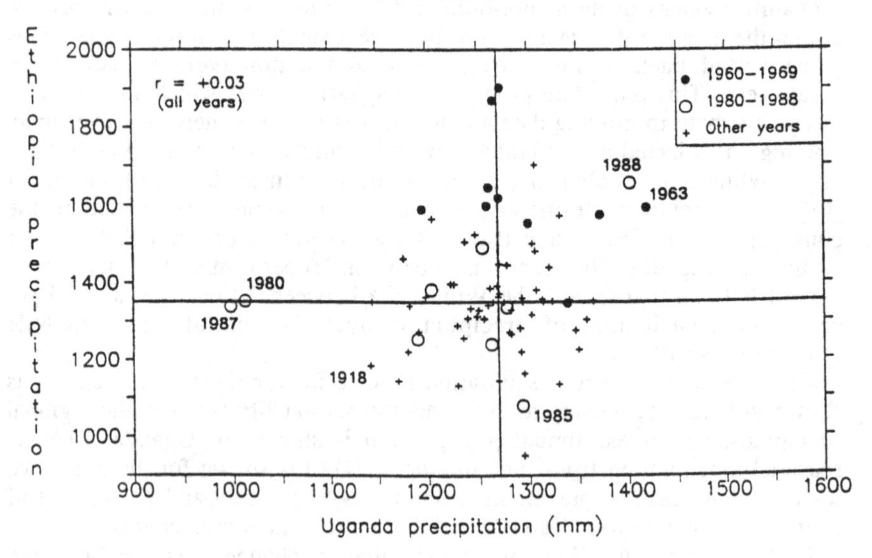

Figure 6: Annual precipitation for Uganda plotted against that for Ethiopia (1914-1988). Years in the 1960s and 1980s are seperatley identified.

1975 and 1980 respectively for the central Sudan, Ethiopia and Uganda (Figure 5). The precipitation peak in the early 1960s over Uganda and the mid-1960s over Ethiopia, contributed to the high levels of Lake Nasser over the first decade after its completion, ensuring that its peak level of 177.41 m above mean sea level was reached in November 1978. The subsequent decade saw falling levels in Lake Nasser which were only reversed in 1988 following a high Blue Nile discharge (Sutcliffe *et al.*, 1989). The very low precipitation input in 1987 over both Uganda and Ethiopia and the high 1988 input over Ethiopia are both evident from Figure 5.

One of the intriguing features of the Nile river system is the contrasting nature of the precipitation regimes which supply the headwaters of the two branches of the Nile. The White Nile is fed from equatorial rains which possess a weak bimodal seasonal distribution, while the Blue Nile is supplied from a seasonally arid regime over the Ethiopian Highlands. The meteorology controlling these two regimes is quite distinct (e.g. Griffiths, 1972) and the correlation between the annual precipitation input to these two catchments is therefore not necessarily good. Figure 6 shows the association between annual precipitation over Ethiopia and Uganda for the period 1914-1988. There is effectively zero correlation between these two regimes implying that there is neither a synchronous (positive correlation) nor asynchronous (negative correlation) tendency in the interannual fluctuations of the two precipitation regimes which contribute to main Nile discharge. In certain years high precipitation input to one catchment may be offset by low input to the other (e.g. 1980), while in other years both catchments may be poorly supplied (e.g. 1918), or well supplied (e.g. 1963). The high Nile discharges of the 1960s are reflected in the high precipitation inputs over both catchments and contrast with the generally poorer precipitation performance during the 1980s (Figure 6).

Future precipitation changes

At least four different approaches to estimating future precipitation over the Nile Basin can be pursued. First, short-term seasonal forecasts of precipitation can be made using global fields of sea surface temperature (SST) anomalies (Parker *et al.*, 1988). This method is being further developed and currently is more relevant for the Sahel zone of central Sudan than for the Ethiopian or Ugandan catchments. Future extensions of this method may, however, include such regions (Ward, N.M., pers. comm.). Second, analysis of the instrumental time series of precipitation may generate a statistical basis for forecasting future precipitation changes. The anecdotal seven year flood/drought cycle of the Nile, or the proposed 18.5 yr lunar cycle in Nile discharge (Currie, 1987) are examples of such an approach. Evidence for robust statistical cycles in the precipitation record over either Ethiopia or Uganda is, however, weak. Furthermore, such statistical forecasts assume a past and future constancy in the boundary conditions which control the climate of the Nile Basin. Such as assumption is increasingly unsustainable under current rates of global environmental change.

Third, historical analogues may be constructed from which future precipitation scenarios for the Nile Basin may be inferred. For example the proposed relationship between warm periods in Europe and high Nile flows (e.g. Hassan, 1981), or the reconstructions of African precipitation during the Holocene pluvial (COHMAP, 1988) and the latter end of the Pleistocene (e.g. CLIMAP, 1976) are examples of such an approach. Difficulties here again include changing boundary conditions which mean that climates under warm conditions in the past may not be directly transferable to warm conditions in the future. A fourth approach to estimating future precipitation over the Nile Basin involves the creation of climate scenarios from General Circulation Model (GCM) experiments. In this section we present some results using this latter approach.

Estimating the future rate and pattern of global warming and its consequences for other climatic parameters such as precipitation is best accomplished at present using GCMs. These are used in experiments which generate two global climates, one representing climate under current (1xCO_2) greenhouse gas concentrations, and the second representing climate under doubled (2xCO_2) concentrations of greenhouse gases. The difference between the 2xCO_2 and 1xCO_2 climates is the estimated greenhouse climate signal. Different GCMs, however, use different spatial resolutions and different physical parameterisations by which the atmosphere is simulated. Their resulting climates are therefore not similar and none of the leading GCMs can be regarded as necessarily better than any other. One method to utilise the simulated greenhouse climates of different GCMs and produce a composite model climate scenario has been developed by Wigley *et al.* (1992). The details of this method are not discussed here, but the following composite GCM scenario for the Nile Basin is derived from the five leading GCM experiments, namely those performed at the Goddard Institute for Space Studies, the Geophysical Fluid Dynamics Laboratory, Oregon State University, the National Centre for Atmospheric Research, and the UK Meteorological Office (see Wigley *et al.*, 1992). The scenario presents the equilibrium change in climate which could result from a doubling of CO_2-equivalent (i.e. from 300 ppmv to 600 ppmv). The date by which such a doubling would occur is uncertain, but probably lies between the years 2030 and 2070 (Bolin *et al*, 1986).

Figure 7 shows the change in mean seasonal air temperature over the Nile Basin for the winter (DJF) and summer (JJA) seasons. Winter temperatures increase by between 3° and 4°C over the Nile Basin, summer temperatures by slightly less at between 2° and 3°C. For summer, the temperature increase is greatest over the lower Nile Basin. Associated with such temperature increases will be an increase in potential evapotranspiration (Etp). Changes in actual evapotranspiration (Eta) will depend on moisture availability, but will undoubtedly lead to greater evaporative water loss from the Nile Basin and consequently reduced Nile discharge. Will this increased evaporative water loss be offset by increased precipitation?

Figure 8 shows the change in seasonal precipitation rate (in mm/day) for the winter (DJF) and summer (JJA) seasons over the whole Nile Basin and Table 2 summarises the projected changes for the three regions analysed previously. Precipitation changes are presented in three ways in Table 2: as seasonal change in total precipitation yield, as a percentage change in existing yield, and as a probability that precipitation will decrease by some unspecified amount. This latter value is calculated from the assumption that the five absolute precipitation changes derived from the GCMs are sampled from a normal distribution, thus enabling a theoretical probability for a given precipitation change to be estimated. This probability estimate in effect provides an indication of the confidence we have in the direction of the projected precipitation change derived from the five GCM experiments. Where the probability is high (low) the models generally agree that precipitation will decrease (increase).

Figure 8 and Table 2 indicate contrasting projections of seasonal precipitation change between equatorial Africa (Uganda) and seasonally arid Africa (Ethiopia and the central Sudan). For Ethiopia and the central Sudan no clear greenhouse precipitation signal emerges from the five experiments. Absolute and relative precipitation change is close to zero and the probability of a precipitation decrease is close to 0.5. There is a hint that summer precipitation over the Blue Nile catchment may decrease slightly and winter precipitation over the central Sudan may increase slightly. This latter result shows the danger of treating GCM scenarios as realistic predictions of climate change; contemporary winter precipitation over the central Sudan is zero, thus projecting a seasonal increase of 9 mm would require a major change in the seasonality of the monsoon circulation and seems an unlikely possibility. The reason for this implausibility is that GCM experiments inadequately simulate present-day climate and generate, for example, winter precipitation in the central Sudan which in reality is nonexistent.

The model signal over Uganda is somewhat clearer. The projected probability of a precipitation decrease is only 0.1 (DJF) and 0.15 (JJA) with estimated absolute precipitation increases of 36mm (+28 per cent) and 63 mm (+18 per cent) respectively. This contrasting result between the White and Blue Nile catchments is not surprising since the meteorology of the two catchments differs substantially, and the instrumental precipitation record over the twentieth century demonstrates zero correlation between Uganda and Ethiopia precipitation (Figure 6).

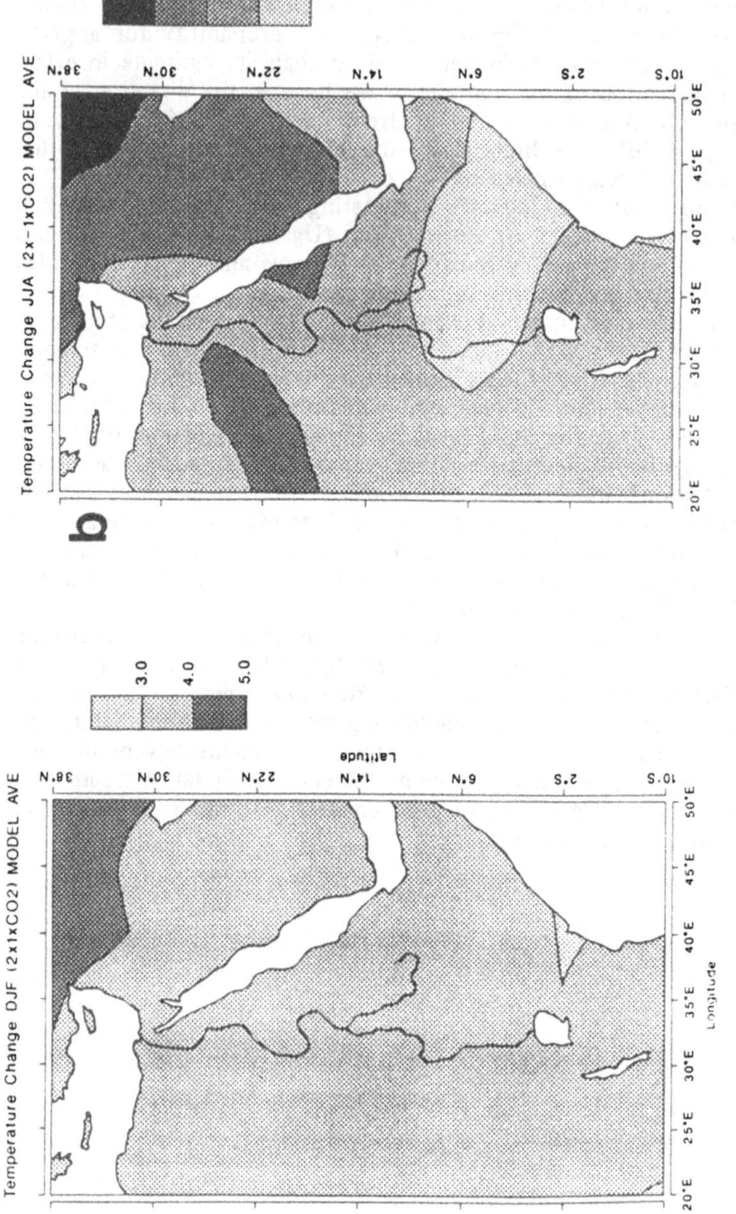

Figure 7: A+B, Model average of the 2x Co_2 - 1x Co_2 change in surface air temperature for the Nile Basin (contour interval is 1^0C).
a) DJF season, b)JJA season

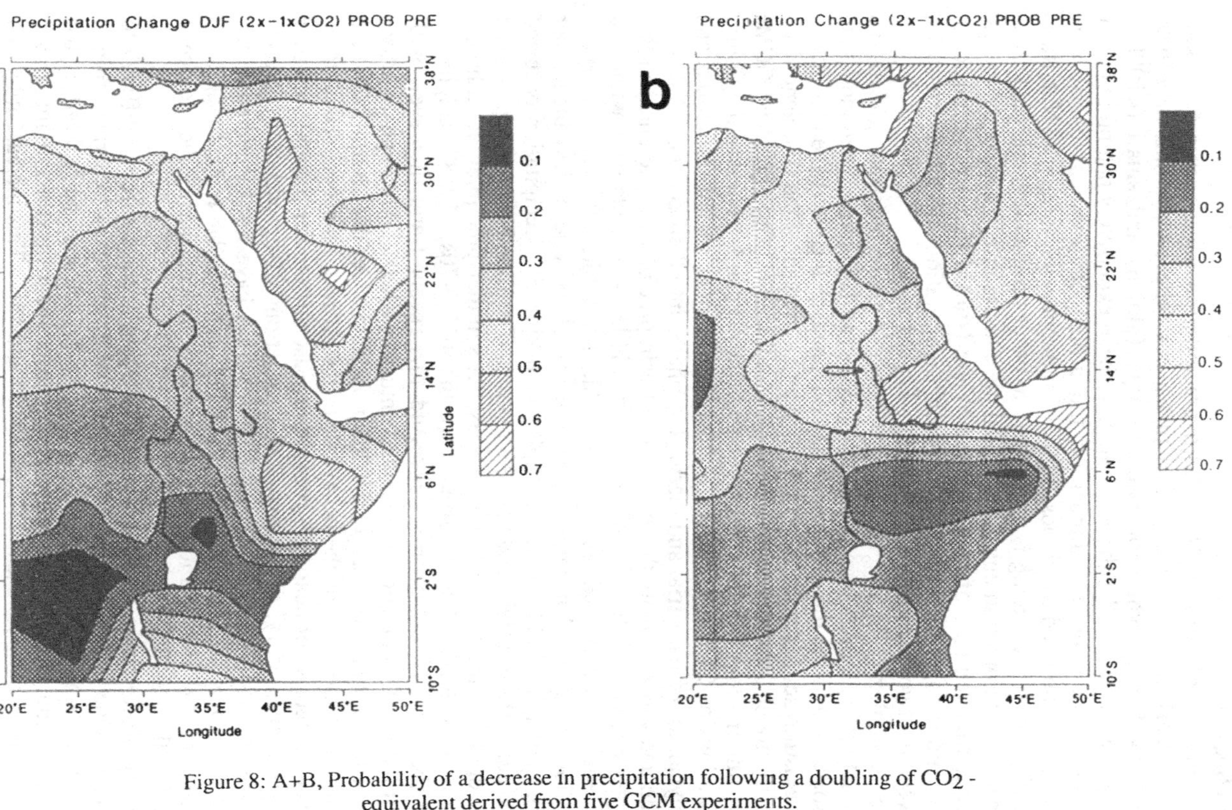

Figure 8: A+B, Probability of a decrease in precipitation following a doubling of CO_2 - equivalent derived from five GCM experiments.

Conclusions

Substantial variations in Nile Basin precipitation have occurred on long (>1000 year) and short (<10 year) timescales. These variations have been forced by factors which operate on large (global) and small (<100km) spacescales. Historically, these factors have been exclusively natural in origin. The progressive human impact on the global environment over recent centuries has now led to the possibility of anthropogenic forcing of Nile Basin precipitation. Such forcing is most likely to occur through regional land-cover changes and through an enhanced global greenhouse effect. Just as the natural precipitation regimes of the Blue and White Niles differ, so the impact of these anthropogenic forcing factors on precipitation are likely to differ between the two catchments.

The complexity of the interactions between these external, global and regional scale forcings, requires a combined ocean-land-atmosphere modelling approach to address the question of future climate change over the Nile Basin. The existing set of atmospheric GCMs may be viewed as the first generation of such models, with the small number of combined ocean-atmosphere GCMs now under development representing the second generation. Future generations of GCMs will need to incorporate higher spatial resolutions and the dynamics of terrestrial and marine biospheres into their conceptualisation.

The analysis of twentieth century Nile Basin precipitation presented in this Chapter has shown that the White and Blue Nile catchments have witnessed different trends in annual precipitation yield. The White Nile has experienced little long-term trend in annual yield (an overall decline of less than 2 per cent), although notable short-term episodes of unusual precipitation occurred in the early 1960s (high) and the 1980s (low). The Blue Nile has experienced an overall decline of 5 per cent in seasonal summer precipitation, although most of this decline has occurred since the mid-1970s. 1988 precipitation over the Blue Nile catchment reversed this recent trend, however, with precipitation 13 per cent above the century average. Although of lesser significance for Nile discharge, precipitation yields over the central Sudan are an important indicator of the behaviour of the African monsoon. Here, there has been a decline of nearly 10 per cent in seasonal summer precipitation over the twentieth century, with most of this reduction having occurred since the mid-1960s. 1988 and 1989 precipitation values have not recovered to the century average. These recent changes in the central Sudan represent up to 20 per cent of the precipitation fluctuations experienced during the Holocene period in this region.

Likely precipitation levels over the Nile Basin in the future are difficult to establish. One method presented here uses a climate scenario derived from a number of GCM experiments. This scenario (which may be realised by the middle of the twenty-first century) suggests mean seasonal temperature increases over the Nile Basin of between 2° and 4°C. This increase might be slightly greater over the lower Nile Basin than over the Upper Nile Basin. There is no clear signal from the composite model scenario about the magnitude or extent of precipitation changes over the Blue Nile catchment or over the central Sudan. Over the White Nile catchment the scenario suggests a high probability (between 0.8 and 0.9) for increased precipitation in both the summer and winter seasons.

With the inevitable increase in potential and actual evapotranspiration which would result from the higher surface air temperatures, combined with the scenario precipitation projections, the current best guess for greenhouse-induced forcing of Nile discharge would be for reduced Blue Nile flows, and constant or slightly increased White Nile flows. Such a suggestion is not wholly incompatible with the observed trends in White and Blue Nile discharge over recent decades. This scenario remains highly uncertain until substantial improvements are made in combined ocean-atmosphere GCMs.

Two implications follow from the analysis presented in this Chapter. First, the magnitude of historical fluctuations in Nile discharge (experienced both in this century and in previous millennia) is unlikely to be lessened in the future. Indeed, the additional climate forcing which will result from increased greenhouse gas concentrations may well increase the amplitude of such fluctuations and may also generate a quite different seasonality in the contributions to main Nile discharge. Second, the greater confidence we have in projections of future temperature than in projections of future precipitation over the Nile Basin implies that increased water loss through increased evapotranspiration should occupy our attention. Either part or all of this water loss may be compensated by increased precipitation yield over the White Nile catchment, but either possibility lends greater emphasis to the need for the completion of the Jonglei Canal and other planned water conservation projects in the southern Sudan in order to lessen the sensitivity of future Nile discharge to evaporation loss.

Table 1: Linear regression analysis for three regional precipitation time series.

Region	Series	Period	Median	Slope	Per cent Change Precipitation over length of series
Uganda	Annual	1901-88	1269mm	-0.27mm/yr	-1.8%
Ethiopia	Seasonal	1912-88 (JJA)	754mm	-0.48mm/yr	-4.9%
Central Sudan	Seasonal	1902-88 (JJA)	312mm	-0.33mm/yr	-9.2%

Table 2: Seasonal changes in precipitation (absolute, relative and probability of a decrease) for a doubling of CO_2-equivalent derived from the composite GCM climate scenario.

	DJF			JJA		
	Absolute Change	Per cent Change	P(Precip decrease)	Absolute Change	Per cent Change	P(Precip decrease)
Uganda	+36 mm	(+28.1%)	0.1	+63.mm	(+18.2%)	0.15
Ethiopia	0 mm	(-)	0.5	-9 mm	(-1.2%)	0.55
Central Sudan	+9 mm	(0 0)	0.35	0 mm	(-)	0.5

Acknowledgements
The work on extending and updating the Global Precipitation set was undertaken as part of contract PECD/7/10/198 by the U.K Department of the Environment. The GCM climate scenario for the Nile Basin was supplied by Ben Santer of the Max Planck Institute, Hamburg and the temperature curves in Figure 4 by Phil Jones.

References
Adamson, D.A., Gasse, F., Street, F.A. and Williams, M.A.J. (1980). Late Quaternary history of the Nile, *Nature,* **288**, pp. 50-55.
Ali, A.M.A. (1989). *Heavy rainfall at Khartoum on 4-5 August 1988: a case study*. Meteorological Magazine, **118**, pp. 229-235.
Ahlcrona, E. (1988). *The impact of climate and man on land transformation in central Sudan.* Sweden, Lund University Press, Lund.
Attia, B.B. and Abulhoda, A. (1990). *The ENSO phenomenon and its impact on River Nile Hydrology.* Cairo, Paper presented at the 'International Seminar on Climatic Fluctuations and Water Resources', December 11-14, 1989.
Berger, A. (1988). Milankovitch theory and climate. *Reviews of Geophysics,* **26**, pp. 624-657.
Bolin, B., Doos, B.R., Jager, J. and Warrick, R.A. (eds.) (1986). The greenhouse effect, climate change and ecosystems. Chichester, *SCOPE*, **29**, J.Wiley and Sons Ltd.
Bradley, R.S., Diaz, H.F., Eischeid, J.K., Jones, P.D., Kelly, P.M. and Goodess, C.M. (1987). Precipitation fluctuations over northern hemisphere land areas since the mid-19th century. *Science,* **237**, pp. 171-175.
Cadet, D. and Nnoli, N. (1987). Water vapour transport over Africa and the Atlantic Ocean during summer 1979. *Qtly. J. of the Royal Met. Soc.,* **113**,

pp. 581-602.
CLIMAP (1976). The surface of the ice-age earth. *Science*, **191**, pp. 1131-1137.
COHMAP (1988). Climatic changes of the last 18,000 years: observations and model simulations. *Science*, **241**, pp. 1043-1052.
Currie, R.G. (1987). On bistable phasing of 18.6 year induced drought and flood in the Nile records since AD 650 . *J. of Climatology*, **7**, pp. 373-390.
Diaz, H.F., Bradley, R.S. and Eischeid, J.K. (1989). Precipitation fluctuations over global land areas since the late 1800s. *J. of Geophys. Res.*, **94** (D1), pp. 1195-1210.
Dickinson, R.E. and Henderson-Sellers, A. (1988). Modelling tropical deforestation: a study of GCM land-surface parameterisations. Qtly *J. of the Royal Met. Soc*, **114** (B), pp. 439-62.
Druyan, L.M. and Koster, R.D. (1989). Sources of Sahel precipitation for simulated drought and rainy seasons. *J. of Climate*, **2**, pp. 1438-1446.
El Tahir, E.A.B. (1987). A feedback mechanism in annual rainfall, central Sudan. *J. of Hydrology*, **110**, pp. 323-334.
Griffiths, J.F. (ed.) (1972). *Climates of Africa*, **10**, Amsterdam, World Survey of Climatology, Elsevier.
Hassan, F.A. (1981). Historical Nile floods and their implications for climatic change. *Science*, **212**, pp. 1142-45.
Howell, P., Lock, M. and Cobb, S. (1988). *The Jonglei Canal: impact and opportunity*. Cambridge University Press.
Hulme, M. (1989). *Analysis of worldwide precipitation records and comparison with model predictions*. Norwich, Report for DoE Contract PECD/7/10/198 (September), Climatic Research Unit.
Hulme, M. and Trilsbach, A. (1989). The August 1988 storm over Khartoum: its climatology and impact, *Weather*, **44**, pp. 82-90.
Janowiak, J.E. (1988). An investigation of interannual rainfall variability in Africa . *J. of Climate*, **1**, pp. 240-255.
JIT (Jonglei Investigation Team) (1954). *The Equatorial Nile Project and its effects in the Anglo- Egyptian Sudan*. 4 Vols. p.529, Khartoum: Sudan Government.
Jones, P.D., Wigley, T.M.L., Folland, C.K. and Parker, D.E. (1988). Spatial patterns in recent worldwide temperature trends. *Climate Monitor*, **16**, pp. 175-185.
Nicholson, S.E. (1988). Land surface-atmosphere interaction: physical processes and surface changes and their impact. *Prog. in Phys. Geog.*,**12**, pp. 36-65.
Normand, C. (1953). Monsoon seasonal forecasting. *Qtly. J. of the Royal Met. Soc.*, **79**, pp. 463-473.
Parker, D.E., Folland, D.K. and Ward, M.N. (1988). Sea-surface temperature anomaly patterns and prediction of seasonal rainfall in the Sahel region of Africa pp. 166-178 in: *Recent Climatic Change* (ed.) Gregory, S., London, Belhaven Press, pp. 328.
Ropelewski, C.F. and Halpert, M.S. (1987). Global and regional scale precipitation patterns associated with the El Ñino/Southern Oscillation . *Mon. Wea. Rev.*, **115**, pp. 1606-1626.

Salati, E., Dall'Olio, A., Matsui, E., and Gat, J.R. (1979). Recycling of water in the Amazon Basin: an isotopic study. *Water Res.* Research, **15**, pp. 1250-1258.
Sutcliffe, J.V. and Parks,Y.P. (1987). Hydrological modelling of the Sudd and Jonglei Canal. *Hydr. Sciences Journal,* **32**, pp. 143-59.
Sutcliffe, J.V., Dugdale, G. and Milford, J.R. (1989). The Sudan floods of 1988 . *Hydr. Sciences Journal*, **34**, pp. 355-364.
Wickens, G.E. (1975). Changes in the climate and vegetation of the Sudan since 20,000 BP. *Boissiera,* **24**, pp. 43-65.
Wickens, G.E. (1982). Paleobotanical speculations and Quaternary environments in the Sudan, pp. 23-50 in (eds.) Williams, M.A.J. and Adamson, D.A. *A land between two Niles*: Rotterdam, Balkema.
Williams, M.A.J. and Faure, H. (eds.) (1980). *The Sahara and The Nile.* Rotterdam, Balkema.
Wigley,T.M.L. (1989). Measurement and prediction of global warming, pp. 85-98 in *Ozone depletion: health and environmental consequences,* Russell-Jones,R. and Wigley,T.M.L. (eds.). Chichester. J.Wiley and Sons Ltd.
Wigley, T.M.L., Santer, B.D., Schlesinger, M.E. and Mitchell, J.F.B. (1992). Developing climate scenarios from equilibrium GCM results (in preparation).

7

Hydrological data requirements for planning Nile management

J. SUTCLIFFE and J. LAZENBY

Introduction

The main requirement for hydrological records in the Nile Basin relates to river flow volumes for irrigation and hydro-electric power production. Long-term flow records exist at many sites throughout the basin, and continuity of these records is essential for future management, particularly in the light of long-term changes in relations between the main tributaries. Long records are required to extend short-term records by correlation, which includes cross-border correlation. Consistent quality is important and is best assessed by analysis of rating curves as water balance comparisons are difficult. The planning of water conservation measures to reduce evaporation losses requires monitoring of wetlands by long-term measurements supplemented by satellite imagery. The need for flood forecasting may also require satellite monitoring.

The aim of this chapter is to outline the hydrological data required for the planning of Nile waters development, and then to review present information to determine how far these requirements are met by records currently being collected. The chapter deals with surface water as opposed to groundwater resources. It treats the whole basin as a unit in terms of scientific knowledge, though the responsibility for data collection is divided between a number of different national authorities, who may have different priorities in terms of water resources management.

The Nile Basin, because of its size and the variety of climate and topography, constitutes one of the most complex of major river basins. A map of the basin is given in Figure 1. Important features of the hydrology are the slow response of the system, especially in the case of the White Nile contribution which is affected by lake storage and losses in the Sudd, and the apparent persistence of periods of high and low flows which have become known as the Hurst phenomenon. Planning of water resources management within the basin therefore requires hydrological records of long duration at the key sites within the system. The losses of river flows by swamp evaporation in the White Nile basin require special attention.

The Nile Basin was treated in the first half of this century as a single unit in terms of hydrological data collection and publication; this resulted in the many volumes of *The Nile Basin*, covering river levels, discharge

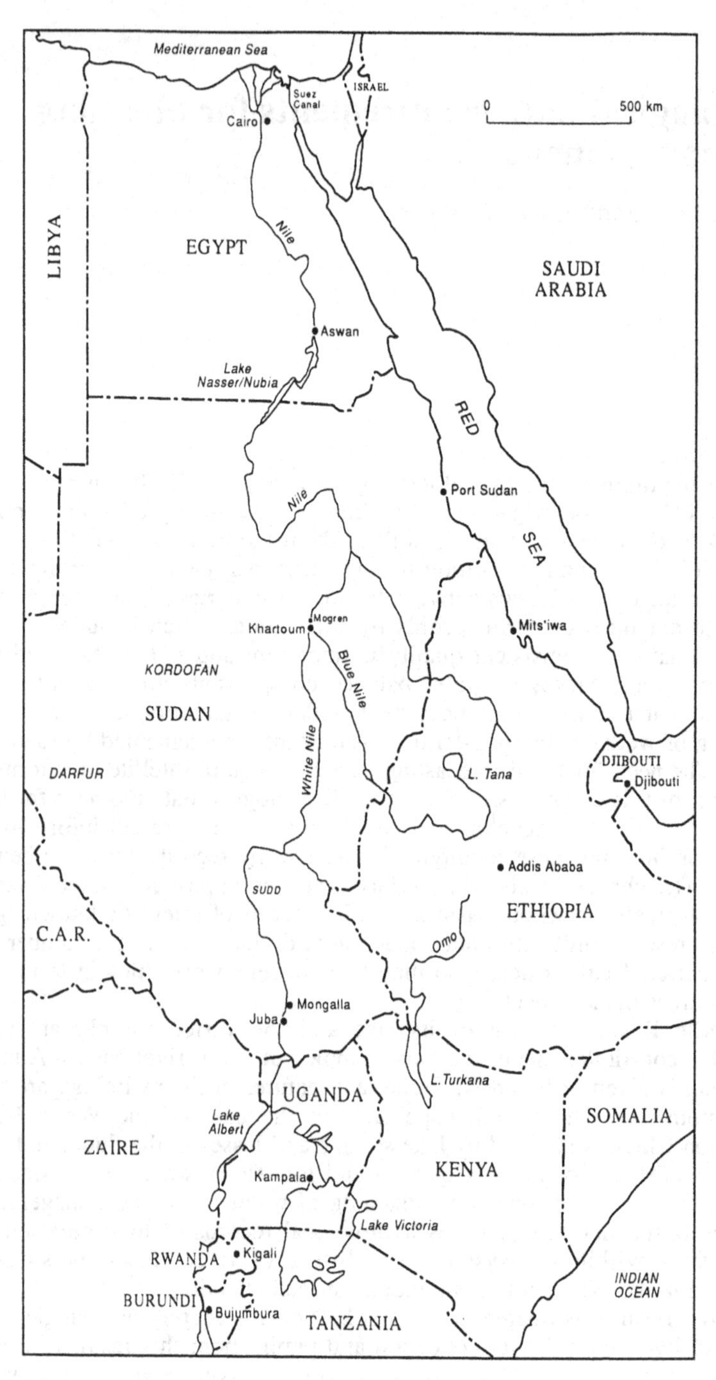

Figure 1.The Nile Basin

measurements and calculated river flows as well as rainfall. This set of volumes (Hurst & Phillips, 1932, et seq) is a great asset to the countries of the Nile concerned, however they may wish to develop the water resources of the area. The optimal development of the resources of the basin will depend on agreement and cooperation between individual countries; the exchange of hydrological data and improved understanding of the behaviour of the complex system are necessary preliminaries to planning. No negotiations on international water rights are possible without an agreed data base, which must be formulated on the maximum time scale.

Hydrological data requirements

The hydrological data required for water resources planning depend to some degree on the major type of development. The development of water resources within the Nile Basin is to a large extent dominated by irrigation and hydro-electric power; water supply for domestic and industrial purposes is by comparison less demanding of water. The volumes of water used in irrigation are relatively large, though a greater risk of failure can be accepted than for domestic or industrial supply.

The use of surface water for irrigation, which is usually required in greater quantities when river flows are low, thus implies the use of reservoir storage. The estimation of reservoir yields through various forms of simulation study depends heavily on annual flow volumes and their variability. In the case of hydro-electric power production, the need to maintain power supplies and operating head generally implies the use of reservoir storage; however, it may be necessary to estimate the power available without storage, so that low flows may be required for statistical analysis. Urban water supply within the Nile Basin is usually derived from direct river abstraction or from groundwater, and therefore statistical analysis of low flows is again relevant.

Thus river flow volumes and low flows are both important for water resources assessment. The need for the maximum length of record is underlined by the natural variations which have occurred in the period of hydrological investigation and over longer time scales. The record of flows at Aswan and Dongola since 1870 (Figure 2) shows periods of persistent high and low flows which both resemble and contrast with changes in the level of Lake Victoria (Figure 3); the comparison is complicated by the different regimes of the White and Blue Niles but this only reinforces the need for long-term records at key sites throughout the basin. The possibility that climate change may add to this natural variability emphasises the need for rainfall and river flow records to monitor the variations resulting from man-made changes.

The need for flood estimates for design and operational forecasting purposes makes high flow measurements necessary, while the importance of sediment and bed load measurements is illustrated by the relatively small storage volumes available in several of the reservoirs and the constraints which potential sedimentation imposes on their operation.

Thus the main requirement is for long-term measurement of the volumes of flow contributed by the different tributaries of the Nile system at key sites

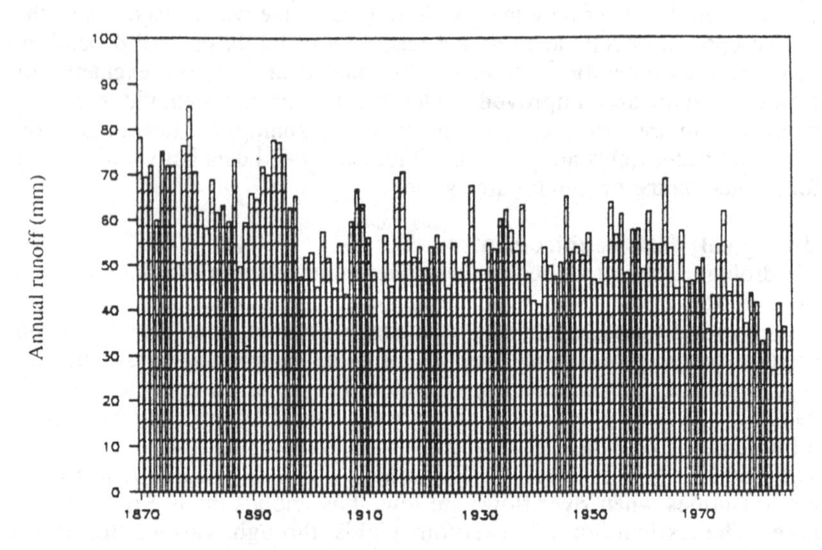

Figure 2: Annual runoff of the main Nile at Dongala, 1870-1987. (mm over the effective over basin). The measured flows at Dongala are supplemented by measurements at Wadi Halfa, Kajnarty and Aswan.

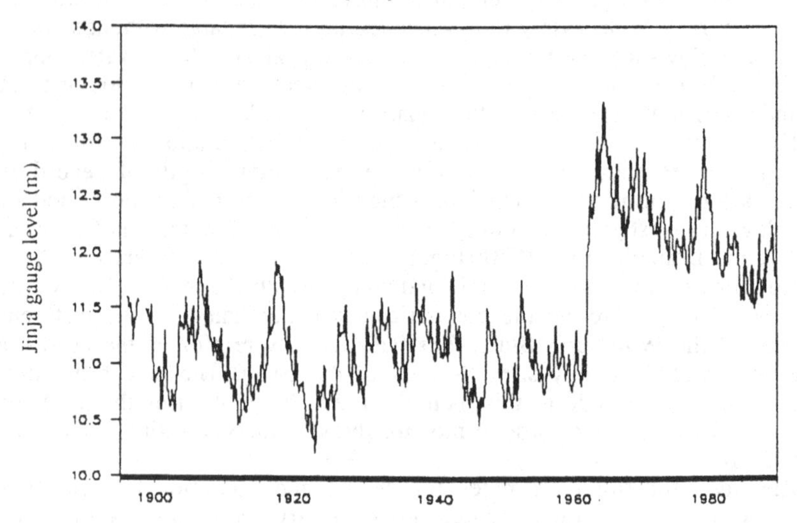

Figure 3: Lake Victoria levels at Jinja, January 1896-January 1990. The early records are derived from the Entebbe gauge.

along their courses, though low and high flows and sediment load are also important. These requirements need support from long-term records of rainfall and evaporation. There are special requirements for detailed hydrological records in the Lake basin, to make the long lake level records useful. An unusual feature of the White Nile basin is the large areas of wetland where river flows spill and are evaporated; there are particular requirements for hydrological data in the Sudd and other areas of potential water-saving projects in the southern Sudan, and these are discussed in the context of these areas.

Need for continuity and consistency

Hydrological records are required to sample variations of river flows both in time and in space, and thus to provide to the water resources planner information on the response of the river system to periods of low rainfall, and to variations of rainfall over different parts of the basin. Because it is unrealistic to expect the existence of a long-term station at every site of interest, it is necessary to plan the combined use of long-term river flow records, long-term rainfall records and short-term flow records to extend the available information by correlation of short-term and long-term records and by regional studies of net rainfall and runoff. The methods of extension depend for their success on the interrelations between stations illustrated by the correlation matrix and on the water balance studies, and in the first instance these depend on the quality of the hydrological records.

For hydrological records to be used in the planning of water resources, they should be as accurate as possible but it is even more important that they should be continuous and consistent. If recent flow records at a project site are to be extended using long-term rainfall or river flow records, or gaps in flow series filled by correlation techniques, bias will be introduced if the recent records at the long-term site have been processed on a different basis from the earlier records, even if the recent records are more accurate as a result of the change. It is essential that reviews of hydrological records are applied consistently to the whole period.

Lake level series frequently predate the establishment of river flow measurements and provide an alternative means of deriving historic series. Where the lake outlet is stable, so that lake levels can be readily converted to outflows, these provide a valuable supplement to river flows and by inference can extend knowledge of past basin rainfall.

Quality control

The quality control of hydrological records is essential if they are to provide reasonable estimates of water resources over a region or during droughts. The appraisal of rainfall and evaporation records should start with the investigation of sites and measuring or estimation techniques. These may change over the years, but double mass comparisons should reveal any significant changes which require detailed investigation. An example of the way in which consistency of records is more important than precision for the study of long-term series will be discussed later in more detail. The water balance of Lake Victoria and its tributaries has been investigated over the years and the coverage of rainfall stations has improved greatly; however, it was easier to model the major rise in the lake level of 1961-1964 by using the small number of long-term gauges around the lake, after adjusting for site changes, than by using the whole improving but non-homogeneous network (Piper *et al.*, 1986).

At sites where river levels have been measured over a reasonable period, the quality of the processed flow records depends on the use of discharge measurements to derive rating curves. The factors affecting the quality of processed flows include the range of gaugings compared with the range of flows at the site, the stability of the hydraulic control and the methods used over the years to derive successive rating curves.

Quality and consistency can be assessed by deriving rating curves by modern computer-based methods from the whole period of records, and by simultaneous study of discharge measurements to monitor progressive movements of bed controls. Because high flows are in general governed by channel controls, the use of a number of years is likely to give more consistent results than the derivation of ratings from individual years. An example of successive ratings and the light they throw on morphological changes is given in Figure 4, where the changes in rating for the Bahr el Jebel at Mongalla are supported by bed level changes revealed by measured cross-sections.

Continuous quality control of hydrological data is preferable, but the review of accumulated records is desirable before major investigations of water resources potential are undertaken. In addition to the review of rating curves, indirect methods of appraisal of hydrological data are useful. Such methods include double-mass analysis and inter-station correlation. The double-mass comparison of flow records at sites on the same river or on adjacent basins should produce an essentially linear relationship, and any changes of slope will draw attention to dates which need investigation. However, changes of slope may be due to natural causes as well as measurement problems. Two examples illustrate this; in Figure 5 outflows from Lake Victoria are compared with flows measured at Mongalla, and the slight change of slope in the early years suggests that the Mongalla flows may have been overestimated in the years before regular gaugings began. In Figure 6, on the other hand, the flows of the White Nile at Malakal and the Blue Nile at Khartoum are compared, and show a real change after the rise in Lake Victoria in 1961-1964. Inter-station correlations, for example of annual flows, can also draw attention to stations where particular investigation is required.

Water balance comparisons are one of the most powerful tools of the hydrologist. Annual or long-term average river flows are compared with corresponding basin rainfall estimates and anomalous values are easily revealed. This approach is not easy to carry out in the Nile Basin because of the complex nature and immense size of the basins of many of the tributaries; the problem is made more difficult by the fact that several of the tributaries have their sources and the greater part of their rainfall in other countries where hydrological records are not so easily available. A recent study of hydrological data in the Nile Basin countries led to the development of

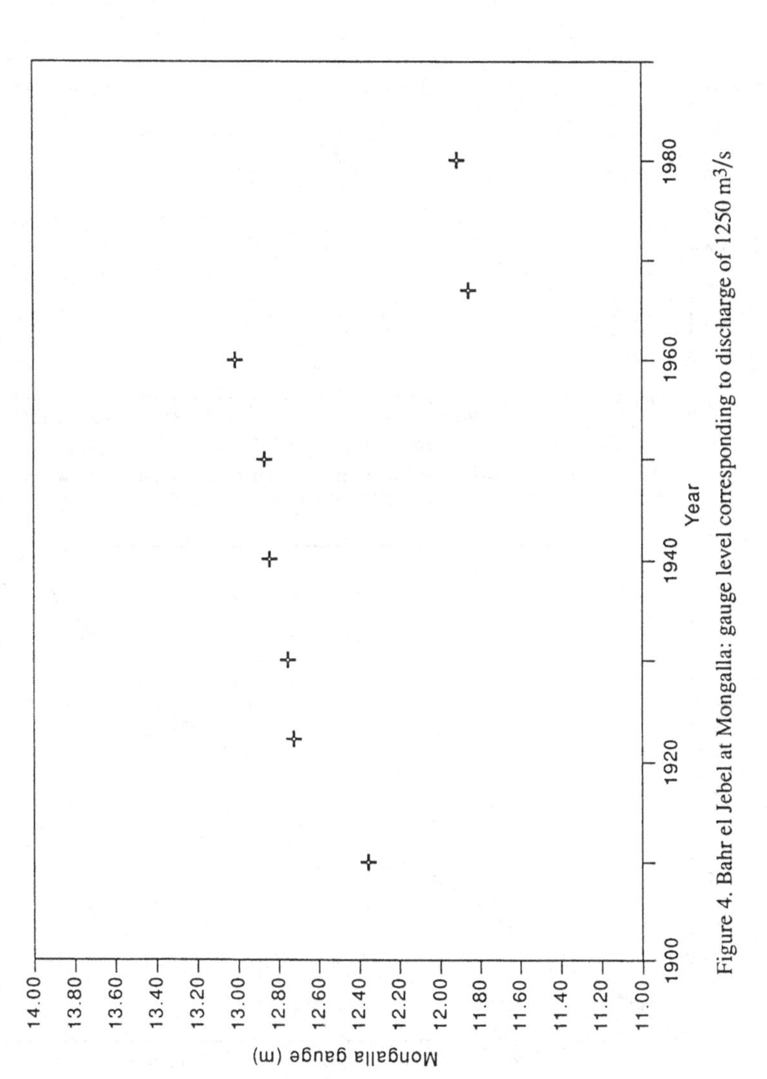

Figure 4. Bahr el Jebel at Mongalla: gauge level corresponding to discharge of 1250 m^3/s

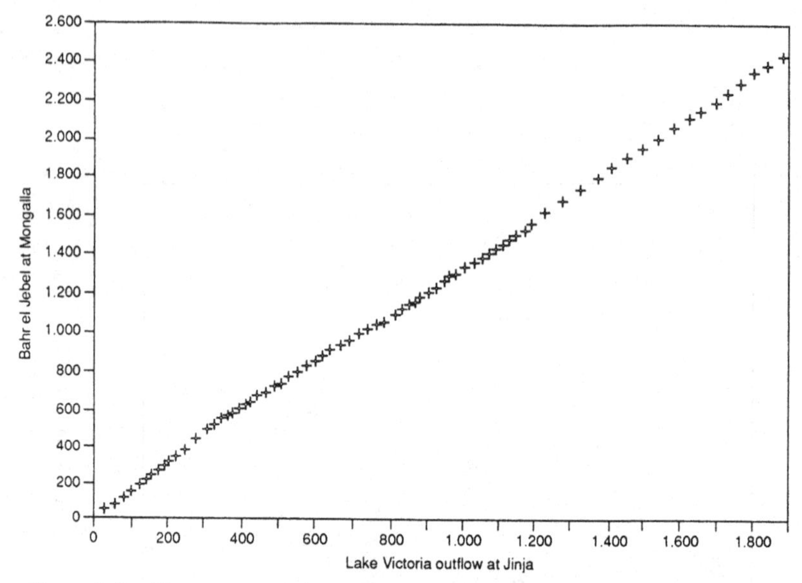

Figure 5: Double mass comparison of Lake Victoria outflows at Jinja and flows of Bahr el Jebel at Mongalla, 1905-1978 (km^3)

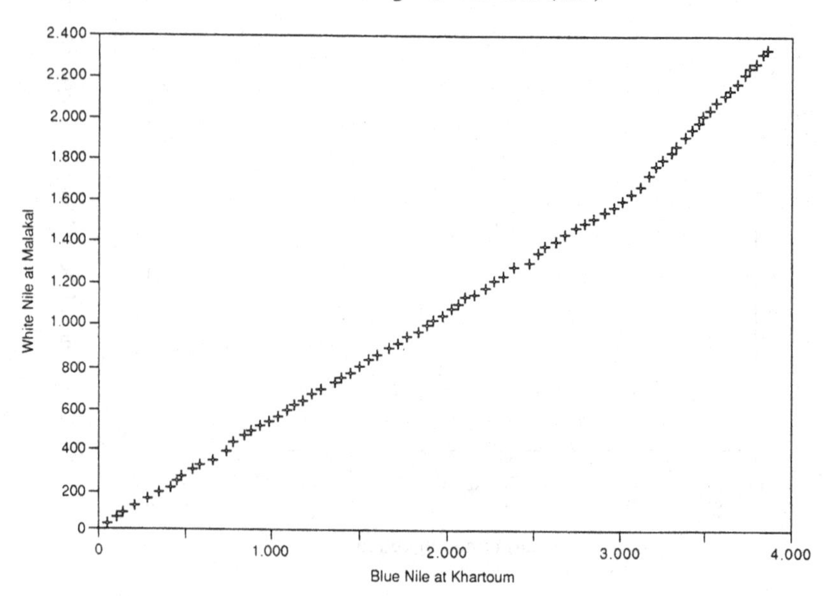

Figure 6: Double mass comparison of flows of White Nile at Malakal and Blue Nile at Khartouom, 1905-1982 (km^3)

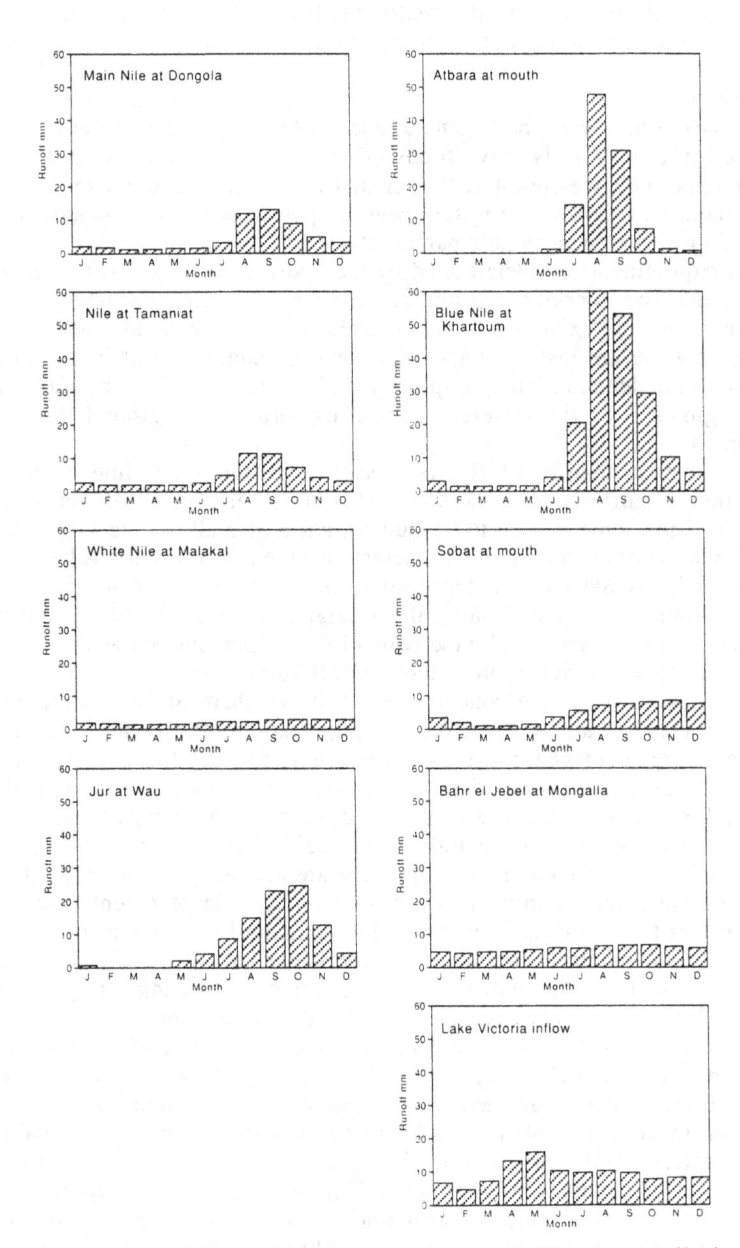

Figure 7: Mean monthly runoff of Nile tributaries, expressed as mm over effective basin.

relatively simple methods of deriving net rainfall from gross rainfall, but the lack of basin-wide rainfall data made it difficult to test the method in the area. Nevertheless, because of the complexity of rainfall regimes and of the changes in rainfall and runoff over the years in different tributaries, there are advantages in the use of net rainfall in water balance studies.

The Nile Basin

The discussion has so far been general and could be applicable to any major river basin; attention is now focussed on the Nile and its particular requirements. The viewpoint will remain hydrological, and related to the records required to assess the water resources potential, but will include brief details of specific projects whose nature affects the need for data.

What requirements are determined by the hydrological peculiarities of the Nile system? The important features are the large amount of storage within the Lake basin, the existence of major wetlands like the Sudd where large amounts of water are lost by evaporation, and the dominance of irrigation in terms of water demand. The complexity and variability of the rainfall and runoff regimes over the different tributaries form a background to these requirements.

An important feature of the lake system is the long time scale of fluctuations in outflows from the lakes. This affects the time scale on which hydrological phenomena need to be studied, but it also allows the long series of lake levels to be used once the mechanism for the lake fluctuations has been understood. Thus the water balance study of the lake system has been an important component of hydrological investigation within the basin, and its importance has been matched by the scale of the programme of measurements and research of the WMO Hydrometeorological Survey.

The wetlands, which are concentrated in the southern Sudan but are also present in Uganda and elsewhere, are important in the context of Nile hydrology as areas of water loss by evaporation. The Sudd in particular has been studied over many years from a hydrological viewpoint because of the proposals for reducing the losses by the construction of the Jonglei Canal. Less attention has been paid to the Bahr el Ghazal tributaries and the Machar Marshes, where significant amounts of water are also evaporated. Difficulties of flow measurement within the wetlands have to a large extent confined measurements to inflows and outflows, and to rainfall measurements around the perimeter, which limits detailed investigation within the wetland. However, the ability to monitor flooded areas using satellite imagery provides a tool which can partly offset the lack of detailed flow measurements.

The importance of irrigation in water use makes irrigation efficiency vital in the context of matching supply and demand. Modern techniques of measuring soil moisture and evaporation above crops in addition to water distribution make it possible to monitor irrigation methods in the field and to test alternative methods of allocating and applying water to improve efficiency. The search for irrigation efficiency requires the continuous measurement of rainfall, potential evaporation and irrigation canal flows on a routine basis for operational purposes in addition to their part in research trials.

Basin hydrology

A knowledge of the complexity and variability of the regime of the different tributaries is important to the consideration of data requirements. A broad knowledge of the basin regime can be derived from the records of the key long-term river flow stations listed in Table 1.

Table 1 Summary of flows at key sites

River	Station	Start date	Area $10^3 km^2$	Mean runoff km^3	mm	SD km^3	CV %
Main Nile	Dongola	1870	1610	89.0	55	34.1	38.3
Atbara	Mouth	1903	113	11.7	103	3.7	32.1
Main Nile	Tamaniat	1911	1470	73.3	50	12.2	16.6
Blue Nile	Khartoum	1900	260	49.7	191	11.2	22.5
White Nile	Malakal	1905	1140	29.6	26	5.1	17.3
Sobat	Mouth	1905	232	13.7	59	2.7	19.5
Bahr el Ghazal	Mouth	1938	274	0.3	1	0.2	58.8
Jur	Wau	1930	49	4.5	91	1.6	36.1
Bahr el Jebel	Mongalla	1905	483	33.1	69	12.2	36.8
Semliki	Bweramule	1940	30	4.2	140	1.1	25.9
Victoria Nile	Jinja	1900	254	25.0	98	9.0	36.1
Kagera	Nyakanyasi	1940	56	6.4	114	2.0	31.4

Notes. The flows at Dongola include earlier records at Aswan, Wadi Halfa and Kajnarty. The basin areas are estimated as contributing. The coefficient of variation (CV) is the standard deviation (SD) divided by the mean.

The rainfall over the basin ranges from virtually zero to over 2000 mm, and its distribution varies from a single season increasing in duration southwards from two to six months in the Sudan and western Ethiopia to two seasons centred about March to May and November to December in the Lake Victoria basin. This varying rainfall pattern is reflected in the seasonal distribution of runoff from the different tributaries; the average monthly runoff at a number of sites, expressed in terms of mm over the effective basin, is illustrated in Figure 7. The flows of the Bahr el Jebel are affected by the storage in Lake Victoria and other lakes, while the flows of the White Nile at Malakal reflect losses and attenuation in the Sudd.

The changes over recent years in rainfall are reflected in the runoff of the different tributaries. In the areas of single season rainfall, over the Sudan and Ethiopia, average annual totals have decreased. Over the Sudan, average rainfall has decreased by about 100 mm between the two periods 1950-1967 and 1968-1986. This change is reflected in the flows of the Blue Nile at Khartoum, which are expressed in terms of mm over the effective basin in Figure 8. Similar changes are shown by the records of the Atbara, Sobat and even the Jur at Wau. In the areas farther south with two rainfall seasons, there is no evidence of a decrease of annual rainfall, and the evidence of the river flows suggests that net rainfall has increased, possibly as a result of changes in seasonal distribution. The flows of the Bahr el Jebel at Mongalla (Figure 9) reflect the increase in rainfall after 1961, damped by the storage in Lake Victoria. This increase in runoff also occurs in the case of the Kagera above Lake Victoria. This divergence of the two rainfall regimes explains the double-mass curve of Figure 6.

Table 2 shows the intercorrelation between the annual flows of key gauging stations, based on the periods of common records. These demonstrate the relative coherence of the northern group of stations on the Atbara, Blue Nile and Main Nile, and the southern group of stations at Malakal and above the Sudd, and the lack of correlation between the two groups.

Table 2 Correlation matrix (R^2) between annual flows

	Atb.	Tam.	Kha.	Mal.	Sob.	Mon.	Jin.	Kag.
Dongola	0.660	0.914	0.844	0.051	0.244	0.003	0.011	0.024
Atbara mouth		0.509	0.633	0.016	0.022	0.070	0.103	0.150
Tamaniat			0.798	0.108	0.307	0.005	0.000	0.000
Khartoum				0.000	0.128	0.063	0.107	0.145
Malakal					0.435	0.709	0.572	0.512
Sobat mouth						0.160	0.074	0.089
Mongalla							0.884	0.746
Jinja								0.800

The issue of climate change and its effect on river flows is important in the Nile Basin because the low runoff coefficients associated with the rainfall regime imply that a small change in rainfall will have a disproportionate effect on runoff. The low runoff is illustrated by the overall average runoff from the contributing basin at Dongola of only 55 mm. This sensitivity of the Nile system and individual tributaries to changes in rainfall is of course the reason for the wide fluctuations which have occurred in river flows in the past.

The range of present climates over the basin is so great that it is questionable whether hydrological modelling is yet appropriate to predict the effect of a given climate change. The natural vegetation over the basin is largely a function of climate and soil characteristics, and thus the vegetation will change as the climate changes, but hydrological models have yet to take full account of this accompanying change. However, studies of average rainfall and runoff over Africa (Sutcliffe & Knott, 1987) suggest that variations in runoff with rainfall from basin to basin are similar over the whole continent and depend like net rainfall on the amount and seasonal distribution of rainfall. Thus empirical relations between rainfall and runoff, coupled with assessments of net rainfall, may be an appropriate tool to forecast the effects of climate change.

The discussion of hydrological data requirements outlined in general terms is now directed to the different parts of the basin in turn.

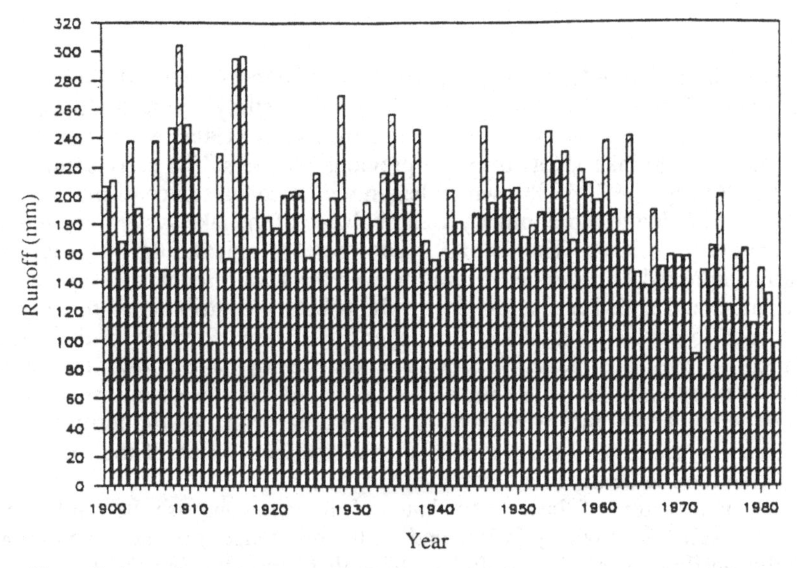

Figure 8: Annual runoff of Blue Nile at Khartoum, 1900-1982, expressed as mm over effective basin.

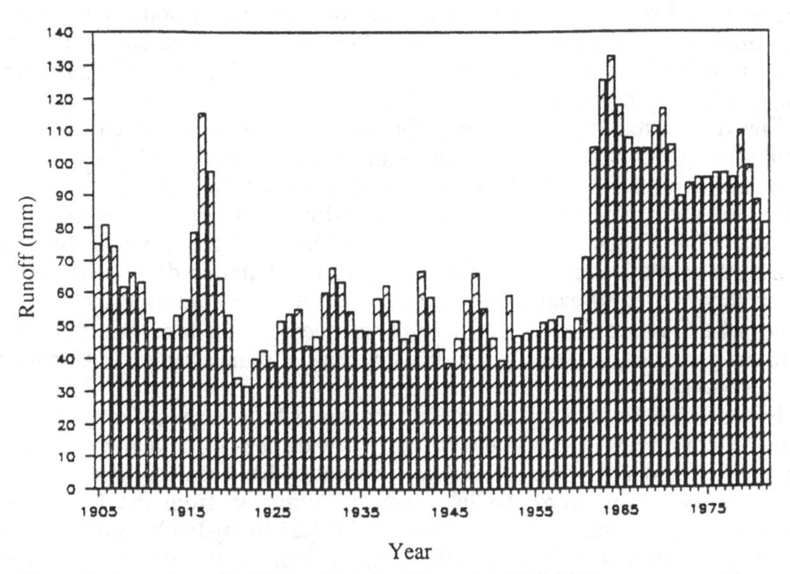

Figure 9: Annual runoff of Bahr el Jebel at Mongalla, 1905-1980 mm expressed in mm over effective basin.

Lake basin

The Lake basin of the Nile is sufficiently distinctive to have been the subject of a number of studies over the years. The earliest study of the hydrological history of Lake Victoria was carried out by Lyons (1906), who traced the rises and falls in lake levels over the previous 50 years. The early studies of Hurst & Phillips (1931,1938), which began with reconnaissance of the Kagera basin in 1926, led to a preliminary water balance of the lake system, including the tributary inflows. The WMO Hydrometeorological Survey provided greatly improved knowledge of rainfall over the lake through measurements on islands, and added measurements of the tributary contributions; these data were used to develop a model of the upper Nile Basin. Reviews of the water balance of the lake (Kite,1981; Piper *et al*.,1986) have shown that it is possible to explain the behaviour of the lake with the present information, if not to predict the future behaviour except in probabilistic terms. It is necessary to understand the water balance of the lake system if the long lake level series are to be used with confidence.

The main factors of the water balance of the lake comprise the rainfall on the lake itself, the tributary inflow and on the other side the lake evaporation and the outflow. The lake rainfall is difficult to monitor, but the long-term average rainfall has been estimated as 1600 mm by WMO, using rainfall measurements with a conceptual model of the rainfall mechanism. In a review (Piper *et al*.,1986) fluctuations from year to year and seasonally have been estimated from eight long-term stations after some errors had been eliminated by double-mass curve analysis. This gave a continuous and reasonably consistent series of lake rainfall.

The main tributary flowing into the lake is the Kagera, which drains the mountains of Rwanda and Burundi, with average annual rainfall up to 1800 mm, and then meanders through a series of lakes and swamps, totalling about 3000 km^2 above and below Rusomo Falls, which delay peak flows by several months. The Kagera is joined near the lake by the Ngono, which drains an area with rainfall over 2000 mm. Other lake tributaries drain a variety of areas: there is significant contribution from the rivers draining the forested slopes of the escarpment in Kenya to the northeast of the lake, but less runoff from the plains of the Serengeti to the southeast and the swamps of Uganda to the northwest.

Evaporation from the lake can be estimated by energy balance methods from meteorological records and is likely to vary less from year to year than the other components; average annual evaporation has been estimated at about 1600 mm. Outflow from the lake can be estimated from the lake levels measured at Jinja and the rating curve established by hydraulic modelling for the natural outfall at the Ripon Falls; since the construction of the Owen Falls dam the discharge has been maintained according to agreement on the basis of the lake level and the 'agreed curve', so that the outflows have been maintained at the natural rates.

All these terms in the water balance are available over the period 1956-1978, with some tributary flows deduced by comparison with direct measurements, and have been extended back to 1925 using basin rainfall

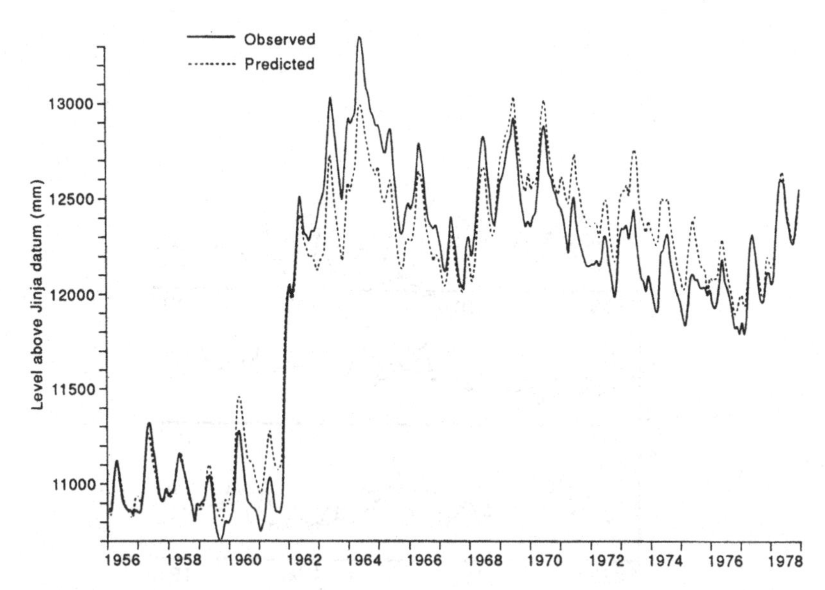

Figure 10: Water balance of Lake Victoria: observed and predicted lake levels, 1956-1978. (Source: Piper *et al.*, 1986)

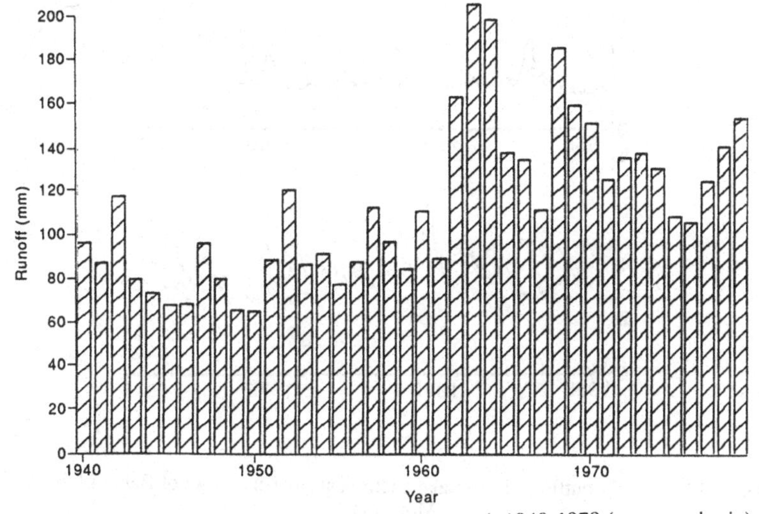

Figure 11: Annual runoff of Kagera at Nyakanyasi, 1940-1979 (mm over basin).

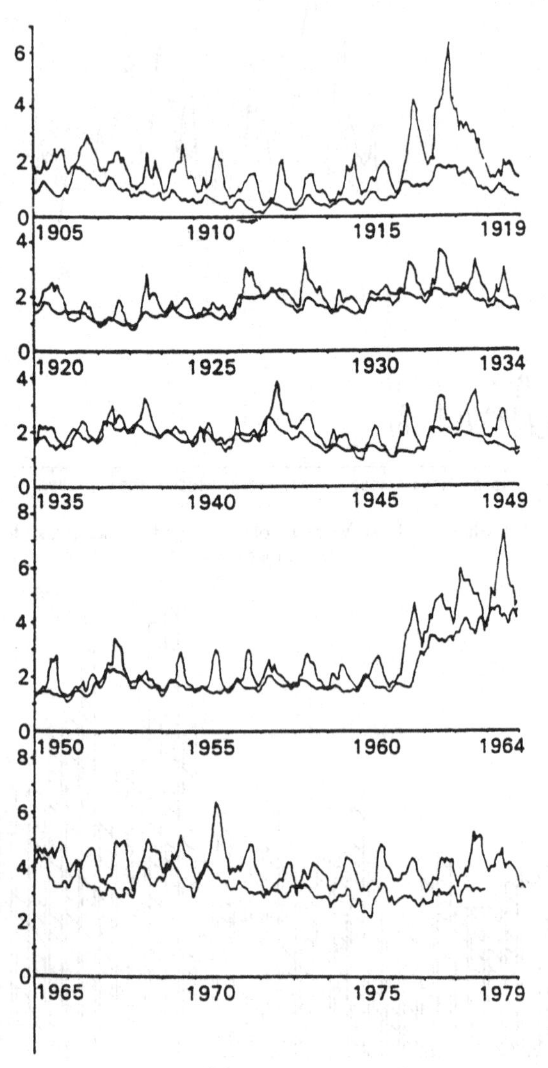

Figure 12: Monthly outflows from Lake Victoria at Jinja and flows of Bahr el Jebel at Mongalla.
(Source: Howell *et al.*, 1988)

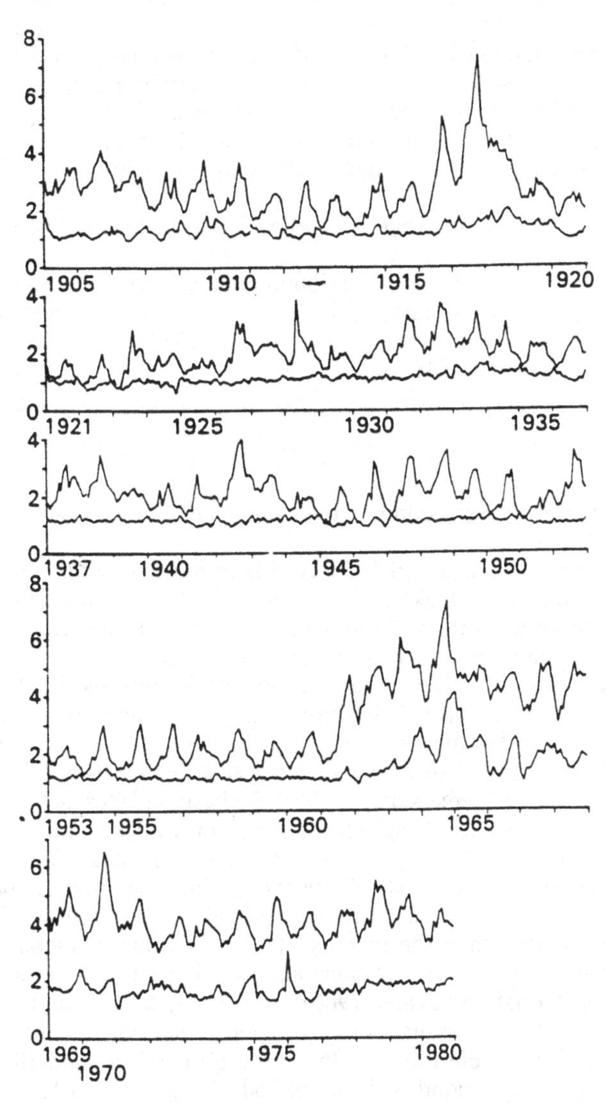

Figure 13: Monthly Mongalla inflows and outflows from the Sudd 1905-1980 (km^3).
(Source: Sutcliffe and Parks, 1987)

measurements and a conceptual model. The main items are summarised in Table 3, and show the relative importance of rainfall and the greater relative variability of tributary inflow. The close correspondence between average rainfall and lake evaporation causes any increase in rainfall to result in a disproportionate increase in tributary inflow and eventually in outflow. The rainfall of 1961-1964 therefore gave rise to a large rise in the lake and doubled the outflows from the lake; there is evidence from the Kagera and the other tributaries that the runoff inflow has remained higher than before 1961, and the outflows have remained high but decreasing slowly.

Table 3 Average balance of Lake Victoria basin, 1956-1978

	Lake rainfall	Tributary inflow	Lake outflow	Lake evaporation
			524	
Mean	1810	343	152	1595
SD	229	106		
			29.0	
CV(%)	12.7	30.9		

Note: each item is expressed in mm over an area of 67,000 km^2

The water balance for the period 1956-1978 is illustrated by Figure 10, where the measured and predicted lake levels are compared; it is evident that it is not necessary to introduce any extraneous factor to explain the rise in the lake. The continued high levels of tributary inflows after the initial rise is illustrated by Figure 11, where the flows of the Kagera have followed the pattern of Lake Victoria rises. It has been argued that land use changes have been responsible for this increase, but the sensitivity of runoff to the amount and seasonal distribution of rainfall seems sufficient explanation. Some persistence is provided by the storage within the basin in lakes and swamps.

Given that the problem of the water balance of the lake can be explained satisfactorily, it is possible to base analysis of the water available from Lake Victoria on the outflows alone, which together with evaporation estimates give an indirect estimate of lake inflow. Because lake level records (Figure 3) cover a longer period than the records of the individual components, they provide a longer data base for statistical analysis. Indeed, it is possible to use the additional information (Lyons, 1906) provided by historic and prehistoric levels; historical information makes it possible to deduce that the high levels of the period since 1961 were mirrored by very high levels around 1878 and in 1892-1895, which correspond with the period of high Aswan flows. On the other hand, there is prehistoric evidence from a cave excavated near Entebbe that the levels of 1964 had not been exceeded by more than 60 cm in the previous 3720+/-120 years (Bishop,1969), which puts a limit on the outflows during this period.

The water balances of Lakes Kioga and Albert (Mobutu Sese Seko) have not been investigated in the same detail, but their contributions to the flows of the Bahr el Jebel are not as significant as those of the Lake Victoria basin.

Nevertheless an understanding of their regimes is necessary to plan data requirements. Comparisons of flows measured at Jinja and Namasagali suggest that the contribution of the Lake Kioga basin is negative, with evaporation exceeding inflow, in years of low Lake Victoria outflows, and positive, with inflow from the basin exceeding evaporation from the lake area, in the years of higher Lake Victoria outflows since 1961. Whereas evaporation exceeds rainfall (about 800 mm) on Lake Albert itself, the contributions of the Semliki tributary have been measured at Bweramule (catchment 30,000 km^2) for a number of years with an average contribution of 4.2 km^3. The net contribution of the lower lake basins and the torrents between Lake Albert and Mongalla can be deduced by comparison of measurements at Jinja and Mongalla, illustrated by Figure 12, and by the double-mass curve of Figure 5.

A wide network of gauges has been established to measure rainfall, evaporation, lake levels and river flows in the Nile basin in Kenya, Uganda, Burundi and Rwanda; this network was started by the Egyptian authorities, established by the national hydrological services, and extended after 1967 by the Hydrometeorological Survey. For example, all the important tributaries into Lake Victoria have been measured, and the main channel of the Nile has been measured at key points along its course. By 1974 some 100 flow gauging and 36 lake stations were established in Uganda alone, but nearly all these gauges were destroyed by 1982, when responsibility for the network was handed back to the Uganda Water Development Department. The network is now being re-established, but there is an urgent need to accelerate this and to process and publish the series of river flows, some of which are available in a variety of sources; a start has been made by processing all available data for 20 gauging stations. From the point of view of management of the Nile, the urgent requirement is to ensure the continuity of the key stations related to major projects.

The possible major projects envisaged for the Lake basin, in addition to the Owen Falls dam, include a hydro-electric project at Rusomo Falls on the Kagera, hydro-electric schemes on the Nile between Lake Victoria and Lake Albert, and a dam below Lake Albert as part of the storage of Nile waters. Adequate long-term river flow records are available at these sites, but there has been a gap in most observations in recent years. Continued maintenance of the network is essential, and in particular the measurements of the Semliki flows require rehabilitation.

The Sudd

The swamps and seasonal grasslands of the Bahr el Jebel floodplain between Mongalla and Malakal have been known as the Sudd since they constituted a 'barrier' to the earlier travellers up the White Nile. The same problems of access have made hydrological investigation difficult, but the early establishment of discharge measurements and flow stations at Mongalla, Malakal and the Sobat mouth, have provided inflow and outflow records (Figure 13) and confirmed the observations of Garstin and Lyons (1906) that about half the Mongalla inflow is lost by evaporation within the Sudd. These observations led to the proposal in 1904 for the construction of a canal to carry a proportion of the flow past the Sudd in a canal, and thus to the plans for the Jonglei Canal which is at present partly completed. The average outflow from the Sudd for the period 1905-1980 was 16.1 km^3, compared with the inflow at Mongalla of 33.0 km^3; however, the latter figure masks a rise from 26.8 km^3 in 1905-1960 to 50.3 km^3 in 1961-1980.

The planning of this conservation measure to save water for other uses led to the establishment of a series of sets of gauging stations across the floodplain to measure flows in the various parallel river channels which were known as 'latitude flows'. These measured flows were used by Butcher (1938) to draw up water balances for the reaches between these latitudes, though the current lack of means of estimating evaporation accurately prevented him from explaining the water balance without postulating considerable spills to the west for which there is little direct evidence.

During the work of the Jonglei Investigation Team (1946-1954), hydrological studies formed an important part of their research into the effect of the seasonal flooding on the environment. Special investigations of sample reaches were concentrated on the southern reach between Juba and Bor, where access was feasible, and in particular on the Aliab valley opposite Bor and the basin north of Mongalla where the soils and vegetation were also studied in detail to deduce the interconnections (Jonglei Investigation Team, 1954, Volume III). Sutcliffe (1957,1974) subsequently used these surveys to clarify the mechanism of flooding by a study of the water balance of the Mongalla-Gemmeiza basin, making use of the measured inflows at Mongalla and outflows at Gemmeiza and Gigging, and to quantify the relations between flooding levels and vegetation.

After the rise in Lake Victoria levels in 1961-1964, the doubled inflows to the Sudd at Mongalla changed the channel network considerably, and diverted a large proportion of the flows at Bor, for example, to the west of the gauging station into the Aliab valley. These changes made it impracticable to measure latitude flows as in the earlier years, and left the only complete measurements those of the inflows at Mongalla and the outflows at Malakal, taking account of the measured contributions of the Sobat. When further studies of the Sudd were commissioned by FAO in 1982, it was possible to use these flow records, together with long-term rainfall series and evaporation estimates, to investigate the water balance of the whole area (Sutcliffe & Parks, 1987) with a simple hydrological model and to deduce a time series of flooded areas, which were compared with a number of estimates based on air photography and satellite imagery. The satellite imagery and the hydrological model both showed that the flooded areas, which averaged about 8000 km^2 before 1961, increased to over 30,000 km^2 in 1964 following the rise in Lake Victoria and have since remained between about 20,000 and 30,000 km^2. It is unlikely to be possible to measure additional flows until lake levels recede, but a combination of inflows and outflows for the whole area, and continued rainfall measurements and satellite imagery, should enable the behaviour of the Sudd to be monitored.

Present conditions in the southern Sudan are not conducive to a programme of hydrological observations, but it is vital that continuity should be preserved as far as possible.

Bahr el Ghazal floodplain

The tributaries of the Bahr el Ghazal derive their runoff from the higher ground of the Congo-Nile divide, but very little of the river flow reaches the White Nile at Lake No. Each of the rivers follows a standard pattern, from an elevated perimeter of rapid runoff with good drainage and some rapids through a zone where the rivers meander between alluvial banks in a defined and widening valley into a zone of unrestricted flooding over clay plains (Southern Development Investigation Team,1954). The headwaters of the main tributaries - the Na'am, Gel, Tonj, Jur and Lol - receive between March and October the highest rainfall in the Sudan of 1200-1400 mm. Their flows have been measured at a series of stations along the main road from Shambe to Wau and Nyamlell which roughly coincides with the boundary between the zone of runoff generation and that of spill and losses; the runoff is concentrated into the months from June to November and averages some 60-100 mm over the basins. The typical runoff regime of the Jur at Wau is illustrated by Figure 7, where the monthly mean runoff is expressed in mm. The flows of the Jur at Wau have decreased in recent years in line with the Blue Nile and Ethiopian tributaries. The total flow of the rivers is about 11 km^3, but only some 0.3 km^3 is measured at the junction of the Bahr el Ghazal and the White Nile, and much of this is spill from the Bahr el Jebel. The balance of the inflow to the Bahr el Ghazal swamps is evaporated, together with rainfall averaging 800-1000 mm.

These wetlands have been estimated as up to 16,600 km^2 in area (see Chapter 12), and it is evident that a large amount of water is lost in evaporation. Proposals have been made (Hurst *et al*., 1966) to divert much of this water through a canal in a conservation project. The basic hydrological data exist for the preliminary study of the area (Chan & Eagleson,1980). However, the resumption of routine rainfall and river flow measurements over the whole basin is essential for further investigation of the hydrology of the area. Satellite imagery could be useful to delineate the areas of evaporation and to confirm where river flows are likely to be at their maximum.

River Sobat and Machar Marshes

The third major area where losses by evaporation occur is the complex between the upper Sobat and the White Nile, in the area known as the Machar Marshes. The main tributary of the Sobat, the Baro, drains some 41,000 km^2 of SW Ethiopia, but high flows spill to the north through Khor Machar into the Machar Marshes. Comparisons of the flows of the Baro measured at Gambela with measurements at a succession of sites downstream towards the junction of the Baro with the Pibor (Hurst, 1950) show clearly that flows above a certain level leave the river either through several known spill channels or as general spill over the river banks. As a result the contributions of the Sobat to the White Nile above Malakal (Figure 7) are truncated in years when high flows occur on the other Ethiopian tributaries like the Blue Nile. There are several tributaries which drain from the Ethiopian headwaters directly to the Machar Marshes like the Yabus and Daga, but there is only significant outflow through the Machar Marshes to the White Nile north of Malakal in exceptional years.

The series of gauging stations on the Baro and Sobat operated in the past have been used to deduce the amounts of spill into the Machar Marshes by difference between successive sites. Alternative estimates of the average spill from the Baro to the Machar Marshes range from 2.8 km^3 (Jonglei Investigation Team, 1954; Hurst *et al*., 1966) to 3.5 km^3 (el Hemry & Eagleson, 1980). Measurement of the flows of the Yabus and Daga were begun in 1950 and estimates of the total contribution of the direct tributaries range from 1.4 km^3 (Hurst *et al.*, 1966) through 1.74 km^3 (Jonglei Investigation Team, 1954) to 2.0 km^3 plus 1.6 km^3 inferred runoff from the plains (el Hemry & Eagleson,1980), with return flow to the White Nile estimated as about 0.12 km^3. With rainfall averaging about 750 mm, it is evident that a significant amount of water is evaporated in the Machar Marshes, compared with the average flow at the Sobat mouth of 13.7 km^3. Various projects to reduce this loss have been discussed in the past (Hurst *et al.*, 1966), including a reservoir on the Baro, embankment, or diversion of the Baro and possibly the eastern tributaries into a canal.

A programme of detailed measurement and investigation is required to establish the hydrological regime of the Machar Marshes and to plan conservation measures. The measurements should include rainfall and evaporation over the area, inflow from the streams draining directly from the Ethiopian headwaters and successive gauging stations to assess spills from the Baro. The remoteness of the marshes and the eastern rivers would make it difficult to maintain rainfall or gauging stations, and there may be a case for weirs and automatic meteorological stations linked with data collection platforms. It may also be difficult to reinstate the gauging stations on the Baro because of their remoteness except with cooperation between the Ethiopian and Sudanese authorities. However, the use of satellite imagery to deduce a time series of evaporating areas might make it possible to estimate the losses by comparison of rainfall and evaporation and thus deduce the inflows by an inverse approach.

White Nile from Malakal to Jebel Aulia

There are no significant tributaries in the reach from Malakal to the Jebel Aulia dam, and the river management problems refer to the control of flows by the reservoir and the forecasting of flows along the reach. Inflow and outflow measurements have been taken over a number of years, and flood routing studies have been carried out to relate flows and levels along the reach to inflows at Malakal. These studies have been based on hydrological routing methods, which rely essentially on relationships between inflow, outflow and storage, and took advantage of earlier studies presenting idealised cross-sections along the White Nile (Jonglei Investigation Team, 1954). The extrapolation of such methods to the higher flows which will occur when the Jonglei Canal and other conservation measures are completed may require hydraulic modelling and the extension of idealised cross-sections and of rating curves to relate flows to levels.

Blue Nile basin, including Rahad and Dinder

The flows of the Blue Nile have been measured at Roseires since 1912, but there is little published information about the water balance of the basin within Ethiopia. The only natural storage within the basin is in Lake Tana, which was investigated during the period 1920-1933 (Hurst *et al*.,1959). The basin draining through the lake is about 13,400 km^2, and the average discharge is estimated at about 4.0 km^3 or 300 mm over the basin. The next points at which flows are known are at el Deim and at Roseires, where the long-term average discharge is 50.0 km^3, which is equivalent to 255 mm over the basin. Thus the outflow from Lake Tana is less than 10 per cent of the flow of the Blue Nile. Although water balance studies of the other Blue Nile tributaries within Ethiopia would be of scientific interest, the main practical purpose of such studies would be for reservoir development, for example for hydro-electric power.

The regime of the Blue Nile may most conveniently be studied from the records at Roseires or Khartoum; the latter has the longer record. The seasonal distribution is illustrated in Figure 7, which shows the concentration of runoff between July and October. The time series of flows, expressed in mm over the basin, is shown in Figure 8; this includes the effect of abstraction upstream but illustrates the variability of the Blue Nile flows compared with the Bahr el Jebel contribution to the White Nile series (Figure 9). This underlines the need not only for flow measurements but also for sediment monitoring as sediment load varies markedly with flow but also over time with land use.

From the point of view of basin development, the main interest in the hydrology of the Blue Nile within Ethiopia is for flood forecasting for reservoir operation and to give warning of possible inundation in Khartoum and in the agricultural areas downstream. Although the gauge at el Deim above Roseires could be used for this purpose, the lead time would be increased if it were possible to incorporate rainfall measurements or estimates into the forecasting system. The problems of establishing raingauges and a communications system in the upper Blue Nile basin make it attractive to use rainfall estimates derived from satellite imagery, and specifically from the durations of cold cloud temperatures below a threshold value, which have proved competitive with traditional raingauge networks in flow forecasting on the River Senegal (Hardy *et al.*, 1989). These techniques proved promising in preliminary trials on the Blue Nile system in studies following the floods of 1988 (Sutcliffe *et al.*, 1989).

The two Blue Nile tributaries within the Sudan are the Dinder and Rahad, which reach the main river above Khartoum. The long-term gauges on these tributaries were located near their confluences with the Blue Nile, but the sites were difficult to calibrate and also less representative than sites which have been established upstream. These flows have been extended by correlation with the Blue Nile flows to give longer series for water resources management studies.

River Atbara and other tributaries

The last major tributary of the Nile is the River Atbara, which has been measured near its mouth since 1903 and in its upper tributaries since 1970. The long-term discharge of the Atbara at its mouth is 11.7 km^3, which is equivalent to about 100 mm over its effective basin; because the basin is farther north than the Blue Nile the runoff is concentrated over a shorter season and the annual variations are greater. The mean monthly runoff expressed as mm over the effective basin is illustrated in Figure 7; the period of runoff is more concentrated than on the Blue Nile, with a high proportion occurring in August and September.

The two main requirements for flow records on the Atbara are for control in the Khashm el Girba reservoir for irrigation, and for flood forecasting. The response of the Atbara is rapid and it will be necessary to obtain a reasonable lead time through rainfall information which, as in the case of the Blue Nile, could require the interpretation of satellite imagery. The monitoring of flow volumes and sediment load requires continuous measurement.

There are a number of ephemeral streams draining the areas adjacent to the Nile north of Khartoum, which may contribute to the flow of the main river after exceptional storms like that of August 1988. There are several gauges established on these wadis which would give an indication of the scale of flows which have resulted from these and other storms.

Main Nile

The progress of the Blue Nile and Atbara floods, superimposed on the contributions of the White Nile, down the Main Nile between Khartoum and Egypt is one of flood wave transmission. Evaporation losses and abstractions for irrigation may have to be taken into account, but forecasts of the volume and timing of the flood wave are needed to give warning of possible inundation of agricultural areas, and to give advanced notice for planning purposes of the volumes to be expected in the Aswan reservoir.

The travel times down the main river have been investigated over many years and empirical relations derived. The response is sufficiently slow that daily levels and derived flows should be adequate to provide the basis for improved forecasting methods. However, it would not be possible to measure the flows at upstream and downstream sites with sufficient precision to estimate local inflows or abstraction by difference; these would have to be estimated directly. For forecasting purposes rainfall estimation using satellite imagery would require supporting by modern methods of transmitting river levels, for instance using data collection platforms.

Aswan flows and climate change

For the planning of water resources management in Egypt, the need is for statistical or deterministic analysis of the magnitude-frequency relationship of future flows, both in the medium term and in the long term. Assisted by the ready availability of published flow records (e.g. Shahin,1985), the problem of the statistical structure of Nile flows has focussed more scientific attention on this than on other river systems. A description of these researches is outside the scope of this discussion, but the long-term persistence of river flows and other natural phenomena revealed by Nile records has been named the 'Hurst effect' (Hurst *et al.*,1965) and has given rise to a large body of research.

This effect means that records for analysis should cover as long a period as possible, and that the long record at Aswan covering the period from 1870 to the present may not provide sufficient evidence to model the changes from the earliest period of high flows to lower flows in the years since 1900 and to higher flows on the White Nile since 1965. The persistence of these fluctuations and the evidence of Lake Victoria levels might suggest a connection with lake storage, but the length of the high and low flow periods makes this unlikely to be the whole explanation. Indeed the persistence shown by the inflows to Lake Victoria shows that the explanations must be sought in the sensitivity of runoff to changes in rainfall and to global climate.

The search for longer records has led to some attention to the long series of annual high and low levels at the Roda gauge in Cairo (see chapter 2), which have been recorded with some gaps since 622 AD and listed by Toussoun (1925). However, the level series have been suspected of changes in unit and reflect aggradation of the river channel over the centuries (Figure 14). Nevertheless the utility of these records is illustrated by the relationship between the annual maxima recorded at the Roda gauge between 1870 and 1921 and the floods measured independently at Aswan during this period (Figure 15); this relationship incidentally appears to refute the doubts which have been expressed about the measured flows at Aswan, for instance by Yevdjevich (1963) who presents the series as an example of inconsistency. Other studies of long-term variations have been based on rainfall and other historical evidence (Nicholson, 1980).

Past natural variations may be reinforced by the effects of secular climate change. The contribution of hydrological records to this study is the provision of reliable and consistent rainfall and river flow series to calibrate hydrological models as a predictive tool, or more simply to provide analogues to convert predictions of rainfall change to estimates of changed water resources (Gleick, 1989).

General observations

The discussion has dealt with the data requirements for the hydrological study of individual tributaries. This section is concerned with the basin as a whole, before the conclusions of the chapter are summarised.

A major contribution to the problems of water resources planning within the Nile Basin would be the monitoring of climate fluctuations in real time using satellite imagery, together with the investigation of possible climate change, and the interpretation of such findings in terms of available river flows.

In general the design of the hydrological network has proved a reasonable basis for investigation of the water resources of the Nile Basin, particularly after the detailed investigations of the Lake Victoria basin. The need is for

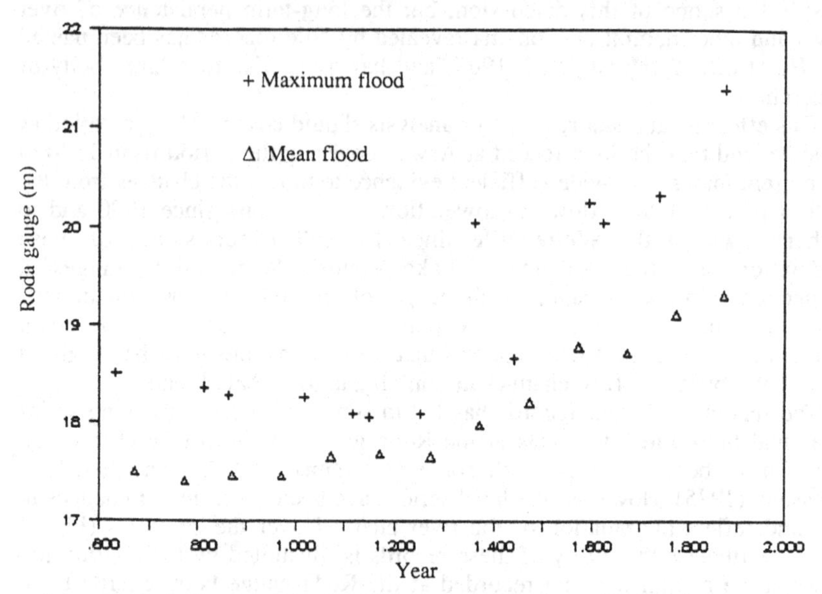

Figure 14: Roda Gauge at Cairo 622-1921: mean and maximum flood levels by centuries. (Source Tousson, 1925)

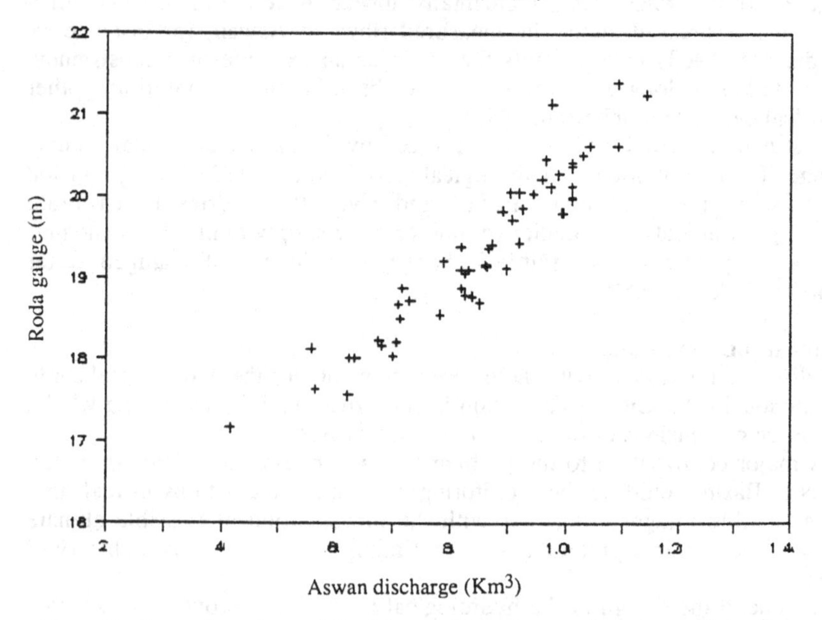

Figure 15: Comparison of annual flood level at Roda Gauge and maximum 10 day flow volume (km^3) at Aswan, 1869-1921.

maintained measurements and for consistent quality control to provide a basis for the extension of short-term records.

The one major gap in the investigation of the hydrological regimes of the basin is the measurement and analysis of erosion and sediment load, which is a severe problem in the maintenance of water control projects. This problem is particularly relevant to the Blue Nile and Atbara tributaries, but regional monitoring and investigation would enable erosion to be linked to such factors as basin size and slope, geology and climate.

Research into the improvement of irrigation efficiency in terms of water use could benefit from the use of modern methods of hydrological measurement; techniques of estimating soil moisture content and abstraction by irrigated crops could be used in conjunction with methods of measuring evaporation above a crop surface in field experiments.

There is a need for the exchange of information on the hydrology of the various rivers of the Nile Basin, irrespective of negotiations on the use of the waters of the basin. Such exchange would enable plans to be made and negotiations undertaken in the light of accurate knowledge. It might be argued that the exchange of hydrological data benefits the downstream countries, who can use the information for forecasting purposes, rather than the upstream states. However, one of the main uses of hydrological records is to provide material for the extension of short-term records by correlation or other means; it is natural for the downstream countries to begin systematic measurement of river flows earlier because of their dependence on these rivers, and thus exchange of data would enable the upstream countries to make best use of their possibly shorter data series.

In addition to the Permanent Joint Technical Commission of Egypt and the Sudan, the Hydrometeorological Survey of the upper Nile lake system set up initially by WMO and UNDP has provided a framework for the exchange of data and scientific expertise between a number of the countries of the Nile basin, in addition to the programme of measurements and research which it undertook. This organisation might provide an example for the extension of scientific exchanges.

Summary of conclusions

It was with reason that Hurst and Phillips (1931) claimed that 'there is no large river in the world upon which such an extensive and accurate system of measurement is carried out'. They warn of the dangers of lack of continuity without scientific control of the programmes, methods and accuracy of all hydrological investigations, but in general this continuity has been maintained in spite of difficulties.

Within the Lake basin, the network has been adequate with 100 gauging stations in Uganda alone in the past; this network requires urgent rehabilitation. Progress is being made, and as far as Nile management is concerned the lake level series can be used to fill some of the gaps in flow records.

The numbers of gauging stations within the Ethiopian basins of the Nile tributaries and measured flows are not available, but long records exist where the tributaries enter the Sudan.

Within the Sudan the river system is relatively sparse but the gauge network has been carefully planned to monitor all the key tributaries, with some 200 river gauges within the basin of which a number date from the early years of this century. In general the need is for the maintenance of continuity and quality control and for the measurement of sediment transport. The wetlands where water conservation measures are planned require measurements of inflows and outflows as well as satellite monitoring of flooded areas.

Acknowledgements
Many of the ideas for this Chapter arose out of a Hydrological Assessment of several of the countries of the Nile Basin, undertaken by Sir Alexander Gibb & Partners in association with the Institute of Hydrology and the British Geological Survey and with the assistance of Reading University, Department of Meteorology, and Professor M.D. Newson of the University of Newcastle-upon-Tyne. This provided the opportunity to discuss the hydrological problems of these countries with those responsible for developing their water resources. The authors are grateful to their colleagues and to a number of people in the countries of the Nile Basin for valuable discussions.

References

Bishop, W.W. (1969). Pleistocene Stratigraphy in Uganda. *Geol. Surv. Uganda. Memoir,* **10**, Entebbe.

Butcher, A.D. (1938). *Sadd Hydraulics* . Cairo,Government Press.

Chan, Siu-On & Eagleson, P.S. (1980). *Water Balance Studies of the Bahr el Ghazal Swamp*. Dept. of Civil Eng., Massachusetts Institute of Technology, Report No. 261.

Gleick, P.H. (1989). Climate change, hydrology, and water resources. *Reviews of Geophysics,* **27**, pp. 329-344.

Hardy, S., Dugdale, G., Milford, J.R. & Sutcliffe, J.V. (1989) The use of satellite derived rainfall estimates as inputs to flow prediction in the River Senegal. *Inter. Assoc. Hydr. Sci.,* Publ. No. 181, pp. 23-30.

el Hemry, I. & Eagleson, P.S. (1980). *Water Balance Estimates of the Machar Marshes*. Dept. of Civil Eng. Massachusetts Institute of Technology, Report No. 260.

Howell, P., Lock J. & Cobb, S. (eds) (1988). *The Jonglei Canal: Impact and Opportunity*. Cambridge University Press.

Hurst, H.E. (1950). The Hydrology of the Sobat and White Nile and the Topography of the Blue Nile and Atbara. *The Nile Basin,Volume VIII.* Cairo, Government Press.

Hurst, H.E. & Black, R.P. (1943). Monthly and Annual Rainfall Totals and Numbers of Rainy Days at Stations in and near the Nile Basin. *The Nile Basin, Volume VI.* Cairo, Government Press.

Hurst, H.E., Black, R.P. & Simaika, Y.M. (1959). The Hydrology of the Blue Nile and Atbara and of the Main Nile to Aswan, with some reference to Projects. *The Nile Basin, Volume IX.* Cairo, Government Press.

Hurst, H.E., Black, R.P. & Simaika, Y.M. (1965). *Long-term Storage: an*

Experimental Study. London, Constable.
Hurst, H.E., Black, R.P. & Simaika, Y.M. (1966). The Major Nile Projects. *The Nile Basin. Volume X*. Cairo, Government Press.
Hurst, H.E. & Phillips, P. (1931). General Description of the Basin. Meteorology, Topography of the White Nile Basin. *The Nile Basin, Volume I*. Cairo. Government Press.
Hurst, H.E. & Phillips, P. (1932). Measured Discharges of the Nile and its Tributaries. *The Nile Basin, Volume II & Supplements*. Cairo, Government Press.
Hurst, H.E. & Phillips, P. (1933). Ten-day mean and Monthly mean Gauge Readings of the Nile and its Tributaries. *The Nile Basin, Volume III & Supplements*. Cairo, Government Press.
Hurst, H.E. & Phillips, P. (1933). Ten-day Mean and Monthly Mean Discharges and its Tributaries. *The Nile Basin, Volume IV & Supplements*, Cairo, Government Press.
Hurst, H.E. & Phillips, P. (1938). The Hydrology of the Lake Plateau and Bahr el Jebel. *The Nile Basin, Volume V*. Cairo, Government Press.
Jonglei Investigation Team (1954). *The Equatorial Nile Project and its effects in the Anglo-Egyptian Sudan*. Report of the Jonglei Investigation Team. Khartoum, Sudan Government.
Kite, G.W. (1981). Recent changes in level of Lake Victoria. *Hydrol. Sci.. Bull*. **26**, pp. 233-243.
Lyons, H.G. (1906). *The Physiography of the River Nile and its Basin*. Cairo, Government Press.
Nicholson, S.E. (1980) Saharan climates in historic times, in *The Sahara and the Nile*. ed. Williams M.A.J. & Faure H. Rotterdam, A. A. Balkema.
Piper, B.S., Plinston, D.T. & Sutcliffe, J.V. (1986) The water balance of Lake Victoria. *Hydrol. Sci. J*.,**31**, pp. 25-37.
Shahin, M.M.A. (1985). Hydrology of the Nile Basin. *Developments in Water Science*, **21**. Amsterdam, Elsevier.
Southern Development Investigation Team [SDIT] (1955) *Natural Resources and Development Potential in the Southern Provinces of the Sudan*. Khartoum, Sudan Government.
Sutcliffe, J.V. (1957). *The Hydrology of the Sudd Region of the upper Nile*. Ph D. thesis. University of Cambridge.
Sutcliffe, J.V. (1974). A hydrological study of the southern Sudd region of the upper Nile. *Hydrol. Sci. Bull*.,**19**, pp. 237-255.
Sutcliffe, J.V., Dugdale, G. & Milford, J.R. (1989). The Sudan floods of 1988. *Hydrol. Sci. J*., **34**, pp. 355-364.
Sutcliffe, J.V. & Knott, D.G. (1987). Historical variations in African water resources. *Inter. Ass. Hydr. Sci. Publ*. **168**, pp. 463-475.
Sutcliffe, J.V. & Parks, Y.P. (1987). Hydrological modelling of the Sudd and Jonglei Canal. *Hydrol. Sci. J*. **32**, pp. 143-159.
Toussoun, Prince Omar (1925). *Memoire sur l'histoire du Nil*. Memoires de l'Institut d'Egypte, Volume 8.
Yevdjevich, V.M. (1963). *Fluctuations of Wet and Dry years: Part I, Research Data Assembly and Mathematical Models*. Hydrology Paper No.1, Colorado State University.

el Zein, S. (1974). An approach to water conservation: projects for reduction of losses in some tributaries of the Nile. *Sudan Engrg Soc. J.*,**21**.

Part III

Future utilisation of Nile waters

8

Future irrigation planning in Egypt

R. STONER

Introduction

The Sudan has yet to utilise its full share of the Nile flows as defined by the 1959 Nile Waters Agreement; Egypt already takes its full allocation. The regional drought during the 1980s has concentrated thinking on methods of improving the efficiency of water use within Egypt. Efficiency requires the maximum re-use of water within the Nile Basin whilst maintaining essential low flows in the Delta to prevent salination. Irrigating areas of the western desert from the Nile offers no potential for re-use of water and is therefore an 'inefficient' use. Enterprising private farmers using groundwater or lifting drainage water point the way to increasing re-use. Management problems must also be addressed. The advent of easy-to-use computer simulations of the irrigation system will permit experienced engineers to try out alternative operational procedures.

The problem

In the first ten years of operation after the completion of the High Dam the level in Lake Nasser continually rose and there was some speculation about the possibility that there was more water available than had been envisaged. The 1980s brought us down to earth quickly and dramatically with the Lake dropping to critically low levels by the end of 1987 and the real question of whether this drought condition was a permanent proposition for the future. There was some respite in 1988 but flows were down again by 1989. Previous chapters have examined these questions in some detail and it is only intended here to recapitulate briefly so as to set the scene for a discussion on what measures might be taken to get the most out of the water that we have.

The Nile Waters Agreement of 1959 assumed the following availability and proposed allocations:

			km^3
Average inflow	84	to the Sudan	18.5
Lake losses	10	to Egypt	55.5
Net availability	74		74.0

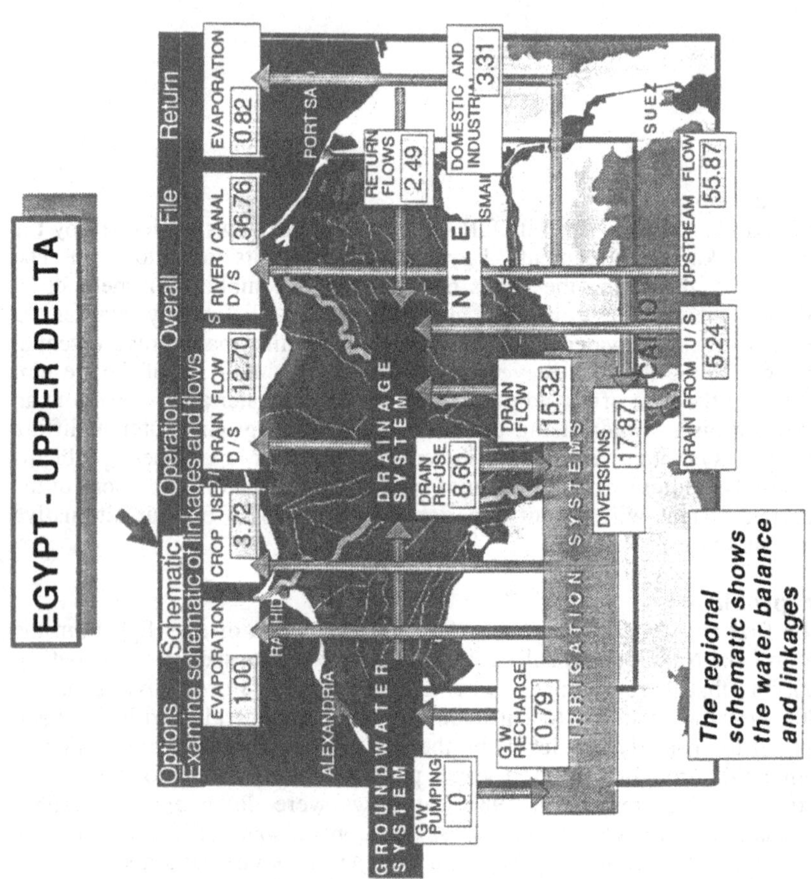

Figure 1: Screen display from Nile simulation model.

Over the years Egypt has taken its full share of 55.5 billion cubic metres (km^3) and sometimes a little more, whereas the Sudan has never got within 5 km^3 of its allocation.

Looking at the actual storage in Lake Nasser from 1968 onwards at the end of July, when the lowest annual level is reached, a fairly steady rise occurred until the late seventies and then the decline, which is still with us, began.

Table 1 Lake Nasser Storage (km^3)

Year	Storage (end July)
1968	29.4
1970	45.4
1972	69.2
1974	67.0
1976	108.4
1978	109.0
1980	103.1
1982	99.1
1984	72.9
1986	53.7
1988	41.4
1989	75.3

A great deal of the rise was attributable to two very good years, the floods of 1974 and 1975. What is interesting from Egypt's point of view is that if the Sudan had taken only 5 km^3 per annum extra during this period the Lake would have fallen to its dead storage level of 30 km^3 by about 1983 and would never have risen above it thereafter, even allowing for the large flood of 1988. Measures for the Sudan which will perhaps enable that country to use its full allocation of 18.5 km^3 in the future are under consideration. If that is indeed achieved and the modest Nile flows do not improve, then Egypt will suffer very serious water shortages.

The table below shows the current water balance as at 1987 and what it might have been with only 50.0 km^3 released as opposed to a full supply of 55.5. The projected km^3 position as at the year 2000 is also given below.

Table 2
Estimated annual water balances
(all figures are in km^3)

Water Supply	Full water	Drought	Full water	Drought
Year	1987	1987	2000	2000
Water Savings	As present	As present	Full	Full
Inflow				
Aswan release	55.5	50.0	55.5	50.0
Outflow				
Edfina	3.5	3.5	0.6	0.6
Canal tails	0.1	0.1	0.1	0.1
Drainage outflow	13.9	13.9	10.9	10.9
Evaporation	2.0	2.0	2.0	2.0
Sub-total	19.5	19.5	13.6	13.6
Water Use				
Municipal & Industrial	2.4	2.4	4.8	4.8
Irrigation	33.6	28.1	37.1	31.6
Sub-total	36.0	30.5	41.9	36.4

The two items that require explanation in the outflow figures are evaporation which represents losses from free water surfaces of canal, rivers and lakes and the outflow from the Rosetta branch of the Nile past Edfina Barrage. The latter is discussed later in this chapter. It now remains to look at ways in which water might be better used and this forms our main discussion.

Some suggested solutions

At its simplest the water balance reduces to incoming and outgoing water. The questions are these: Can we augment water coming into the basin and can we make better use of that going out? The scope for the former is limited and constrained by political and environmental factors but we would suggest that there are great opportunities for better use of water before it is eventually lost to the basin. And it is all lost, whether it is discharged to the sea or evaporated (given that year-on-year changes in storage have no effect in the long term). Some drainage to the sea is needed even if the inefficiencies in operation could be entirely eliminated. It is required to carry away salts, at least as much as is introduced for the Nile each year, and it is also needed to carry away pollutants introduced by domestic and industrial waste even if these can be minimised by suitable treatment processes. All of the remaining water is lost to evaporation. Ideally the whole of this water should be passed through a crop on its way out into the atmosphere and, in fact, much of it is. But that does not necessarily mean that it is used most efficiently. And here we are not talking about farm efficiencies or even canal command efficiencies which have no real significance in a closed basin. Such efficiencies think in terms of beneficial use (i.e crop evapo-transpiration) compared with water delivered. If, as in the Egyptian context, water apparently lost is picked up in the groundwater or in drainage and is re-used, then these efficiency concepts have no significance in relation to water conservation although they may have in regard to energy expended. The efficiency that is of most concern here is not the return to water delivered but the return per unit of water evaporated. In that context the growing of rice and sugar cane, which both use a great deal of water, especially in the hottest climate of Upper Egypt, could be questioned.

There is a number of ways in which both losses to the sea might be reduced and better use of evaporated water might be made. Most of them involve improved management practices at all levels.

Increasing the water available

Works upstream in the southern Sudan could substantially increase the water available to Lake Nasser by draining the marsh lands of the Sudd. A start was made on the Jonglei Phase 1 project which would have yielded for Egypt up to an extra 1.9 km^3. However, the local inhabitants objected very strongly and the work had to be abandoned. Three other possible projects have been identified: Jonglei II, Machar Marshes and Ghazal Canals which altogether could yield as much as an additional 14 km^3 per annum. If, in accordance with the 1959 Agreement 50 per cent were to benefit Egypt, then the total additional water would be 4.7 km^3 net taking the amount available from 55.5 to 60.2 km^3.

There are very considerable ecological and social problems involved in these projects and at the moment they appear to be far into the future if they are ever built at all (see Chapter 12).

The agricultural sector

There is little doubt that the most dramatic potential improvements in the use of irrigation water lie in agriculture itself. Egyptian agro-climatic conditions are particularly suitable for irrigated agriculture and the technology already exists for substantial increases in crop yields. It has been estimated that the yield of most crops could be increased by one-third and those of cotton and sugar cane by as much as 70%, while maize, sorghum and groundnut yields could be doubled. Projects are in place for disseminating research technology to farmers. More effort in this direction will yield quick results. However, freedom of choice by the farmers in what they grow is somewhat limited by government intervention setting crop quotas, compulsory purchase at low prices, and control of credit. To a large extent an historical cheap food policy which benefits the urban sector at the expense of the rural areas has been at the root of the problem The government recognises the problem and a current policy of removing these distortions will provide incentives and flexibility to producers.

There is still need for further research and while the advent of water from the High Dam has assured two crops a year, there is every possibility that by using shorter term varieties a third crop can be introduced. By judicious crop selection this might be achieved without additional water. Indeed some enterprising farmers already achieve this.

The scope for agricultural improvement is enormous and the return to investment in the sector is high. It should therefore have the priority it deserves.

Land

By far the best land for irrigated agriculture is in the Delta. It has the added benefit that it enjoys a milder climate than other areas of Egypt. This means that evapo-transpiration is lower and thus more crop per unit of water can be grown. Two things are then surprising: first that urban encroachment on to such land was allowed to go unchecked for many years and second that production is limited by a shortage of water, particularly at the tails of canals at some periods of the year. The first matter has been addressed by government but there is still some question of whether the measures introduced have been wholly effective. The second matter will be discussed later.

Given that water is the scarce resource and that the needs in the Delta should be given priority, it is particularly difficult to understand the various projects in the Western Desert where sand of the texture of granulated sugar is being irrigated with no hope of the resulting drainage water being returned to the basin. A number of these projects have failed but still valuable water is being lost to the system in this way.

Within the Delta, to the north, a number of reclamation projects on salt-affected land have been taken up. Where these have been developed by private farmers there has been considerable success but still there are large tracts of land retained for government farms and on these progress has been very slow. Throughout the world governments are poor farmers and Egypt is no exception. Such land should be transferred to the private sector to accelerate its development.

Improvements to the irrigation system

Tertiary canals (*meskas*) in Egypt generally run below ground level and the farmers lift the water to irrigate the fields. A rotation system is operated with so many days 'on' followed by a similar number 'off '. Wide variations occur but for rice in the Delta the rotation now practised is five days on, five off. As ever, farmers are reluctant to irrigate at night which does cause water to escape at night at the tails. Much of the system is under-capacity due to channel deterioration, restrictive and inappropriate cross structures, and poor control.

For the future one single measure could effect a substantial improvement; that is to introduce a continuous flow system. This would virtually reduce to half the flows in presently 'rotated canals'. It would enable the present surplus capacity to be used for night storage thus enabling daytime only irrigation whilst not preventing night irrigation. Occasional closures at periods of low demand would still be required for maintenance and new construction.

The aim must be to simplify operation such that the system can function with the minimum attention by operating staff. This will be achieved by reducing the number of regulators to a minimum, removing or enlarging structures that presently restrict flows and cause head losses, making provision for automatic control and, in appropriate circumstances, remote control.

There are some parts of the system where gravity rather than lift irrigation is practised. Needless to say these areas are very prodigal in their use of water. Wherever possible these should be replaced by the conventional low level *meska* and converted to lift irrigation.

In the system at large there is little or no measurement or control of discharge except at major command level. It is fundamental to successful future irrigation that measurement and regulation of discharges should be extended to the lower levels of the system. It is impractical and perhaps counter productive to attempt to do this down to *meska* level and therefore the lowest level proposed is at the head of branch canals and at cross regulators. Undershot gates will continue to be used for control on large canals which can be calibrated more reliably, are more suitable for mechanical operation and would thus more readily fit into an automatic and remotely controlled operating regime.

Other proposals for main and branch canals include remodelling of some canals to ensure adequate capacity, both for conveyance and night storage; construction of adequate new bridges, aqueducts and culverts; and improvement or replacement of tail escapes; improvement of canal banks and access roads; replacement of *meska* offtakes and provision of pump sumps for direct lift irrigation from these canals. Pilot projects for remote control, communications and telemetry are proposed prior to extending this form of operation to other areas.

Similar measures are also required at *meska* level, but in addition *meskas* should be lined or culverted through villages where presently they are often used to discharge effluents and to dump rubbish. Proper pump sumps are also required, as are reconstruction of leaking drain crossings and of banks and the provision of adequate access.

Drainage re-use

The re-use of drainage water is already widely practised whether it be an individual farmer pumping from a drain or major drainage pump stations delivering water back into a main canal. Indeed, it is this practice that raises the overall system efficiency to about 65 per cent at present. Three types of drains are recognised; those carrying largely fresh water which can be used freely either directly on the land or returned to the river or the canal system; those, particularly towards the north of the Delta, which pick up saline groundwater and can only be used after controlled mixing with canal water; and lastly, those that are heavily polluted with domestic and industrial waste.

Ultimately the drainage to the sea should be minimised to that level which is necessary to carry excess salts from the Nile Valley. If, for example, all water with more than 2500 ppm of dissolved solids were the only water rejected to the sea, then the maximum achievable system efficiency would be something over 80 per cent, i.e 20 per cent rejects, and it is that sort of level that should be aimed at.

As far as possible drainage water should be used as soon as it is feasible, that is, in the area where it is produced, rather than discharge it all downstream thus making it more difficult to manage efficiently further down the system. The emphasis should therefore be on smaller re-use schemes scattered widely throughout Egypt, and normally these would amount to pumping drainage water into canals.

The problem of using sewage-polluted drains has yet to be tackled but even on the notorious Bahr el Bakr drain, which carries sewage effluent from Cairo, some farmers are irrigating food crops directly and with considerable success. They are then passing clear drainage water from their fields back to the drain. While this practice cannot be encouraged from the public health viewpoint, it may point the way to other possible uses, perhaps for forestry, tree crops or reed beds. In the longer term, sewage treatment will improve the situation but there is no doubt that the problem will be with us for a long time to come.

Careful attention to the re-use of drainage water provides the single largest scope for increasing the availability of irrigation water and it should receive the closest possible attention.

Winter losses from the river to the sea

The last barrage on the Nile is at Edfina and flow past this structure is all lost to the sea. On average this amounts to some 5 km^3 per annum occurring in the months from October to March. Part of this (about 1.2 km^3) is due to stability problems with Esna barrage which has a restricted head differential across it and also to the requirement of an adequate head over the lock sill at Nag Hammati for navigation purposes. (Work is in hand to address both of these problems). The remainder is either released for power generation purposes or for maintaining irrigation levels during the January/February canal closure period. During recent years of severe shortage of water at Aswan the power requirement has been dropped, saving possibly 1.5 km^3. Now, if the canal closure period were to be staggered over a longer period it may be that enough water would be available to meet the navigation need and the losses through Edfina could be eliminated altogether. This matter requires early detailed examination.

Groundwater

Before the construction of the High Dam groundwater was used to some extent to supplement irrigation in the winter but since that time has fallen into decline. Nationally this is unfortunate because the storage available in the groundwater (perhaps 400 km^3) is considerably greater than the capacity of Lake Nasser (130 km^3). Further, like Lake Nasser, it is recharged annually from the Nile via the irrigation system but unlike the Lake is not subject to the same high losses. Judicious use of groundwater can help all farmers in times of shortage particularly those at the canal tails. It can also provide a measure of drainage and consequently reduce the flows to the drains, i.e provide on-the-spot drainage re-use. Virtually the whole of the irrigated area is undertaken by a suitable aquifer and all except the northern third of the Delta contains good quality water.

Three types of development proposed are: informal private use at *meska* level; private exploitation canal at command level; and also public development at this level.

Experience throughout the world suggests that private development should be encouraged wherever this is feasible; it is at once cheaper initially and more efficiently managed than government operations. The best option here is to allow and encourage enterprising farmers to develop groundwater wherever possible, and generally this will be all that is necessary in areas with shallow watertables. Wells of this kind produce 20 to 100 m^3/hour and irrigate about 15 feddans. A number of farmers already operate such wells.

Private development at canal command level is aimed at where there is a serious water shortage. The technology proposed is exactly the same as that at *meska* level. Large numbers of wells may be required and will need development in properly organised projects which will address problems such as credit availability, cost recovery, motivation of farmers and the capability of local contractors. A scheme of this type has already been prepared and is ready for implementation.

Public development would be confined to those areas where private individuals would not be able to exploit the groundwater, such as those areas with deep watertables. Wells would be of large capacity (300 to 500 m^3/hour) pumping directly into canals. Again, a scheme has been prepared for this type of operation.

Groundwater will provide a most useful additional storage capacity and the schemes under consideration should be implemented and then replicated across the entire system as soon as some useful lessons have been learned.

Management of the system

Management at all levels requires upgrading and a programme of training both within Egypt and overseas has been undertaken to this end. At system level the management problem is very complicated and, in fact, is scarcely possible by traditional methods, especially in conditions of water shortage and need for making the most of the resource in terms of efficiency. Historic computer programmes have been developed but because of their complexity they have not been widely understood and thus they could be operated by only a few experts. It is not only in Egypt that this sort of problem arises and because managements consider that they are losing control to a few computer experts who, they feel, do not understand their problems, the business is treated with suspicion and even hostility.

In the last few years the advent of cheap computer graphics that can be run on ordinary microcomputers points the way to overcoming some of these problems. Such a programme has been developed for the Nile system that absolutely anyone can operate without any knowledge of computers. It simulates every element from the operation of Lake Nasser to the drainage outflow to the sea and gives the operator a wide choice to make changes to the inputs, to run the model with alternative river flow sequences and essentially to ask the question of 'what if ?' All the results are presented on the screen although printed copies of such results can also be produced, more or less, instantly. The figure shows a typical screen for the Upper Delta water balance illustrating the various interconnections and demonstrates how the results of a most complex analysis may be simply presented. A number of very powerful programs are linked together to drive the model; these remain invisible to the manual operator but are readily accessible to those who are interested in the details.

References

Attia, F. and Kefruy K. (1984). *Groundwater extraction and use in irrigation agriculture in Old Lands in the Nile Valley.* Cairo, Research Institute for Groundwater.

Sir M. MacDonald and Partners (1979). *Rehabilitation and improvement of water delivery in Old Lands.* Cairo, Egyptian Ministry of Irrigation.

Stoner, R. F. (1984). *Nile below Aswan. 5 year plan investment review in land reclamation.*

Water Master Plan (1983). Cairo, Ministry of Irrigation.

9

Water resources planning in the Sudan

D. KNOTT AND R. HEWETT

Introduction

The flows of the Nile represent the major water resource for the Sudan. Indeed their importance can be best summed up in the ancient Latin dictum "aut Nilus, aut nihil". At present some 17 km^3 out of a total entitlement of 20.55 km^3 are allocated to irrigation, storage and hydro-electric projects. Future demands will far exceed the availability of supply and will act as a severe constraint to development in the Sudan. This situation calls for co-ordinated planning and conservation measures to increase the yield of the Nile and geographical development that makes best use of the available resource.

The development of the waters of the Nile, which is the principal surface water resource of the Sudan, has been the major economic activity of that country over the last 70 years. It has involved the construction of four major dams and the implementation of over four million feddans of irrigation. In recent years water resource development has been the subject of a number of planning and feasibility studies, culminating in the Master Plan presented in the Nile Waters Study of 1979.

The purpose of this chapter is to review the current status of water resources planning and in particular to focus on the growing imbalance between future demands for water and the availability of supplies. Figures quoted are mainly from the Nile Waters Study which outlines the main development options.

Climate and land resources

The vast land area of the Sudan (2.5 million km^2) extends from the Tropic of Cancer to the Equator. The climate, described in some detail in Chapter 6, is largely influenced by the ITCZ, which separates a southerly humid wind blowing from the Atlantic and a northerly dry wind from the Eurasian land mass. Hence the northern and central regions have a hot dry continental climate, while the south has a wet tropical climate. July, August and September are the peak months of the rainy season, but the amount of rainfall varies widely. Rainfall decreases steadily from the south to the north, from 1600 mm per year to almost zero, with isohyets roughly parallel to the lines of latitude. Evaporation rates increase in the reverse direction ranging from

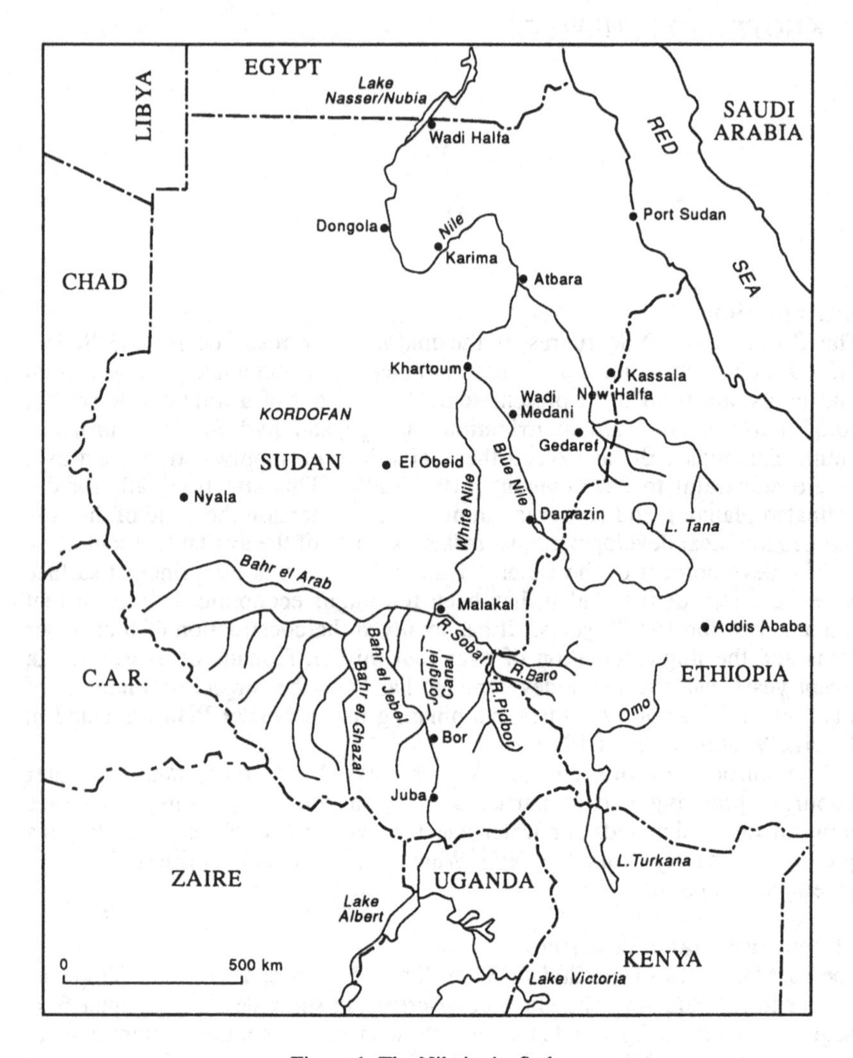

Figure 1: The Nile in the Sudan

1600 mm per year in the south to 2600 mm in the north. Temperatures are high throughout the country, averaging about 29°.

The land areas along the Nile in the Sudan can be broadly divided into three agro-climatic zones, separated by the 200 mm and 800 mm isohyets:

1. In the north a desert zone, sparsely populated and with little development potential. Irrigation is confined to ribbon development along the Nile where cereal, vegetable and fruit crops are grown, entirely dependent on the flows of the Nile for their source of water.

2. A savanna zone of the central clay plains between the 200 and 800 mm isohyets, where the predominant soils are deeply cracking clays. The area supports both rainfed and irrigated agriculture, the main rainfed crop being sorghum and the principal irrigated crops, cotton and wheat. To the west lie the Qoz sands where the natural vegetation is open woodland with grass. Here the population is largely nomadic pastoralist.

3. The Southern Zone. This area, which embraces all the tributaries that form the White Nile, has two main land forms, the southern clay plains and the ironstone plateau. The former is characterized by swamps which expand seasonally as the Nile overflows in the flood. The vegetation, which supports abundant wild life, is typically papyrus and grasses. The area is little developed and the main agricultural activity is the rearing of livestock, accounting for 50% of all cattle in the Sudan (See also Chapter 12). The ironstone plateau supports subsistence rainfed agriculture, based on a mixed cropping of cereals, vegetables, cassava and some livestock.

The variability of rainfall is more pronounced in the north than in the south. Thus for example, in the desert region one single storm may equal 20 times the average annual rainfall, while in the central region a storm may equal 2 to 3 times the annual average. In recent years there has been a decline in rainfall which has been more marked in the more northerly areas as illustrated in Table 1.

Table 1 Decline in rainfall (mm) (1900 - 1980)

Station	Average Annual Rainfall (mm)		Decline
	1901-1965	1965-80	%
Karima	42	24	42
Kassala	338	256	24
Wad Medani	386	306	21
Kosti	514	437	15
Kadugli	778	670	14
Juba	1025	943	8
Wau	1166	1050	10
Yambio	1482	1452	2

Source: Sudan Meteorological Dept.

Surface water

It is convenient for planning purposes to consider the Nile system in the Sudan as five discrete sub-basins.

1. The complex system of tributaries and swamps upstream of Malakal. These tributaries, the Bahr el Jebel (which divides into the Bahr el Zeraf), the Bahr el Ghazal, and the River Sobat combine to form the White Nile. Owing to the attenuating effect of the swamps, the flows into the White Nile are quite regular and exhibit only small monthly variation.

2. The White Nile, some 800 km in length from Malakal to Khartoum, where the average annual flow including the Sobat is 26 km^3. This reach is characterized by a very flat gradient, descending only 12 m, a slope of less than 1:50,000.

3. The Blue Nile Basin, comprising the Blue Nile itself, some 700 km in length with a drop of 120 m from the Ethiopian border to Khartoum, and its two principal tributaries, the Dinder and Rahad. The annual average contribution of this basin to flows at Khartoum is 51 km^3. In contrast to the White Nile, the flows are highly seasonal since the basin derives nearly all its flows from rainfall in the Ethiopian Highlands.

4. The Main Nile, formed by the confluence of the White Nile and the Blue Nile, which flows 1500 km from Khartoum to Lake Nubia. The river flows through a series of cataracts with a total drop in level of 250 m. The average annual flow of 77 km^3 exhibits the combined characteristics of its two tributaries, with the seasonal pattern of the Blue Nile superimposed on the regular flow of the White Nile.

5. The Atbara, the last major tributary in the Sudan, which joins the Main Nile 300 km downstream of Khartoum, contributing 12 km^3 per year. Since the Atbara also derives its flow from the Ethiopian Highlands, the pattern of flow is similar to that of the Blue Nile. The seasonal effect is, however, more pronounced and the river is virtually dry for six months of the year.

The total annual flow at the Egyptian border, before allowing for any significant abstraction, is taken historically at 84 km^3 (1905-1959). Under the Nile Waters Agreement of 1959 the Sudan is allocated 18.5 km^3, which after adjustment for losses is taken as equivalent to 20.55 m^3 measured at Sennar, a more convenient point of reference. As well as the net consumption for irrigation development, any incremental evaporation losses in storage projects must be accounted for in the Sudan's share of water.

Groundwater

By comparison with the vast surface water resources of the Nile, groundwater resources are small. Nevertheless the resources, primarily located in the Umm Rwaba deposits and Nubian Sandstone, are estimated to have a potential yield of 2 km^3/ year.

Current usage is mainly from boreholes and shallow wells situated along the banks of the Nile, its tributaries and other wadis. In most areas groundwater is used primarily for water supply, but particularly in the Northern Province groundwater is also used for irrigation, either as a supplement to the Nile or as the principal source. Total current usage in the Sudan is estimated at 0.3 km^3/year from some 10,000 boreholes and 7700 shallow wells, though this is likely to be an underestimate. Figures for areas recharged from the Nile Basin are not available separately.

Existing development

Irrigation

The total area currently irrigated in the Sudan is estimated at 4.5 million feddans (1.9 million ha). Almost all this irrigated land lies in the central plain along the Nile and its tributaries.

Irrigation in the Sudan is dominated by the Gezira-Managil, the largest singly managed irrigation scheme in the world, totalling about 2.1 million feddans (880,000 ha). Essentially the Gezira-Managil scheme is gravity fed from the Sennar reservoir, but elsewhere the bulk of irrigation is by pumping.

The main exceptions to this are the New Halfa scheme on the Atbara, where irrigation is supplied by gravity from the Khashm el Girba reservoir, and areas along the banks of the Nile where flood basin or recession agriculture is practised.

The principal crops grown under irrigation on the Blue and White Nile and the Atbara are cotton, wheat, groundnuts and sorghum, and in certain areas vegetables. Sugar is also an important crop in the central Sudan, notably at Kenana. Farther north wheat is grown increasingly, and in the Northern Province fruit and fodder are included in the cropping patterns. Cropping intensities vary widely from around 20 per cent on some of the White Nile pump schemes to 100 per cent in Managil.

Table 2 shows the estimated current areas and corresponding water commitments from the Nile and its main tributaries. Actual water use is currently less than these figures as some of the pump schemes have fallen into disuse. The latest estimate of current utilization of water is just over 14 km^3.(See also Chapter 3).

Table 2 - Irrigated areas and water allocation

	Net Cultivable Area (000 feddans)	Annual Water Commitment (km^3)
Upstream of Malakal	20	0.05
White Nile		
Pump Schemes	494	
Sugar Projects	146	
Total	640	1.90
Blue Nile		
Gezira/Managil	2080	
Rahad	300	
Pump Schemes	310	
Sugar	70	
Other	220	
Total	2980	11.8
Atabara - New Halfa	450	1.62
Main Nile	350	1.20
Reservoir Evaporation		0.53
Total	**4440**	**17.10**

Source: Nile Waters Study

The ability of the canal system to meet peak irrigation water demands is a particular problem arising from the combined effects of drought and floods. In periods of drought a chain of difficulties is set up. Low rainfall and high evapo-transpiration create higher water demands, while late sowing of crops beyond the optimum date and sowing several crops at the same time leads to an exaggerated peak. The consequent higher than normal canal flows during the flood season result in greater sediment deposition in the canals, so reducing the canal capacity. Hence the period of watering extends well beyond the flood season, increasing the demand on storage, and leading to shortages at the end of the season. This can result in a situation where crop losses in a single year can wipe out the returns of the five previous years.

Storage projects

There are four main storage projects in the Sudan. On the White Nile, the Jebel Aulia dam, completed in 1937 some 30 km upstream of Khartoum, impounds a reservoir volume of 2.5 km^3. This storage was originally intended for the regulation of flows to Egypt storing White Nile flows for release when the White Nile drops away, but its role has been superseded by the construction of the Aswan High Dam. On the Blue Nile there are two storage projects, the Roseires reservoir completed in 1966 with a storage capacity of

2.4 km^3 and the Sennar reservoir built in 1925 with a capacity of 0.7 km^3. Both these reservoirs, which supply regulated flow to the irrigation schemes on the Blue Nile, are affected by siltation from the high sediment load in the river during the flood season. The Khashm el Girba reservoir on the Atbara was also completed in 1966. It had an original capacity of 1.3 km^3 and supplies water for the New Halfa irrigation area. Siltation has progressively reduced the storage capacity so that it is now estimated at less than one third of the original figure. There are consequent severe reductions in the area irrigated.

Since the Jebel Aulia and Sennar dams were completed before 1959, losses from the reservoirs are already taken into account in the water regime assumed for the Nile Waters Agreement and hence do not count as a part of the Sudan's share of water.

Hydro-electric facilities

There are hydro-electric facilities at three of the main dams with installed capacities as follows:-

Roseires	250 MW
Sennar	15 MW
Khashm el Girba	12 MW

These hydro-electric stations have been the mainstay of the Sudanese power system for the last 25 years, and have generated most of the electricity supplied to the grid.

Water supply

The Nile is the source of water supply for all the urban centres located along its banks, notably Khartoum, Juba, Malakal, Kosti, Kassala, Atbara and Wad Medani. Quality rather than quantity of water is the main problem. Special measures are required to deal with the high silt content of the water.

Flood protection

Except for local protection to river banks and some minor groynes and protection works at pumpstations and other strategic points, no major flood protection works have been carried out on the Nile. In fact to restrain the river in times of flood is such a major task that it cannot generally be justified for protection of any but the most important areas.

There is some evidence of a rising river bed in certain areas, but observations to date are not yet sufficiently numerous or over a long enough period to draw any firm conclusions.

Irrigation development

Plans for the development of irrigation in the Sudan are concerned with the rehabilitation and modernization of existing schemes, intensification of cropping, and the development of new irrigation schemes. The framework for irrigation development is comprehensively set out in the Master Plan of the Nile Waters Study (1979). The main opportunities for increasing irrigation in the Sudan are summarized below.

Major projects on the Blue and White Niles
The earlier Blue Nile Waters Study (1978) evaluated a potentially irrigable area of 2.85 million feddans (1.2 million ha) both to the east and the west of the Blue Nile from Roseires to Khartoum. After a screening process this was reduced to four major projects:

	Irrigated Area (feddans)	Annual Water Requirement (km^3)
Rahad II	300,000	1.0
Kenana III	300,000	1.0
Kenana II	300,000	0.9
Gezira Intensifiction	270,000	1.1
Total	1,170,000	4.0

The Rahad II project would be supplied by pump stations and canals taking its water from the Blue Nile. The Kenana projects could be supplied either by gravity from the Roseires reservoir or by pumping downstream. The Gezira project involves an increase in intensity on the existing area from 75 per cent to 100 per cent. Associated with these projects would be the heightening of the Roseires dam to increase the volume of water stored and provide more seasonal regulation of flows. A 10 m heightening of the dam would increase the reservoir capacity by 4.2 km^3 to 6.6 km^3. Evaporation losses would also increase by 0.45 km^3 per year.

The Nile Waters Study carried out a similar exercise for irrigation from the White Nile. Two projects were evaluated, Jebelein-Renk and Renk-Gelhak, with a total irrigated area of just over 400,000 feddans (170,000 ha) and associated water requirement of 1.25 km^3. Since these projects would be supplied by pumping from the White Nile no additional storage facilities would be required.

Smaller pump schemes
Two studies have recently been carried out for the modernization of the smaller pump schemes on the Blue and the White Niles. Although concerned principally with rehabilitation of the existing irrigation schemes, the studies identified possible extension areas of 140,000 feddans (60,000 ha) which would require additional water of 0.5 km^3.

The Upper Atbara project
Studies have also been carried out for the Upper Atbara irrigation project. This involves the construction of a reservoir with an initial storage capacity of 1.5 km^3 at Rumela on the Setit, the main tributary of the Atbara. The irrigated area would be of the order of 300,000 feddans (120,000 ha) and the

water requirement 2.0 km^3 per year. This water requirement is higher per unit area than that shown for the Blue Nile and White Nile Schemes and arises largely from different assumptions on irrigation efficiency.

Irrigation upstream of Malakal
The potential for irrigation development in this region is largely undefined although some possibilities have been identified. There are prospects for large scale gravity irrigation to the north of the Sobat and east of the White Nile. Further potential exists in the floodplain to the east of the Jonglei canal (but see Chapter 12), and to the west of the Bahr el Jebel. There is also scope for the extension of supplementary irrigation for tea and coffee in the upland areas adjacent to the southern border.

The total area could exceed 250,000 feddans for which a provision of water amounting to 0.5 km^3 would be required. It may be noted that due to the more favourable climatic conditions, the unit demand for irrigation water is significantly less in this region than in areas to the north.

Irrigation on the Main Nile
Irrigation development on the Main Nile would be mainly concerned with increasing intensities on existing areas from 100 per cent to 150 per cent. This would lead to an increase in water requirements from 1.2 to 1.6 km^3 per year with a consequent increase in storage requirements.

Hydro-electric development

Upstream of Malakal
Some potential exists for the development of hydro-electric projects on the Bahr-el-Jebel between Nimule and Juba. In this reach, which includes some rapids, the river with an average flow of 29 km^3 drops some 160 m in level. Two main sites, at Fola rapids and Beddan rapids, have been identified. At each site 30 m of head could be developed, justifying an installed capacity of between 200 and 300 MW. Evaporation losses in the associated reservoirs would not be significant. The nature of the terrain north of Juba as far as Malakal is flat and marshy and not suitable for the development of hydro-electric projects.

The White Nile
Between Malakal and Khartoum, the gradient of the river is very flat and consequently this reach is also unsuitable for hydro-electric development. Proposals have been considered for adding a 25 MW power station at Jebel Aulia. However the low hydraulic head, which is reduced to zero during the flood season as a result of backwater effects from the Blue Nile at Khartoum, severely effects the economic viability of the project. Furthermore, the need to operate the reservoir in a hydro-electric mode would tend to increase evaporation losses.

The Blue Nile
The two principal sites on the Blue Nile at Sennar and Roseires have already been developed. However, there is still the possibility of increasing the

installed capacity at Sennar, where only 15 per cent of the total water passing the dam is utilized for hydro-electric generation. The economic viability of increasing the installed capacity from 15 MW to 45 MW has already been established, and the project would have no adverse environmental effects.

The heightening of the Roseires dam discussed in the previous section would also increase the amount of energy generated by the existing power station.

The Main Nile

The main potential for hydro-electric development in the Sudan lies on the Main Nile from Khartoum to Lake Nubia. A number of sites, mainly sited at the downstream end of the cataracts, have been investigated as listed in Table 3 below. At all these sites evaporation losses would be significant in the context of the Sudan's allocation of water. A feasibility study is now being carried out for the Merowe project on the fourth cataract. Several alternative configurations for this project have been considered, varying from one large dam with a significant element of reservoir storage, to three smaller dams, which would be essentially run-of-river schemes. The latter approach would allow staged development and involve lower evaporation losses, but would constrain the pattern of seasonal energy generation.

Table 3 - Potential Hydro-electric Projects on the Main Nile

Project	**Installed Capacity** (MW)	**Evaporation Losses** (km^3/yr)
Sabaloka	120	0.15
Shereik	240	0.5
Shirri Island	450	0.32
Low Merowe	600	0.8
High Merowe	750	1.75
Dal	600	2.4

Source: Nile Waters Study

Other development objectives

Water Supply

Though the requirements for urban water supply are small at present, they are likely to become significant in the future. Given the current rate of population growth in the Sudan, an urban population in the next century in excess of 25 million people can be envisaged. Normal criteria for urban water demands will result in a total abstraction approaching 2.0 km^3 per year with an associated net consumptive use of 0.6 km^3. Provision for this level of demand will have to be allowed for in future planning of water allocation.

Future flood protection

The existing and proposed dams on the Nile and its tributaries within the Sudan have almost no effect on flood alleviation, as any attempt to store water during the main flood season will quickly silt up the reservoirs. Only some relatively minor local protection studies and works are planned.

However, a flood early warning system is currently being considered and work on the configuration and implementation of the system is expected to commence shortly. This will provide several days of flood warning to most parts of the Sudan and provide some monitoring of longer term changes.

Navigation

Navigation on the Nile is and will remain an important mode of transportation within the Sudan. However, the consumptive use of water for this purpose is negligible and consideration of navigation does not feature in the overall water balance. It could, however, affect the operation pattern of reservoirs on the Nile, and it may be necessary to include navigation facilities in the major structures.

Conservation projects

It is evident from the assessment of water requirements set out in the previous sections that future demand far exceeds available flows. One possible solution to the problem is to undertake projects to increase the yield of the Nile. It is estimated that some 45 km^3 of water is lost annually by evaporation and transpiration in the swamp areas of Southern Sudan. These losses are distributed as follows:-

	km^3/yr
Bahr el Jebel basin	14
Bahr el Ghazal basin	12
Sobat and Machar basin	19
Total	45

Source: Control and Use of Nile Waters in the Sudan. Ministry of Irrigation and H.E.E. (1975)

Measures to reduce these losses have been under consideration for many years. Four main proposals have been put forward. The Jonglei Phase I project involves the construction of a canal of 20 to 25 M. m^3/d capacity to by-pass the swamps of the Bahr el Jebel and Bahr el Zeraf. Work on the canal was started in 1976 but was suspended owing to security problems. When completed the project is estimated to increase the annual flows in the White Nile by 4.75 km^3 (3.8 km^3 net). Jonglei Phase II would involve enlargement or duplication of the canal and the provision of storage in the equatorial lakes, and add a further 4.25 km^3.

A third proposal would aim at reducing evaporation losses in the Machar marshes by construction of flood embankments along the river Baro and a diversion canal from the Baro to the White Nile. A fourth proposal to reduce losses in the Bahr el Ghazal would consist of a series of storage reservoirs and diversion channels. The estimated reduction in evaporation losses are 4.0 km^3

and 7.0 km^3 respectively.

The total estimated saving of water from these measures could amount to some 20 km^3 of which half would be allocated to the Sudan. The figures, however, are very tentative and more analysis is required to develop firmer estimates. It should also be added that the cost of these works would be substantial and the environmental problems formidable (see Chapter 12).

Conclusions

This brief review of water resources planning in the Sudan clearly demonstrates that the Sudan is facing a growing problem of water shortage. The allocations under the Nile Waters Agreement imposes a constraint on the development of irrigation and hydro-electric facilities, and priorities for use will have to be established.

Conservation projects offer the only means of significantly increasing water availability. Further studies of these projects, and in particular of the environmental aspects will need to be undertaken. In the future co-ordinated planning, involving a continuous dialogue between all the riparian countries, will be essential for the development of the Nile Basin.

References

Coyne et Bellier, Sir Alexander Gibb & Partners, Hunting Technical Services, Sir M. MacDonald and Partners. (1978). *Blue Nile Waters Study,* Volume 1 - Main Report.

Coyne et Bellier, Sir Alexander Gibb & Partners, Hunting Technical Services, Sir M. MacDonald and Partners. (1979). *Nile Waters Study*, Volume 1 - Main Report.

SOGREAH. (1982). *Upper Atbara Feasibility Study*. Final Report.

Sir Alexander Gibb & Partners,(1987). Merz and McLellan, *Updating of the Feasibility Study for the Heightening of Roseires Dam.* Final Report - Volume 1.

Sir Alexander Gibb & Partners. (1989). *White Nile Pump Schemes Modernization Study*. Final Report.

Sir Alexander Gibb & Partners (1990). *Blue Nile Pump Schemes Modernization Study*. Final Report.

10

Irrigation and hydro-power potential and water needs in Uganda- an overview

B. KABANDA and P. KAHANGIRE

Introduction

This brief contribution to the debate on the utilization of the Nile aims to highlight the technical information available on Uganda's water requirements in relation to that river.

Uganda, with its total land area of 236,810 km^2, lies wholly in the White Nile Basin. The country's hydrology is dominated by the River Nile and its associated lake system. Open water covers 36,278 km^2 or 15 per cent of the land, and another 5,183 km^2 or 2.2 per cent is covered by swamps. To Uganda, therefore, the Nile is not only a source of water for all sectors of the economy, but the environment upon which everything depends.

Socio-economic factors

Uganda's economy, like that of most third-world countries, is heavily dependent on the export of raw materials - mainly coffee - for which prices on the world market have been decreasing over the past decade, while it imports finished goods and services at ever increasing prices. The population, currently estimated at sixteen million, is increasing at a high rate close to four per cent per annum which is putting great pressure on the land and other natural resources, including water.

In addition to the irrigation potential, the Nile system offers Uganda a great source of hydro-electric power for its domestic needs and for export. Although Uganda is less rich in mineral deposits such as gold and petroleum, it has a large natural resource in water, and it is on this that its future socio-economic development will depend.

It is clear that for the Ugandan economy to survive, the country must increase agricultural production for export and food self-sufficiency, and diversify its export base. The available land must be optimally exploited, using the best farming techniques, and irrigation should be developed and used where appropriate to increase food production. Water should also be considered as one of the available natural resources which could be exploited to broaden the export base for economic and social survival and development.

Water resources and water resource monitoring

Surface water

Uganda has a good climate with average temperatures of around 23-27°C, and long-term annual rainfall ranging from 1600 mm near Lake Victoria, to 800 mm in the north-eastern part of the country and 700 mm around Lake Albert. As a result, Uganda has in general fairly well distributed resources of surface water (see Figure 1).

Contrary to the general belief, water could in fact be a constraint on Uganda's development. These are the main reasons:

i) Most of the water sources in Uganda other than the Nile are of limited quantity, and many are seasonal. This means that there are limited sites where large development projects such as hydro-power and irrigation schemes could be constructed.

ii) Over the whole country, annual rainfall is nearly matched by potential open water evaporation. Figures 1 and 2 indicate that long-term average annual rainfall varies between 700 and 1600 mm, while open water evaporation varies between 1200 and 2900 mm over the country. Over the main lakes, evaporation equals or exceeds rainfall, and drought periods are becoming longer and more severe. This may lead to actual nation-wide water shortages.

iii) Water is distributed unequally over the country so that some areas like Karamoja are semi-arid. This leads to localized natural water shortages.

iv) All the major water bodies in Uganda are shared with neighbouring countries. This clearly imposes political limitations on their use and necessitates negotiations over their control and allocation.

Groundwater

Geological conditions determine that accessible, groundwater resources are limited in quantity. The development of groundwater will continue to be mainly for small-scale domestic use.

Hydrology

The hydrology of the country is dominated by the Nile system. Hydrological observations on the Nile within Uganda were initiated in 1898 by the Egyptian Irrigation Department, of the colonial administration. A national Hydrological Surveys Department was set up in 1947 and it was succeeded by the Water Development Department.

The country can be divided into eight drainage basins as below:

L. Victoria sub-basin	59858 sq km
L. Edward sub-basin	18624 sq km
Victoria Nile sub-basin	26769 sq km
L. Albert sub-basin	18223 sq km
L. Kyoga sub-basin	57669 sq km
River Aswa sub-basin	26868 sq km
Albert Nile sub-basin	20004 sq km
L. Turkana sub basin	4299 sq km

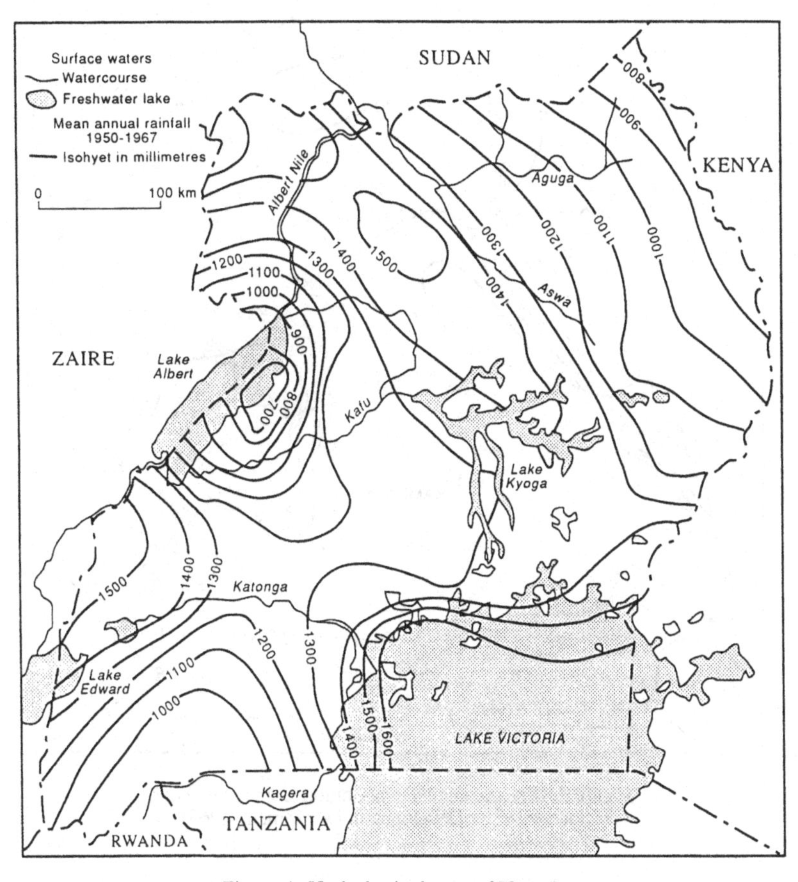

Figure 1: Hydrological map of Uganda.
(Source: Water Development Department, 1990)

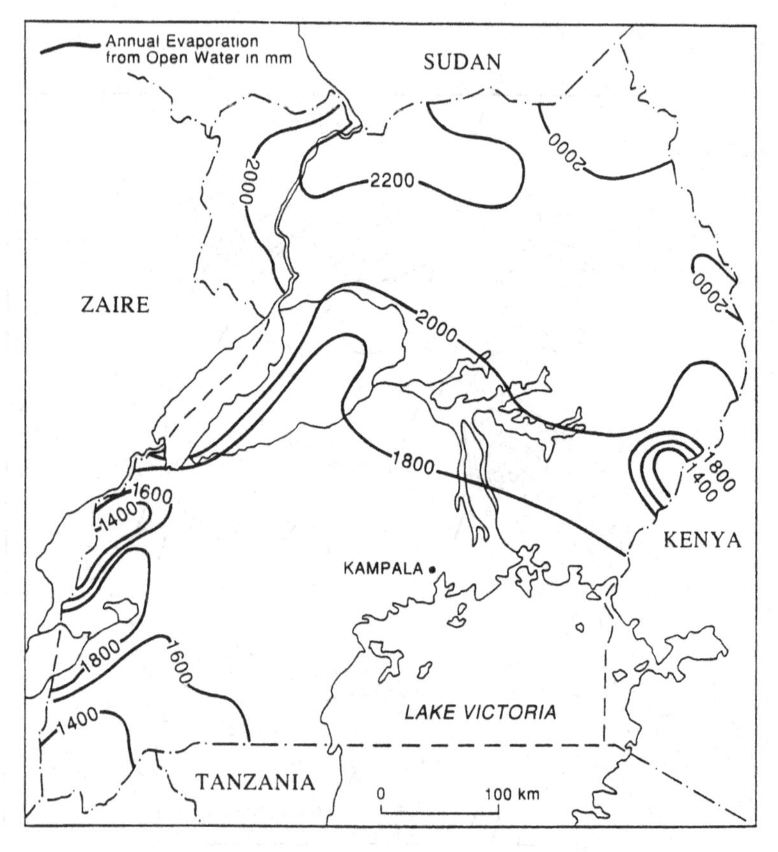

Figure 2: Uganda, annual evaporation from open water.
(Source: Water Development Department, 1965)

The hydrological observation network expanded gradually and by 1979 there were 142 stations of various categories and instrumentation in the country. Unfortunately almost all of the equipment was destroyed in the disruptions of the civil wars of the 1970s and 1980s. Government efforts to rehabilitate the stations has been piecemeal and constrained by lack of funds. As a result the network has not been fully rehabilitated to date. The original network was fairly well designed, spread over the entire country with most emphasis on the Nile system.

The UNDP/World Bank Sub-Saharan Hydrological Assessment project (1989) confirmed that the hydrological service had collapsed and a lot of external support (equipment, technical assistance and training) would be needed to revive it.

The sound network created by the former East African Meteorological Department modernized and intensified by the Hydromet Survey Project within the project area was also destroyed during the period of turmoil and there are now less than twenty first and second order meteorological stations operational and reporting reliably consistently.

There is no single groundwater monitoring station in the country although groundwater development projects for rural water supply have increased. Water quality records have also lapsed although some monitoring of groundwater has been maintained. The useful water quality records initiated by the Hydromet Survey Project have also been discontinued

Simulation and forecasting activities

Uganda as a participating country in the Hydromet Survey Project has had access to the *Mathematical Model of the Upper Nile Basin* which uses the Sacramento model for catchment modelling. A water quality and environmental impact model has also been developed and Uganda had personnel trained on the development and use of the model. The water quality model has been tested on Lake Victoria for the various model components.

The mathematical model was calibrated on six control catchments, the three lakes and four river reaches on the River Nile. The model was also applied to the River Aswa catchment with good success. No forecasting models have been developed and used in Uganda although the techniques are useful for hydropower operations and flood warning for streams in mountainous areas such as Mubuku/Sebwe and the flood prone river Katonga. Because of the lack of adequate computer facilities and trained staff, Uganda has not been able to utilize effectively the models developed under the Hydromet Survey Project for water resources planning and management. The Hydromet Survey Project also has degenerated and is not continuing with the exercise of improving the models or assisting countries develop and sustain the capability to enable them to take full advantage of the capacity to model and deploy the massive data collected.

In the area of meteorology there has been little simulation and forecasting models activity. The meteorological service does however participate in a drought monitoring and food security project which will utilize various simulation systems.

Water demand

General considerations

Uganda at present lacks a comprehensive national master plan for water, and has therefore not identified its water requirements for various sectors of the economy, and for alternative planning scenarios. What is presented here is based on previous piecemeal studies. The lack of reliable data also hinders the development of reliable national long-term projections of water use and demand.

Domestic and industrial use

The quantities of water needed by these sectors are relatively small. The Hydromet Project Studies (1982) estimated that the mean annual demand will grow from 5 cumecs to 40 cumecs (1.26 cubic kilometres) between 1980 and 2000 (Figure 4).

Irrigation water requirements

Although irrigation is not widely practised, and the current water use for irrigation is very low, the agricultural sector will eventually be the largest water user in Uganda. A UNDP report (1989) estimated that only 2000 hectares are currently irrigated.

For a long time, irrigation in Uganda will be supplementary; either to increase acreage or to grow crops more intensively. The existing studies of water demand have considered soils and topographic suitability, crop varieties and water availability, leaving out areas whose irrigation will require heavy pumping costs. This means that large areas of land such as Karamoja which could be turned green, have not been considered while assessing the demand until now for irrigation water in Uganda, due to the prohibitive costs of water delivery. In the long run, these areas should be considered so as to reach the ultimate irrigation potential of Uganda. Even using these conservative estimates, however, it is evident that water availability will limit irrigated agriculture in Uganda.

According to the 1982 Hydromet studies, Uganda's irrigated area was expected to reach 241,500 hectares (Figure 3) by 2000. The mean annual demand for irrigation water from the Nile, which was negligible in 1980 and still is so today, could reach 85 cumecs (2.7 km^3) by 2000 (Figure 4). It is estimated that the present total water consumption for domestic, industrial and irrigation uses is 70 cumecs (2.21 km^3 per annum) and the total annual demand will approach three km^3 by 2000.

A government survey of water requirements as applied to irrigation and swamp reclamation was carried out in 1954-5. A follow-up study in 1970, twelve years before the Hydromet studies, estimated Uganda's water requirements for irrigation by 2000 to be much smaller than Hydromets's projected potential of 2.7 km^3, at only 0.3 km^3 per annum. However, it is estimated that this would rise to almost 5.5 km^3 per annum by 2050. In a recent FAO report (1987) Uganda's irrigation potential was estimated at over 400,000 hectares, which area would require about 6 km^3 of water per annum.

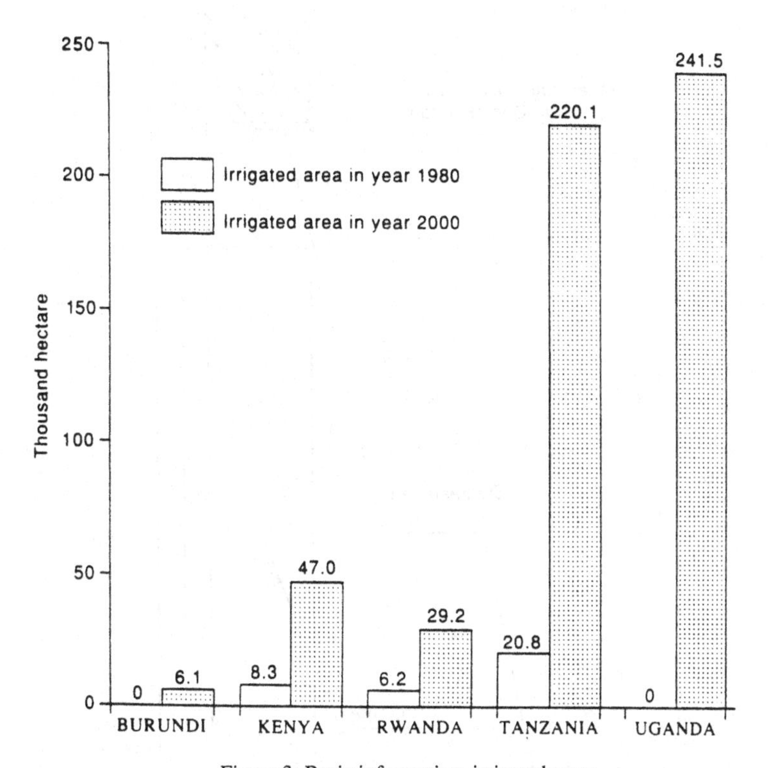

Figure 3: Basic information, irrigated areas.

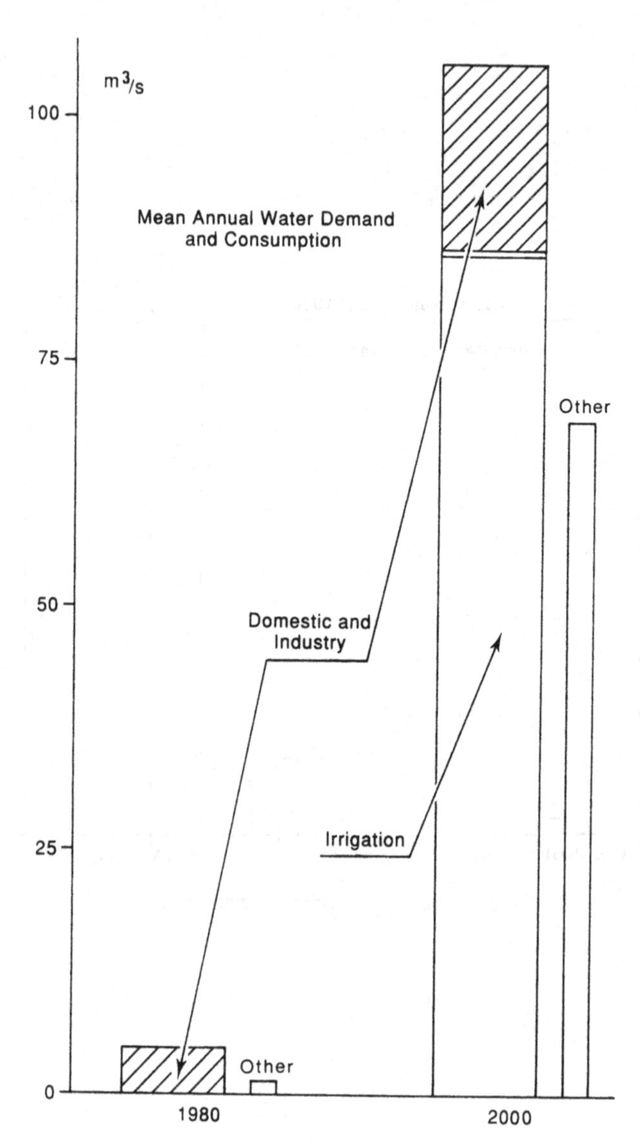

Figure 4: Uganda, mean annual water demand and consumption.

Hydro-electric power

As stated earlier, Uganda has great potential for the development of hydro-electric power on the Nile, particularly between Lakes Kyoga and Albert (Mobutu Sese Seko). Such development would require a guaranteed Nile flow adequate for power generation. Therefore, although the generation of power does not consume water, Uganda will require a fairly constant flow in the Nile through its territory, in order to carry out its projected strategy for developing hydro-power. Potential sites include Bujagali, Karuma, Ayago-Nile and Murchison Falls with a total potential installed capacity of as much as 1694 MW.

It is considered that the design flow for the next power station at Murchison Falls should be about 640 cumecs, based on the long-term average Nile flow as recorded at Jinja since the 1960s. The hydro-electric power potential of the Nile is not fully harnessed, and a design flow of 1000 cumecs could be considered for future schemes on the Nile, assuming that the post-1961 flow is sustained. If the potential of the Nile in this were fully exploited, hydro-electric power could bring to Uganda increased foreign currency earnings through exports of power.

Integrated development of the Nile Basin

The integrated development of the Nile Basin is necessary to achieve the maximum benefit to all the countries within it. Support should be given towards this aim, on the basis of:

i) All countries participating in the development as equal partners.

ii) Each country's right to use the Nile waters in its territory should be recognized, as long as such use does not cause appreciable harm to other riparian users. Each country considers the waters in its territory to be first and foremost its own natural resource.

iii) An integrated basin-wide development plan should take into account the natural occurrence and location of the resource, and the potential for development and use of the Nile waters in each country. Development should aim to maximize the benefits for as many other riparian states as possible. For example, hydro-electric power development in Uganda could benefit several other riparian countries.

iv) The general objective of development being the maintenance and enhancement of the overall social and economic development of all the people of the Nile Basin.

v) Co-operation being based on clearly identified objectives and benefits accruing to each participating country.

Conclusions and recommendations

i) Uganda has a great potential for the development of irrigation and hydro-electric power, but in the long run water availability could be a constraint to the country's sustainable socio-economic development unless there is adequate planning in advance.

ii) Only inconclusive information is available concerning Uganda's future water demand from the Nile. However it is estimated that Uganda will require six km^3 per annum by 2050 for irrigation, and its total water requirements will be about eight km^3 per annum by that year. It is therefore probable that Uganda's future water requirements from the Nile will eventually be of the order of ten km^3 per annum, or 317 cumecs by 2100.

iii) The lack of an integrated national water resources master plan makes it difficult for Uganda to state with certainty its future water demand in general, and that from the Nile system in particular.

iv) All efforts towards the integrated development of the Nile Basin for the mutual benefit of all the riparian countries should be supported.

Acknowledgements

Uganda has greatly benefited from previous water use studies carried out under the auspices of the British Overseas Development Administration and it is hoped that this and other agencies will assist engineers and others in Uganda to draw up a much-needed master plan.

References

WMO/UNDP Hydrometeorological Survey of the Catchments of Lakes Victoria, Kyoga and Mobutu Sese Seko (1982) (RAF/73/001). Project Findings and Recommendations, Geneva.

World Bank/UNDP (1989). Sub-Saharan Hydrological Assessment (IGADD countries). Final Report. Uganda.

Uganda Government,(1955). Water Resources Surveys of Uganda applied to Irrigation and Swamp Reclamation. (Unpublished report).

Ministry of Overseas Development, Land Resources Division (1970). A Survey of the Water Requirements of Uganda, (unpublished). England.

UNDP (1989). Nile Basin Integrated Development. Fact-Finding Mission report, RAF/86/003-RAB/86014.

Uganda Government (1965). Ministry of Mineral and Water Resources. Hydro-meteorological Records from areas of potential Agricultural Development in Uganda.

FAO (1987). Irrigation and Water Resources Potential for Africa, Rome.

11

The integrated development of Nile Basin waters

Z. ABATE

Introduction

Integrated river basin development should be perceived as as set of anthropogenic activities which takes place in an interlinked and complex ecosystem. The Nile Basin is one such system which represents a very large and highly valued hydrological unit.

This chapter discusses the Nile Basin as an ecological entity and it is assumed that all nine co-basin states could equitably share, utilize and manage natural resources for the mutual benefit and welfare of the basin's population. To this end, the chapter outlines strategic and historically controversial issues that must be addressed, and provides a framework for meaningful regional cooperation and coordination. It discusses issues of water allocation, priority research areas, and gives specific recommendations to support and achieve sustainable use of water.

According to O'Riordan and More (1971, p.194) a water resource system is an integrated complex of interlinked hydrological and socio-economic variables operating together within a well-defined area, commonly a drainage-basin unit. Such systems are composed of interlinked sub-systems, which are simpler on account of the smaller number and greater simplicity of components, their more restricted areal coverage, and their more restricted aims. The whole drainage basin is a sound basis for the integrated development of resources. However, the water needs of modern states often demand the creation of a comprehensive super-system , in which, for example, a socio-economic policy is imposed on a variety of drainage-basin units (each perhaps having very different hydrological characteristics), or in which one very large and uniform hydrological unit is developed under a number of differing socio-economic bases (usually national). In some super-systems the complexity stems from the need to manage diverse hydrological circumstances towards some integrated socio-economic goal (e.g. the California Water Plan), whereas in others the complexity arises from attempts to operate a vast hydrological unit by means of a number of differing institutional bases (e.g.

Note: The views expressed in this chapter are the professional opinion of the author; and are not those of the Ethiopian Government.

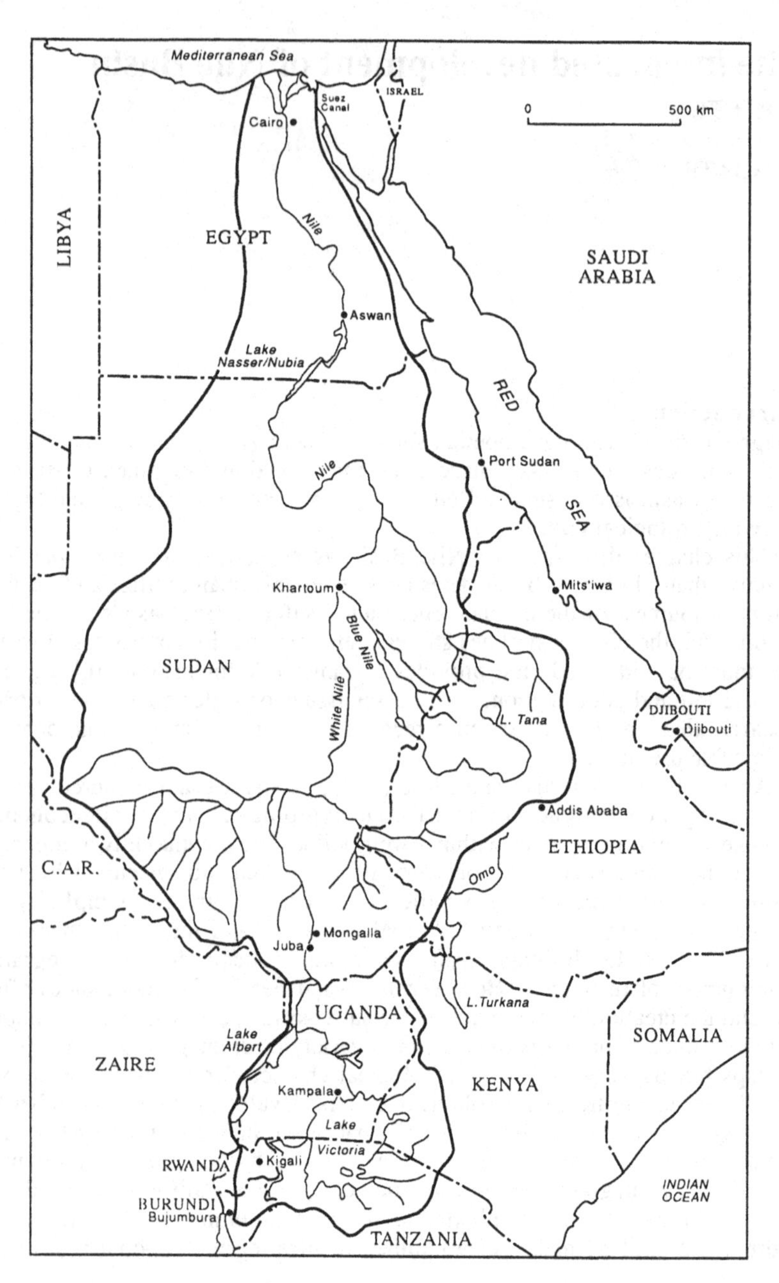

Figure 1. The Nile Basin

the Mekong River Plan).

This chapter treats the Nile Basin as a whole and recommends all sovereign states be brought into equal membership of a community to utilize equitably land and water resources and to protect the physical, biological and social environment with a goal of the sustainable use of the resources. It attempts to point out the needs for such developments as irrigation, hydro-power generation, flood diversion and control, flow augmentation, groundwater, fisheries, water transportation, including the differential effects on the physical, social, and political environment of the states in promoting development activities tailored to an integrated action programme.

The Nile in Ethiopia: an undeveloped power and irrigation potential

The inventory of dams at the beginning of the 1990s shows how limited to date has been the development of the vast hydro-power and agricultural potential of Ethiopia. 6 per cent, or 3.7 million hectares of Ethiopia's potential arable area is considered to be irrigable. This is approximately the same area as Egypt claims is irrigable within its borders, although to date Egypt has only managed to irrigate about 2.5 million hectares. Egypt is already finding it difficult to provide reliable water although improved water management will enable some increase in agricultural use in Egypt. The implication that Ethiopia might utilize water to irrigate even half of the 3.7 million irrigable hectares is very significant to downstream users. That it has an estimated average of 112 cubic kilometres of water annually (compare the nominal resources of Egypt 55.5 cubic kilometres per year and for the Sudan 18.5 cubic kilometres per year) means that Ethiopia has the possibility to command sufficient water to irrigate a much greater area than currently irrigated in Egypt and the Sudan together.

Installed hydro-power, at 400 MW, is only a fraction of national potential and as a result Ethiopia has one of the lowest per capita electrical energy usages in the world (Abate 1991). While the majority of the existing installed capacity and that currently being constructed are not located on the Nile tributaries, 80 per cent of the future major schemes are located in the Nile basin tributaries.

Table 1
Hydro-power in Ethiopia

Scheme	Year commissioned	Energy production (GWh/year)	
		Average	Secure*
Abu Samuel	1939	1.5	-
Tis Abay	1953	68.0	55.0
Koka	1960	110.0	80.0
Awash II	1966	165.0	120.0
Awash III	1971	165.0	120.0
Fincha with Amarti	1973	617.0	613.0
Melka Wakana	1989	560.0	440.0
Sor	1990	60.0	48.0
Total		**1746.5**	**1476**

*Secure indicates the power available in low flow years

Table 2
Planned hydro-power projects in Ethiopia

	Planned commissioning	Energy Production (GWh/Year) Average	Secure*
Gilgel Gibe	1991-1993	864.0	670.0
Chemoga Yeda	1998-2015	3031.0	2526.0
Upper Beles		1617.0	1100.0
Halel/Werabessa		1475.0	1180.0
Aleltu		3550.0	3484.0
Total		**10537.0**	**8960.0**

Source: Abate (1991)
*Secure indicates the power available in low flow years.

Ethiopia's hydropower development plans are based on the assumption that she will become a major exporter of power, with the majority of the energy being generated within the Nile catchment.

Table 3
Planned hydro-power in Ethiopia

Year	Population (millions)	Demand for electric energy (1000 GWh) Domestic & other use	For export	Total demand for electrical energy (1000GWh)	Per cent of total energy
1990	50	5.5		5.5	3.4%
2000	67	7.4	6.0	13.4	8.4
2010	91	10.0	25.0	35.0	21.9
2020	122	13.4	40.0	53.4	33.3
2030	161	17.7	55.0	72.7	45.4
2040	215	23.7	70.0	93.7	58.6

Source: Abate (1991)

The sites of the major hydro-power projects are mainly in Ethiopia's Nile Basin.

Table 4
Priority sites for major hydro-power projects

Name of Basin/River	Technical potential (1000 GWh/year)	Anticipated completion (year)
Abbay/Abbay	7.42	2000
Omo-Ghibe/Omo	*2.26*	*2000*
Abbay/Abbay (Mendaiya)	8.75	2010
Abbay/Didesa	2.20	2010
Baro-Akobo/Bir	6.00	2010
Abbay/Abbay (Mabil)	8.40	2010
Abbay/Guder	3.70	2020
Abbay/Abbay	5.10	2020
Abbay/Abbay (Karadobi)	9.60	2030
Abbay/Beshilo	1.10	2030
Genale-Dawa/Genale	*4.40*	*2030*
Baro-Akobo/Baro	4.60	2030
Abbay/Beles	3.30	2040
Baro-Akobo/Geba	6.30	2040
Omo-Ghibe/Ghibe	*11.90*	*2040*
Total	**85.03**	

Source: Abate (1991)

Note: the projects shown in *italics* are non-Nile catchment projects. They total 18,560 GWh/year or 21.8 per cent of the total proposed capacity. Over half of this would come on stream over half a century in the future.

The Ethiopian authorities have not yet decided on their strategy for the development of water for agriculture. At the outset it is recognized by the Government (Abate, 1991, p.62) that water is not a free commodity and that a charging system would have to be installed when water is used in large quantities in the agricultural sector in future. It is also recognized that there are different approaches to the use of water and that the high-technology approach for water distribution is expensive requiring between US$10,000 to US$15,000 (per hectare) to develop. Small scale irrigation projects require much less finance and much more quickly achieve a pay-back on investment. And because increases in yield of such staples as wheat resulting from the allocation of irrigation water to agriculture are unlikely to provide economic returns, nor contributions to the national food economy of strategic importance, the Ethiopian authorities are unlikely to devote scarce investment resources to large scale irrigation projects where both initial and running costs are high. The small scale irrigation project is likely to be the preferred development strategy.

Ethiopia's agricultural and irrigation strategies are likely to emphasize crops which would benefit from the optimum growing conditions in the high temperature regions of the North-Western lowlands and the cultivation of sugar cane is considered to be an economical option (Abate, 1991).

Environmental problems and issues

Almost all the countries in the Nile Basin face a number of environmental problems such as deforestation, soil erosion, sedimentation, the lack of appropriate institutions, and the lack of adequate financial resources and trained manpower for environmental management and protection. These problems coupled with poverty and high population growth pose a serious threat to the resource base and the life support system of the basin's environment. Furthermore, the basin is tropical and ridden with vectors of human and animal disease such as malaria, bilharzia and rinderpest. Stagnating lakes and ponds are numerous, which in many parts make much of the basin unhealthy for human habitation.

It is therefore important that all the basin states pool their efforts to protect the environment and avert ecological disasters such as drought, famine and disease, desertification, and floods. The first and necessary step is to know in sufficient detail the state and quality of the existing environment of the Basin. It is only when a meaningful and comprehensive strategy and plan of action is developed and implemented in an integrated manner that the problems of the environment can be tackled effectively. It should be stressed that the basin's environment should be seen as one ecological entity and that its sustainability requires concerted action by all basin states. Action programmes have to be drawn up to tame the region to make it habitable for human settlement and livestock husbandry.

Recurrent drought

Drought is a relatively frequent event in parts of the basin. It certainly is not so infrequent that prudent managers would not have contingency plans ready for when it occurs. During the past fifteen years persistent drought has gripped Ethiopia and the Sudan. Nowadays, crop failure due to erratic and unreliable rainfall has become a permanent feature for some groups.

In the absence of a standing inter-country institution, each country has had to constitute a special task force of national officials to monitor drought and advise the governments of the remedial measures necessary. There is a need to develop a regional policy for the management of the consequences of drought and programmes comparable to the historic efforts of epidemic control. The programme should deal with both pervasive, and episodic drought. If the dire climatic predictions associated with the greenhouse effect should come to pass, the governments of the region should be prepared to respond to the predicted damaging trends. In this context improved interregional coordination is essential. Moreover, networking with sub-regional observers for drought reporting purposes might provide the basis for a larger water resource information and extension system. The need to assemble data related to drought conditions cannot help but underscore other needs for timely and accurate information and resource assessments.

General poverty

Among the basin states, seven are least-developed countries with per capita annual income ranging from 120 to 310 US dollars (World Bank, World Atlas, 1991). The population of the co-basin states is expected to reach 812 million from the present 246 million, of which about 140 million live in the basin itself, by the year 2040. (World population prospects, UN, New York, 1986).

The relationship between population growth and economic development is much disputed, but the following five issues are of major importance to the Nile Basin.

1. Given the current co-basin's population growth, the question to be asked is whether they will be able to maintain a reasonable degree of food self-sufficiency, or whether they will become increasingly dependent on food imports.

2. The relationship between population growth and poverty is also important. If the growth of the economy is not sufficient to absorb and renumerate rapidly expanding labour forces, poverty will not be reduced even if average per capita GNP rises.

3. The countries of the region have a very low financial base and are internationally indebted; undertaking comprehensive global development by themselves is not possible.

4. Given the burden of internatonal debt and the rapid population growth rate as high as 3.7 per cent in some co-basin states, whether they can attain any meaningful economic development is a major concern.

5. Factors such as suspicion, political instability and internal hostilities constrain practical economic development effort.

Strategic issues and constraints for integrated develoment

Meaningful regional cooperation

Water is a local, national, and regional resource, and although water occurs hydrologically in definable geographical units, within these units its management is invariably subject to fragmented political jurisdictions. Efforts to deal with water geographically encounter strong resistance from bureaucracies that are functionally organized for diverse purposes. While conceptually appealing, most of the attempts at water management at the river basin level have been failures because nations are not organized on a hydrological basis. Political organizations reflect other interests and are determined by boundaries or based on legal, political and economic factors. Property rights are a good example. In considering the best spatial organization for water development, the fact that water planning regions do not have any particular social and political validity should be borne in mind.

Development needs and aspirations in the basin

For most of the Nile Basin countries the prime development imperative is to raise food for the millions of people residing within them; and each country makes food self-sufficiency a major policy goal; at the same time in all countries there are serious employment and purchasing power problems. Some of the countries enjoy adequate food production, but a substantial part of the population does not have sustained and remunerative employment to be able to purchase adequate diets. All of the countries seek a major intensification of agricultural and other economic activity to offer productive employment.

To build industrial and service sectors, the overall economy of each country requires increased quantities of food per capita as well as low cost food with which to feed the work force. This therefore requires an efficient high input agriculture making intensive and effective use of land and human resources. This in turn, implies investment in inputs like tools, high yielding seeds, fertilizers, pesticides, rural energy supply and the important provision of irrigation water. In these circumstances it is natural that at the national political level each co-basin state aspires to affirm and intensify its sovereignty and autonomy.

Irrigation and agricultural development

The agricultural production of the Basin is governed by alternating periods of water surpluses and deficiencies. The rapidly increasing density of population per unit of agricultural land may tend to stimulate irrigation investment. Not all of this investment will have direct positive effects on agricultural production but it will provide rural employment. Farming in the future may have to use all potential arable tracts, and increases in food production will have to come from increased yields and multiple cropping. Irrigation makes dry season crops possible, and it increases wet season yields by establishing a more favourable water regime for crops. As important, it induces farmers to invest in other inputs such as high-yielding seeds, fertilizers and additional labour by reducing the risk of loss of these investments through drought.

Inter-governmental relations

It is manifest that an essential preliminary to any international cooperation in the Nile Basin is the establishment of sound relations between the riparians and especially between those that have significant opportunities to reciprocate. The hydrological factor varies in significance in riparian relations according to the geography of the respective countries. Ethiopia's Nile tributaries providing such a substantial proportion of the total flow in the system is in a unique position and the agencies responsible for water affairs are actively investigating means of advancing cooperation.

The countries of the regions through their effective powers should foster relations between co-basin water agencies and formulate, reform, and clarify water policies. Among these countries, some of the agencies have risen to a higher level of competence and responsibility while some of their counterparts are only at a formative stage.

Nile Basin-wide planning and the question of multiple jurisdictions
The multiplicity of nations represents the single most powerful obstacle to basin-wide development planning. Governmental jurisdictions represented in the Nile Basin are numerous. All nine nations have many water ministries, commissions, and regulatory and administrative agencies. These political boundaries amount to more than a lifeless grid invisibly subdividing the region. They are repositories of social traditions, values, economic interests, and political power. Planners who overlook these political boundaries will imperil their plans.

Nearly all the governments of Nile countries are based on a single party system and have a unitary form of state. At the regional level, there are multilateral or bilateral groupings like the Hydrometeorological Survey (Hydromet) Project of Lakes Victoria, Kioga and Albert (Mobutu Sese Seko), UNDUGU and the Permanent Joint Technical Commission (PJTC) of Egypt and the Sudan which are set up to bring agreement on differentiated and often particular national interests of countries or signatories.

Historically, some of the basin countries have tried to lay claim to as much of the natural resources bases of the Nile River Basin as possible on behalf of their citizens, and few issues have united the citizens within each of those nations more than the issue of water resources and economic development. Such deeply felt national interests will lead to arduous negotiations over the future allocation of water. On the other hand, the pluralism evident in the basin could yield to an incremental evolution of well founded approaches and institutions based on sound economic and ecological principles. Before significant progress can be made there must be a prolonged study of the conditions and costs associated with institutional change within the basin and the process would be accelerated if appropriate incentives were introduced by competent agencies in the international community.

The problem of establishing and achieving priorities
The existing heavy demands on Nile waters appear to commit the river's resources totally. Nevertheless upper basin nations have rights to develop their water resources.and such upstream water storage and consumption will affect lower basin users. A system of planning has to be evolved to enable logical responses to shifts in national and international circumstances and especially to the changing role of water resources in the inevitably dynamic economies in which the environment is only a part. Meanwhile the governments of Nile Basin states have obligations to their citizens and have to obey their national laws, and at the same time have to abide by internationally accepted principles on the use of shared water resources.

The existing dynamic interplay of interests within a complex framework of laws, rules and traditions is not neccessarily the best starting point for water resource planning and revised principles will have to be recognized. Some typical constraints on planning currently are:

1 The level of utilization of the Nile waters by the co-basin states varies

with their respective socio-economic advancement. For example, Egypt has extended its irrigated area to 2.8 million hectares with the water delivered since the 1960s from the Aswan Dam. Egypt vigorously utilizes 55.5 km^3 of the Nile Waters. According to the master plan, Egypt has a projected demand of 65.5 km^3 annually for her socio-economic development.

After Egypt, the Sudan stands high in the utilization of the water resources of the Nile. The Sudan has developed 1.8 million ha of irrigated agriculture consuming about 13 km^3 per year. With the projected development of 1.5 million irrigated hectares, her annual demand of water could exceed 30 km^3.

Conversely, the upper riparian states that include Tanzania, Rwanda, Burundi, Kenya, Uganda, and Zaire utilize only 0.05 km^3 of the Nile waters. Despite the fact that Kenya, Tanzania, and Uganda have a need for food security and employment generation, their efforts to harness the Nile waters to date have been minor. Ironically, Ethiopia contributing 72 km^3 which is about 86 per cent of Nile waters, utilizes only 0.6 km^3 despite the fact that the country is recurrently hit by devastating droughts and famines.

2 Surface water is scarce in the dry season, and important disputes could arise about entitlement to the share of water among some states. There is a long standing suspicion between upper and lower basin states about the sharing of water resources. Each upper and lower basin state would maintain its claim for surface water by pointing to its current uses of water and its future needs and by asserting that it has no alternative sources. Therefore, all feasible means have to be explored and fairly evaluated to relax the competition for limited water. The most fundamental constraint which is aggravating the water shortage is the increased rate of population growth. The slowing of population growth is the single most important measure which would improve each country's prospects to feed and employ future populations.

Meanwhile the nature of existing institutions has severely deterred the decision making elites of the Basin countries from coming to terms with the need to cooperate rather than press on with the development of immediately accessible water.

3 Indices of environmental constraint have proved to be a useful analytical tool by simplifying the complexity of the environment enabling insights and facilitating the assignment of priorities between programmes and policies, as well as in evaluating the performance of such programmes and policies. But before they can be adopted and used as planning tools the decision makers have to be persuaded of their relevance.

The need for regional cooperation

The rational utilization and optimum development of international water resources requires cooperative and concerted action among the basin states on the basis of defined sets of rules and procedures. One of the current practices to resolve differences and settle disputes arising from the utilization of basin waters among states is to draw up agreements. The utilization of boundary and transboundary rivers often involves a complex framework of management in adddition to the definition of boundaries. Therefore, there is a need to consult, negotiate, cooperate, and use judicial and dispute resolution institutions. The need for a framework that bestows a spirit of cooperation, and settles issues through discussion and negotiation is gaining acceptance in the Nile Basin. Social necessities such as peace, non-interference in inter-state affairs and social stability among co-basin states will bring about in due course constructive cooperation and development. The setting up of such cooperative arrangements is to be encouraged and welcomed by the international community to the extent that they advance the mutual interest of all the basin states.

Nile waters allocation: important issues

There are many uses of water and water behaves as several different commodities at the same time; and, depending upon how water is to be used, it may also belong to one of several markets. For example a demand curve for water in the Middle East is different from that of water rich tropical Zaire. In less fortunate arid zones consumers are willing to pay very high prices for water for residential, commercial, and agricultual purposes.

Water is just one of many resource inputs which should be used in concert to achieve the best economic and social outcomes. Its value is different among the nine nations in the basin with respect to their uses as irrigation water, hydro-electric power, navigation, water-based recreation, as well as for industrial and municipal purposes.

Ideally Nile waters should be divided in such a way that the allocation of a quantum supply to one individual nation can improve the well-being of that nation without decreasing the well-being of the other eight nations. Should the allocation decisions make some nations better off and some worse off, a compensation principle has to be worked out. In other words, the gainers would have fully to compensate the losers. The beneficiaries have to be clearly defined as well as the losers, and any damages sustained or opportunities foregone by the losers should be quantified. The estimation of gains and losses should be made on the basis of the value of economic criteria. The issue is that if the gains to the gainers are greater than the losses to the losers and the latter are adequately compensated then the overall welfare will be advanced. Any basin-wide allocation policy would have to address the issue of establishing minimum acceptable flows for each sub-basin state as well as other principles concerning seasonal flows, pollution and environmental sustainability and amenity.

General recommendations

On-going surveys to enable evaluation and the preparation of national water master plans

The emphasis in the preceding discussion has been on the need for an interdisciplinary approach to maintaining and developing further the renewable natural resources of the region. The following recommendations echo a number of the conclusions reached in earlier chapters. They are repeated here in order to establish the sound scientific principles upon which they are based as well as the conventional professional practice which is the basis of the Ethiopian approach to Nile waters development.

Prepare national water master plans

The first step towards the preparation is the assembly of integrated survey data to provide an inventory of the water resources of each country. All the co-basin states have to prepare comprehensive national water master plans in order to estimate its present and future needs. The master plans will indicate the required agricultural production for the existing and prospective increase in population and corresponding water requirements. It will also forecast the requirements of power for various purposes such as pumping for irrigation, reclamation and drainage, as well as for industries and municipal needs.

Conduct adequate water studies on the river basins

The present status of knowledge relating to the availability of surface and groundwater resources of the Nile basin is not a sufficient basis to attempt a comprehensive quantitative appraisal to enable sound basin-wide water allocation and management. A detailed water balance should be calculated for the major tributaries. There is a great need for precision in the collection of such data in Ethiopia and some other riparians in order to establish with confidence the extent of the resource base and the different demands placed on it. Such data are important as an input for a reliable planning effort aimed at the eventual acceptance of projects and their legitimization.

Study sediment balances

The serious issue of soil erosion has to be addressed. Soil erosion is not only danaging to the countries where it occurs but it also has consequences downstream. In order to aid the technical appraisal of these processes estimates have to be made of reservoir sedimentation, the costs of river training works and of soil erosion control measures. Sediment balances have to be calculated for the mountain and plain elements of the region. Part of the study should carefully examine the role of forests and other land uses in relation to sediment loads.

Introduce low-flow augmentation approaches

A number of methods have to be considered for each nation in the Basin to make more water available in the dry season, by means of upstream storage, underground storage, and storage in the plains. All methods have to be identified which could augment low-flow and mitigate food shortages.

Interbasin transfers should be only considered after detailed technical feasibility studies have been undertaken to assess environmental impacts, the costs of cross-drainage works, and the possible social conflicts which could arise from the expropriation of land and water from traditional users.

Intensify efforts to achieve inter-regional power development and use

Only a small part of the potential capacity for hydro-power generation is presently developed in the Nile Basin. It is therefore, necessary to augment the systematic and comprehensive studies of the development of hydro-electric potential which is an important component of multi-purpose water resource development in the Nile Basin. Another important aspect in this regard is to develop programmes of rural electrification for small pumps, especially for lifting water from rivers for small-scale irrigated agriculture. Agricultural processing will also make use of hydro-energy.

Develop fisheries

Protein malnutrition is a serious problem in the region. Fisheries could be one of many major sources of protein nutrition and for family incomes in several areas of the Basin states. A careful area-by-area evaluation of the contribution of agriculture and fishing should be the basis for decisions on how much water including land should be allocated to fisheries and how much to irrigation in the dry season.

Develop transportation network in the Nile Basin

Transportation is an important component of the daily life of the people inhabiting the Basin. Transportation affects a variety of human activities and enables population to fulfil social needs and economic goals. Roads, navigation channels, rail and air transportation. will all be required. In general, a regional transportation framework could be superior to national frameworks because it would allow unique problems in the area to be solved by the people in the region who are familiar with them. The environmental and social impacts of such regional transportation projects would have to be evaluated.

Broaden the analysis of engineering projects to include their environmental impacts

Each environmental impact study represents a systematic means of anticipating the consequences of action. Investigations should be made to anticipate the potential impacts of developing surface water collection by collector drains from swamps, especially the Sudd, irrigation schemes and any hydro-power generation projects on the regional climate, as well as on traditional ways of life, vegetation, wildlife and the whole terrestrial aquatic and atmospheric ecosystem.

Institutionalize water management research

With the growth of the populations of the nine co-basin states, the traditional approach to estimating future demand takes the current amounts of water supplied for each type of use (be it for domestic supply, industrial and commercial, irrigation, hydro-electricity generation, transportation, fishing and aquaculture, waste disposal, cooling purposes, and water-based recreation needs) and increases the demand figures in proportion to the projected levels of population and economic growth. The projections should also take into account future improvements in efficiency, shifts in economic preferences as well as consumer preferences and changes in industrial demand. Data on current sectoral use are essential if an appropriate and ecologically and economically efficient allocation of water is to be achieved.

Once water has been allocated, it has then to be managed and water management in the highly water dependent Nile Basin is a complex and multifaceted challenge. A broad and integrated approach should be taken. It is to be expected, especially by Egypt and the Sudan, that combinations of interventions and adaptations should be shaped over a number of multi-year planning cycles. Research in water saving irrigation technologies have to be intensified; and on-farm water management has to be improved.

As the water supply available is inadequate for all types of use, means should be developed to enable the efficient re-use of water. Also crops that need less water and which are quick maturing have to be developed.

Research and design: the institutional, political and legislative imperatives for the basin's water policy

The institutional, legislative, social and political imperatives are inescapable but are at the same time extremely difficult to specify because they concern many issues at once. Water policies must be sensitive to the distinct regional concerns and regional differences. Regional institutions should be used to facilitate coalition and consensus-building, undertake conflict resolution, offer technical and information services, establish mechanisms to carry out areas of agreement, enable the employment of personnel inter-governmentally, and build professional working relationships.

Social and political stability can be achieved when equitable principles are the basis of resource allocation. No co-basin state or society should bear a disproportionate cost in obtaining the basic water required for survival or for meeting environmental quality requirements.

A cooperative development framework embracing all the Nile nations should be developed to integrate all the national master plans, analyze them, and work out ways and means to maximize the respective national welfares.

Nile Basin-wide planning objectives should take into account the spectrum of national goals across the nine states. They could range from the development of new waters, national income maximization, income distribution, enhancing and controlling the natural environment and allocating water effectively between various sectors such as agriculture to industry, and where necessary reallocating it. There should be studies of the influence of pricing policies on water use. Also studies of the amounts of water demanded at each price for the important water-using industries and by municipal uses should be conducted in order to determine the value of supplies that might be provided under an efficient system.

In the final analysis, a meaningful cooperative alliance needs to be established, which will only be durable when Governments and peoples of the Nile Basin see that the rewards of cooperative association would improve the chances of providing adequate and secure food supplies, harmonious relations based on peace, stability and a sustainable environment. Any binding treaties should be equipped with processes and procedures for the settlement of disputes, and be self-adjusting to the evolving natural realities of the region and the social needs of its peoples in the future.

References

Chorley, R.J. (1971). *Introduction to Geographical Hydrology*. U.K, Methuen.

Crawford, A. B. and Peterson, D.F. (1974). *Environmental Management in the Colorado River Basin*. Utah State, Logan, Utah.

Freeman, David M. and Brown, Perry (1974). Concepts of Carrying Capacity and Planning in Complex Ecological Systems in *Environmental Management in the Colorado River Basin* by Crawford, A.B. and Peterson, D.F.. Utah State, Logan, Utah.

Garretson A. H., *et al.* (ed.). (1964). *The Law of International Drainage Basins*. New York, Dobbs Ferry.

Howell P., Lock, M. and Cobb, S. (eds) (1988). *The Jonglei Canal: Impact and Opportunity*. Cambridge University Press.

O'Riordan, T. and More, Rosemary J. (1971). *Choice in Water Use*. Bungay Suffolk, The Chaucer Press Ltd.

World Bank,(1988). *World Atlas*.

UN, (1986). *World population prospects*. (1986). New York.

12

The control of the swamps of the Southern Sudan: drainage schemes, local effects and environmental constraints on remedial development in the flood region.

P. HOWELL and M. LOCK

Introduction

Much of the current debate on the Nile and the distribution of its waters has centred on the recent shortfalls in Blue Nile flows owing to poor rainfall in the Ethiopian catchment and the possibility of this being a long-term or even permanent climatic feature. This and growing demand occasioned by population increase in Egypt and the Sudan naturally focuses attention on the potential of catchments other than that of the Blue Nile system and with which there is not necessarily any climatic correlation. High levels in Lake Victoria and increased discharges in the White Nile from the East African catchment during the last three decades and the possibility of increased rainfall in that area (see Chapter 6) therefore tend to reinforce interest in as yet very tentative schemes to reduce losses from evaporation and transpiration in the Sudd and neighbouring wetlands. These include the current Jonglei Canal Stage I now halted by the present civil war, and the projected Jonglei Canal Stage II, which will double the canal capacity.

There are also major canalisation schemes planned for two other swamplands where huge losses by evapo-transpiration occur, but neither of which draws its water from the White Nile catchment in East Africa. The first we refer to as the South-Western area, which includes the Bahr el Ghazal basin. This receives water from the Nile-Congo divide, virtually none of which ever reaches the Nile, and from rainfall which has no correlation with that from which the Bahr el Jebel and the Sudd derives. There appears to be some correlation with Blue Nile rainfall as will be seen from Chapter 13. The second is the Machar Marshes, a vast area of wetland in which much water is lost by spill from the Baro, the main tributary of the river Sobat, as well as numerous torrents flowing into the plains from the Ethiopian foothills (JIT, 1954. Vol III p. 971 *et seq*). Here again, rainfall has no correlation with that of the upper White Nile catchment but with the Blue Nile system farther north. The overall title of Jonglei II is sometimes used incorrectly to describe all these schemes collectively; we prefer to keep them separate, especially as they would derive their water from totally different sources.

The Flood Region (Figure 1)
The whole area, including the Sudd, was at one time classified as the 'Flood Region' (JIT, 1954), which receives water from direct rainfall and from rivers and watercourses which originate from higher ground; from the East African catchment in the case of the Bahr el Jebel, as well as streams which find their way across the eastern plains in a north-westerly direction, runoff from the Imatong and adjacent mountain ranges farther south. Much of this water is diverted in a north-easterly direction by the rivers Veveno, Lotilla and Geni and joins the Pibor river, which in turn joins the Sobat.

The south-western part of the Flood Region is fed from the rivers of the Nile-Congo Divide but, except in the extreme south of the region, little water draining from the west reaches the Nile. North of the Sobat, the Machar Marshes, which can be regarded as part of the Flood Region are, as described above, fed by spill from the Baro and runoff in the form of relatively small seasonal streams from the Ethiopian foothills. Here again little of this water reaches the Nile.

Large or small, the rivers entering the Flood Region have hydrological features in common, succinctly summarised in the report (1955) of the Southern Development Investigation Team (SDIT p.5) which is quoted here.

In spite of the apparent diversity of types of streams in the area, a regular pattern can be distinguished in each river from south to north, east to north-west and west to north-east. In this context the SDIT described the mechanics of what they referred to as the 'standard river' as follows:

> **"First stage: the area of rapid runoff**
> The standard river begins in the elevated perimeter where the area of rainfall is greatest. Owing to the relief of its catchment in this area, drainage is good and some five to ten per cent of the rainfall falling on the catchment drains away in numerous small streams which unite to form the river. Rapids occur at this stage. Owing to rapid runoff from most of the collecting grounds, the individual streams reflect very closely the incidence of rainfall over their catchments and are very erratic in flow. As the river flows from the collecting area the extreme peaks and troughs are damped to a certain extent, but nevertheless throughout the rainy season the discharge varies considerably. Over most of the collecting area there is some soil erosion, especially at the onset of the rains before a grass cover is established; for this reason the river leaving the collecting area carries a considerable amount of silt. After the rains there is a small residual flow of clear water throughout the dry season from the heavily afforested parts of the catchment, or from springs.
>
> **Second stage: the defined valley floodplains**
> The standard river leaves the elevated collecting grounds in a defined but widening valley, runoff from whose slopes still contributes to the flow, and begins the second stage of its course. Its longitudinal slopes, and hence velocity, decrease. The coarser suspended particles in the water are deposited and the river channel is not sufficient to carry the high discharges. A floodplain results, whose width is determined by the higher land of the valley sides and which is inundated when high discharges occur in the river. The main channel is very unstable. It meanders through its floodplain, and by bank erosion continually changes its course. A

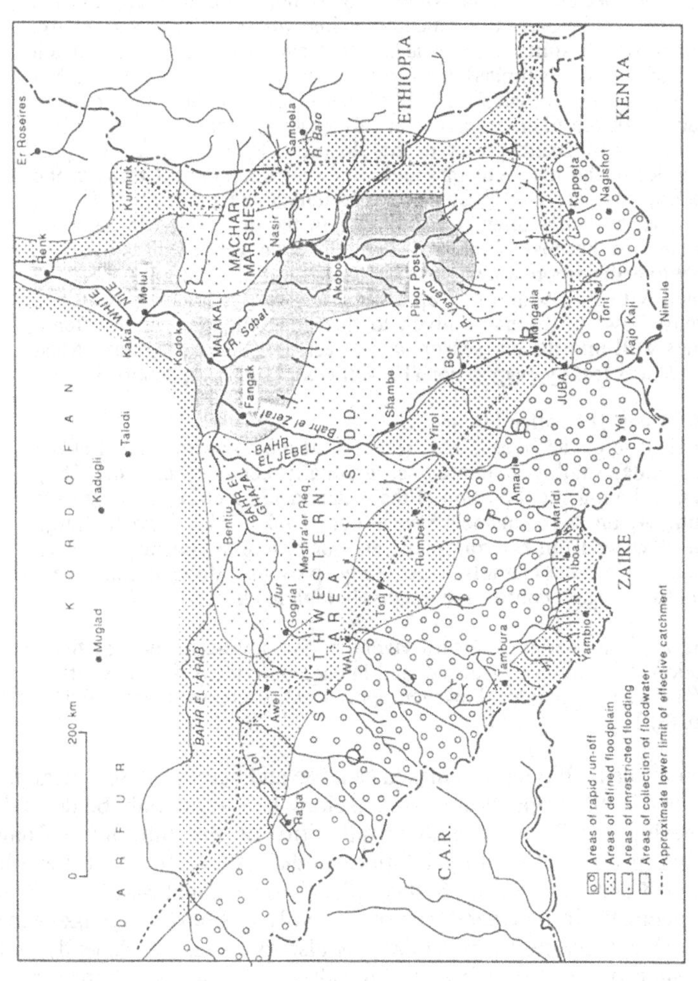

Figure 1. Hydrology of the flood region. Source: SDIT, 1955.

marked feature of the river in this stage, which is of great importance for the whole of the rest of the course, is the alluvial bank alongside the channel and the deposition of suspended material in the water. The bed of the river is of sand, in continually shifting banks, whose coarseness decreases progressively downstream. Considerable reduction of the dry season flow takes place in the bed through seepage.

Third stage: the unrestricted flooding area
The high land which limits the width of the floodplain now disappears and the river enters the great impermeable plains, whose slope is very slight. The water which spills out of the channel can now travel for a much greater distance and is limited in its spread only by the vegetation, which is generally composed of such thick grass that it forms a hydraulic gradient. The current in the main stream is at this stage so slack that vegetation, either floating or fixed, grows in the channel. In places definite spill-channels leave the main channel and wind and interconnect through the floodplain, so that it is difficult to distinguish the main channel. In depressions the water often remains through the dry season in papyrus-locked lagoons, but mainly the area is completely covered with tall grasses rising above the flood-water. Excess local rainfall falling on the area, combined with spill-water, sets up a surge and creeps for long distances. 'Creeping flow' is the name given to this phenomenon, which often continues well into the dry season and occasionally the next rains. Most of this water is lost by evaporation (Howell *et al.*,(1988), Appendix 3).

Fourth stage: the collecting area
The final stage of the river is now reached. At the downstream end of the swamps wide, shallow watercourses gradually become defined. They collect and drain a small portion of the water standing on or creeping over the plain. Again these channels have alluvial banks which prevent a large amount of water from entering them, and they are consequently not very efficient drainage channels. As they progress downstream their floodplains are again confined by relatively high land.

Such is a picture of a 'standard river' in the southern Sudan; the description is, of course, idealised, and in each individual river the details vary. The basic principles, however, apply to all southern rivers."(SDIT, 1955, p.48) [5]

Realisation of the ambitious schemes to canalise and harness these waters, most of which are lost by evaporation and transpiration, will not only be dependent on engineering feasibility and the availability of the substantial finance required, but also on satisfactory solutions to the state of civil war prevailing between north and south in the Sudan. The Jonglei Canal Stage II will also require agreement from Uganda since it involves storage in Lake Albert. Moreover, acceptance by the inhabitants is also essential and depends on the effects of such projects on the hydrological regime and their impact on the ecological conditions upon which the local economies depend. If there are adverse effects, remedial development measures and alternative livelihood projects must be found that are proved to be both technically feasible and economically viable. The difficulties to be found in this connection will be apparent later.

There are already far too many over-optimistic assumptions about the development possibilities and economic potential of this part of the southern Sudan. Our main purpose here is briefly to summarise the predicted local effects of Jonglei Stage I since the area involved - the Bahr el Jebel swamps and hinterland - has been the subject of closer investigation than other parts of the southern Sudan. We also intend to demonstrate the most obvious yet not generally known environmental constraints on development measures which might otherwise serve to alleviate detrimental effects.

Part of the region related more directly to the Bahr el Jebel and the White Nile system was surveyed between 1946 and 1953 by the Jonglei Investigation, a multi-disciplinary team concerned with the effects of the then Equatorial Nile Project, which was part of Egyptian plans for *the Future Conservation of the Nile* (Hurst *et al.*, 1946). This plan included a diversion canal which differed from the current Jonglei Canal Stage I, being larger in capacity (55 million m^3 per day compared with 25), requiring over-year storage in Lake Albert and a system of operation which involved seasonal variations in discharge to meet Egyptian irrigation requirements, now taken care of by the Aswan High Dam. Investigation, surveys, experiments and trials were conducted in this connection from 1946 to 1954 (JIT, 1954. Collins, R.O. 1990, pp. 198-246).

The survey was extended to other parts of the southern Sudan in 1953/54 (SDIT, 1955) by the Southern Development Investigation Team, though work on this was cut short by independence and the first civil war which began in 1955. This lasted until the Addis Ababa Accord of 1972 when peace was restored, a degree of autonomy granted to the Southern Region, and the way made clear for the implementation of Jonglei Stage I, a project quite different in design and operation from that envisaged earlier.

The second source of information is the various surveys and experimental works carried out in the late 1970s and early 1980s by the Dutch Consultants (ILACO/BADA), particularly their unsuccessful trials in mechanised crop production, UNDP and FAO projects, and more especially the range and swamp ecology surveys of consultants, the Mefit-Babtie Group, carried out between 1979 and 1983 (Mefit-Babtie, 1983). All these endeavours (Howell *et al.*, 1988, Ch. 19: Collins, R.O. 1990. p.385), which might have continued and been extended, ended abruptly on the outbreak of civil war once more in the autumn of 1983. No further research has been possible since then.

We propose to refer first, necessarily in very summary form, to the background environmental information and conclusions on the forecast impact of the ill-fated Jonglei Canal which was halted in 1983 some 260 km from base and 100 km short of completion. From this it is possible to predict something of the impact of Jonglei II, though the hydrological effects will be dependent on the method of operation of the Albert Dam and are therefore not directly comparable. We then consider the possible implications of the Machar and Bahr al Ghazal drainage schemes, though based on far less data on their engineering details and on the hydrology and ecology of the area.

Topography

The Jonglei Investigation Team (JIT 1954) classified land types within this region as: "high land", relatively flood-free, better drained areas where human habitation is normally possible at the height of the rains. This is something of a euphemism since this land is rarely more than a metre or two above the surrounding "intermediate land" - the vast areas of land marginally below this level which are subject to periods of flooding from rainfall in the wet season and almost total drought in the dry. The third category given was "swamp", divided into *toich*-land, a common Nilotic word meaning the floodplains seasonally inundated from the rivers and tributaries; and *sudd* - permanent swamp, almost always under water or waterlogged and characterised mainly by *Cyperus papyrus* and *Typha domingensis*.

The JIT stressed the pitfalls of so broad a definition but pointed out that "the main features upon which this classification is made are physiographic, edaphic, and hydrological. The lack of slope (often less than 10 cm/km), the heavy impermeable soils, the comparatively heavy though variable rainfall (between 600 and 900 mm), and with drainage channels generally insufficient to carry it, mean that nearly the whole region is subject to a greater or lesser degree of flooding in the rainy season".

The Mefit-Babtie Team (1983) used different terminology based on vegetation type. Most of the JIT's high land falls within areas mapped by Mefit-Babtie as woodland. The JIT's intermediate land agrees almost exactly with Mefit-Babtie's 'rain-flooded grassland'. *Toich*-land is referred to by the latter as 'river-flooded grassland' and swamp or *sudd* as 'permanent swamp'. Within their major categories, Mefit-Babtie also distinguished a number of vegetation types on the basis of major plant species.

Generally speaking, with considerable local variation, particularly west of the Nile, these land and vegetation characteristics are found in the areas likely to be affected by the current though uncompleted Jonglei Canal Project, as well as those that would be affected by the future large-scale plans for further drainage works in the Flood Region. They are important because all such schemes will have the effect of altering ecological conditions which create the different vegetation types, and as will be seen later, this will affect the current land use practised by the people.

There have been many misconceptions about the nature of this region. As an extreme example, writing in 1976, W. D. Hopper states that:

> "The southern half of the Sudan is potentially one of the richest farming regions in the world, with the soil, sunlight and water resources to produce enormous quantities of food - as much, perhaps, as the entire world now produces! The water is useless today; the headwaters of the While Nile, blocked in their northward flow, spill out over the land to form great swamps. To unlock the promise of the Southern Sudan those swamps would have to be drained, a rural infrastructure put in place, and the nomadic cattle raisers of the region somehow turned into sedentary farmers. The capital costs of such an undertaking would be as large as the promise... yet the potential is real and untapped, and as world food shortages persist such a reserve can no longer be neglected" (Garang, M. de, 1981. pp. 197-205; Hopper, D., 1976, pp. 197-205)

There are, it is true, areas of marginally greater agricultural development potential mainly in the south-western parts of the southern Sudan, though none has the potential of, for example, many parts of Uganda. But one assumes that since the author refers to the spilling of water and drainage, he had in mind the Flood Region, the subject of our study. Similar over-optimistic views have been expressed by those only marginally acquainted with the region, particularly in regard to the huge areas of intermediate or rain-flooded grassland [1]. These, grown green during the rainy season, give the impression of huge grazing resources; closer inspection, as we shall see, shows that they have only limited value after burning, at the beginning and at the end of the rains.

Jonglei Canal Stage I

The Jonglei Canal Stage I was originally intended to follow the direct line from Jonglei, an obscure Dinka village at the edge of the swamps, to the junction of the river Sobat with the White Nile. Subsequent survey showed that its head would be better located at Bor, extending its length to 360 km. The canal would be operated on the 'run of the river', there being no upstream control other than the head works to govern the relative flows through it or the natural channels of the river. It was to have a maximum carrying capacity of 30 million m^3 /day, and operational targets of 20-25 million m^3/day. The benefit downstream was calculated to be 4.7 km^3, or 3.8 km^3 as measured at Aswan after allowing for transmission losses, which, under the Nile Waters Agreement of 1959, would be divided equally between Egypt and the Sudan who would also share the costs.

The Jonglei Area

By Jonglei Area, which falls wholly within the Flood Region, we mean the area likely to be directly affected by the hydrological and hence ecological changes caused by the canal and its operation, effects which extend both east and west of the Nile, as well as physical obstruction to the seasonal west-east migration of people and livestock living close to the canal itself. We also talk of the canal zone - the area some 30 km on either side of the line of the canal - where people may have the benefit of perennial water, but may suffer the need to relocate their permanent dwellings and their shifting cultivations as well as suffering dislocation to their transhumant pastoral economy.

Climate

Weather recording stations are few and far between in the Jonglei Area, as is the case with the Flood Region as a whole, and only rainfall can be considered well covered. The climate of the region is markedly seasonal, with a 5-7 month rainy season alternating with a 5-7 month dry season during which no rain falls at all from roughly November to March.

Rainfall is associated with the Inter-Tropical Convergence Zone (ITCZ). In the north of the area two transits of the ITCZ merge into one so the wet season is pretty well continuous. Farther south, there tends to be a break in the rains between the northward and southward transits of the ITCZ. Thus at Bor (6° 10' N) the wettest month is separated from the next wettest by an interval of at least one month in 70 per cent of years, while at Malakal (9° 33' N) such an interval occurs only in 30 per cent of years. This break in the rains, which is unpredictable in both timing and duration, is often very harmful to crop production and explains the apparent paradox that the people of the wetter Bor area, for example, frequently have to import cereals from outside the area, while farther north, around Renk, the rains, although less in quantity, are much more reliable and productive.

Soils

Soils (Howell, *et al.*, 1988, Appendix I) have developed on recent alluvium, a fine-grained and well-weathered material. Gravel and stones are absent except in the vicinity of the Zeraf hills in the north; calcium carbonate concretions (calcrete) are locally abundant in some areas. The soils contain much coarse sand and clay, with relatively little of the intermediate size classes. This gives them a high bulk density and can render them extremely impermeable. During the dry season they become very hard and crack deeply, but the first heavy rains cause swelling and resealing of the cracks to give an impermeable but sticky and slippery surface.

A feature of soils of this type is often the presence of *gilgai*, a term used to describe regular patterns of low mounds and hollows which in the Flood Region have an amplitude of around 50 cm and a wavelength of 10-20 m. The mechanism of their formation is poorly understood; they pose major problems if land is to be levelled for irrigation; they cause problems for the local farmer because crops mature at different rates in different parts of the pattern; and they add to the difficulties of road construction, a particular constraint to development in the Flood Region. They are reported to reform after levelling if seasonal wetting and drying continues. All major plant nutrients, and particularly phosphorus and nitrogen, are in short supply in these soils. The calcium magnesium ratio is low resulting in poor structural properties (Euroconsult, 1981).

Sheet Flooding

The entire Flood Region is extremely flat, rising more steeply round its perimeter. Figure 2 shows slopes along selected transects across the central part of the Jonglei Area.

Given that the soils are impermeable and much of the rainfall comes as heavy storms in which several centimetres may fall in a restricted locality in a very short time, sheet flooding (referred to as "creeping flow" by the JIT) is characteristic of the rain-flooded grasslands of the region (Howell *et al.*,1988, Appendix 3). Such floods advance slowly across the plains without following clearly defined channels, their flow rate being further retarded by the resistance of the vegetation. In most cases creeping flow is only a few centimetres deep, though depths of 50 to 70 cm have been reported. It seems likely that such depths are only reached in shallow but ill-defined channels and depressions that flank them which can be detected from the air in the rainy season throughout the area and particularly the grassland plains east of the

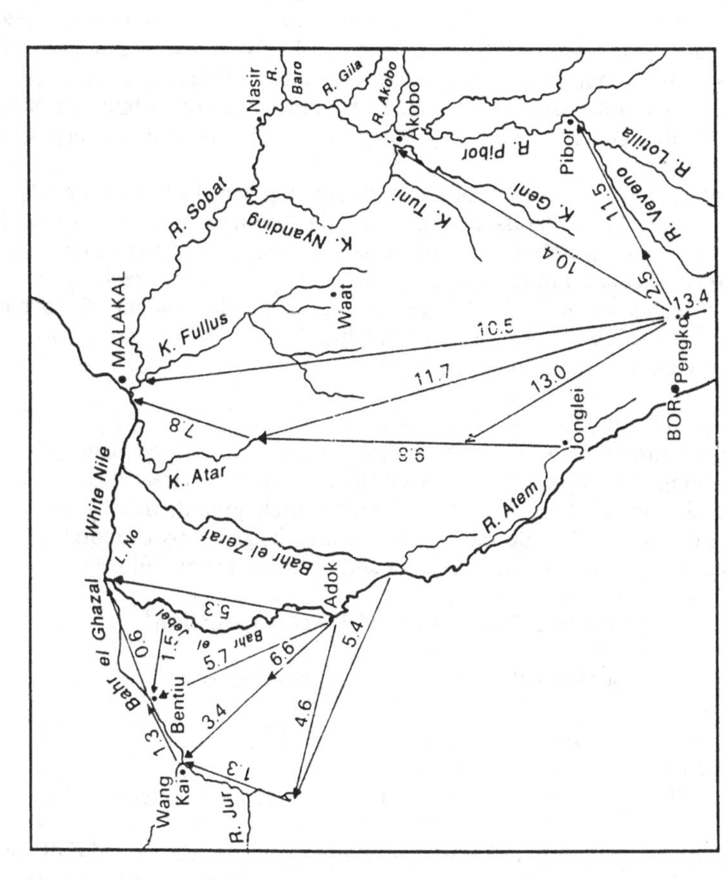

Figure 2: Ground slopes in the Jonglei area. Figures denote centimetres per kilometre. (Source JIT, 1954)

Nile (JIT, 1954, Fig. D27). Some such floods undoubtedly have connections with either permanent or seasonal channels since they are accompanied by fish, in particular lung fish (*Protopterus aethiopicus*) which can breathe air and aestivate in burrows during the dry season, and catfish (*Clarias* spp), which can also breathe air and survive in the foulest of pools.

Sheet flooding in the grasslands east of the Nile is sometimes said to derive from spill from the Nile in the vicinity of Gemmeiza, though this is in fact precluded by topography since the Nile here flows in a shallow but steep-sided trough until just north of Bor so that spill eastwards is not possible. Northward flow from runoff from the Imatongs and adjacent mountains in the south is mainly intercepted and drained by north-easterly tributaries of the Pibor river, though some water may escape northwards in years of very heavy rainfall.

Creeping flow is a further obstacle to development in the region. It can inundate crops, cause hardship to people and livestock and wash away the rudimentary roads which have to be remade annually. On the other hand it serves to redistribute rainfall and contributes to groundwater recharge in the beds of seasonal streams. It also serves to increase the amount of moisture retained in the soil when the rains cease, which is beneficial to grazing grasses that regrow after burning in the early dry season.

Hydrology of the Bahr el Jebel swamps

The Jonglei Area is largely dominated by the hydrology of the Bahr el Jebel. On the average about half the inflow originating from the Great Lakes of East Africa and a number of seasonal torrents which join it, mainly between Nimule and Juba, spills into the swamps and is dispersed by evaporation and transpiration. Figures for recent inflows and outflows are as follows:

Table 1 Mean Annual Discharge in km^3

Period	at Mongalla	at tail of swamps	% loss
1905-1960	26.8	14.2	47.0
1905-1980	33.0	16.1	51.2
1961-1980	50.3 (88% increase)	21.4 (50.7% increase)	57.5

These figures reflect the increased discharges, mainly due to rainfall over Lake Victoria, especially in the early 1960s, which have been a feature of the second half of the century.

In effect the swamps act as a buffer, so that higher inflows lead to higher losses and downstream flow is evened out. Thus an 88 per cent increase in inflow leads only to a 50 per cent increase in outflow. This 'buffering' effect expresses itself as an increase in swamp area - a 130 per cent increase.

The river-flooded grasslands would not be expected to increase as much as permanent swamp; they are an expression of the annual fluctuations in river flow, whose amplitude has not increased greatly, but now fluctuates around a higher mean.

Table 2 Bahr el Jebel Floodplain. Areas in km^2 (Bor-Malakal)

	1952	1980	Increase
Permanent Swamp	2,700	16,200	13,500
Seasonally river-flooded grassland	10,400	13,600	3,200
Total	13,100	29,800	16,700 (130 %)

The obvious conclusion from these figures is that increases in discharges from above the Sudd result in only relatively small increases in outflow (+10.5 per cent during the period 1961-1980 over the period 1905-1960) although the increase in terms of actual annual volume of water, c.7 km^3 (1961-80), was a significant gain downstream. Instead, the area of permanent swamp increases enormously, while there is a substantial though more modest extension of river-flooded grasslands in the floodplains. Similar levels of discharge, probably accompanied by very heavy local rainfall, occurred in 1917 with extensive flooding which was damaging to people of the area, but the inflow had dropped away substantially by 1919, so the floods were not sustained. By contrast the floods of the early 1960s, which reached a peak in 1964, were long drawn out and damage to human interests was not only massive but sustained over a much longer period.

Areas of flooding and consequent vegetation change (see Figures 3 and 4) were estimated by the use of a simple hydrological model based on measurements or estimates of inflow into the Sudd, outflow, rainfall on the swamp and evaporation from the flooded area (Sutcliffe and Parks, 1982 and 1987).

No such calculations in quantitative terms for the Machar Marshes have been made, though the JIT noted (from periodic reconnaissance flights even over a relatively short period of 5 years) that expansion and contraction occurred in the flooded area. Data for the South-Western area and the Bahr el Ghazal basin are given in Chapter 13, but that area was not affected by increases in flows in the period 1960-1980 comparable to those affecting the Bahr el Jebel and the Jonglei Area. As an essential preliminary, plans for drainage in the latter two areas must be preceded by surveys and calculations based on similar models if quantitative estimates of reduced swampland and river-flooded grassland as the result of canalisation are to be made. As will be seen below, diminution of river-flooded grasslands can have a profound effect on the pastoral sector of the local economies.

The rural economy of the Jonglei Area

The Jonglei Area, and indeed very nearly the whole of the Flood Region, is occupied by Nilotes - Nuer, Dinka and Shilluk. All practise a mixed economy - animal husbandry combined with crop production - in an environment that is "unpredictable and capricious".

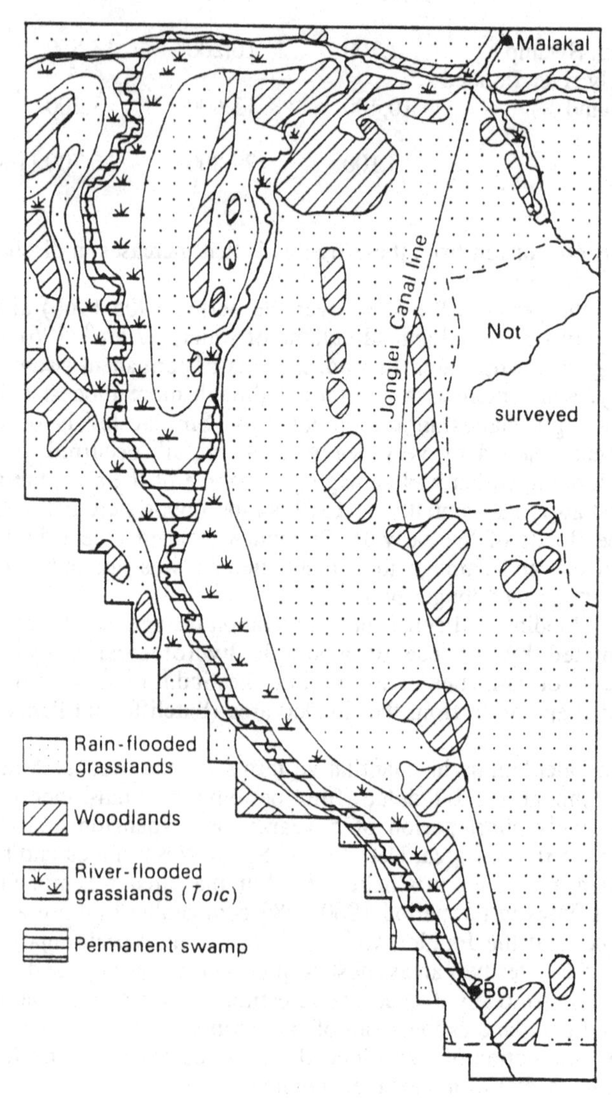

Figure 3: Vegetation in the Sudd area in 1952 (From JIT, 1954).
(Source: Howell *et al*, 1988)

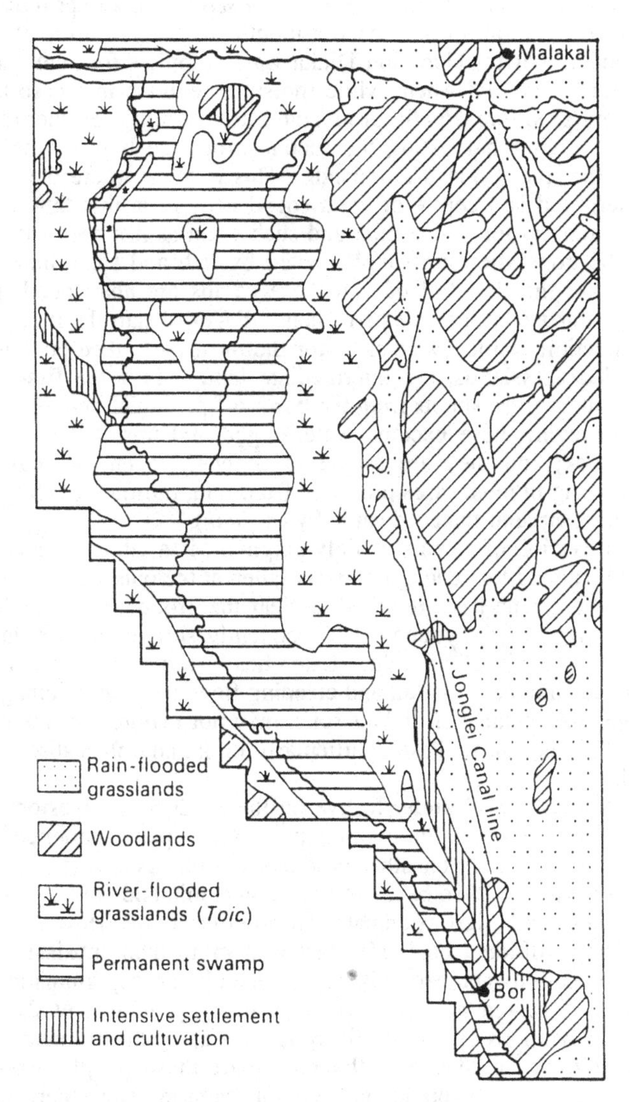

Figure 4: Vegetation in the Jonglei area 1983. From ground and air surveys and LANDSAT imagery.(Source: Howell *et al*, 1988)

The distribution of the land types described above mean that topographical features "combine with climate to restrict the range of agricultural opportunity" (Howell, *et al.*, 1988, p.225 *et seq.*). The abrupt transition from usually heavy rainfall during the wet months of the year to total drought for the rest mean that for Nuer and Dinka water supplies for man and livestock dry out fairly rapidly. Then, while moisture remains in the soil, the herds survive on green regrowth produced after burning of the rain-flooded grasslands. After harvesting the second crop, the bulk of the population moves to the river-flooded grasslands. This is *toich* where the cattle-camps are located, and is a centre of great social activity at this time of year, and economic activity too, since milk and such grain as has been carried there to support the population are supplemented by fish and some hunting of wild animals[2]. At the first fall of rain these camps are abandoned, part of the population hurries back to the permanent villages to start the first processes of crop production, while livestock is for the most part taken once more to the rain-flooded grasslands, the intermediate land, where the first rains again produce young green shoots from the burnt-off perennial grasses. After a few weeks, however, these become waterlogged and unsuitable for cattle and smaller stock, and they are forced to concentrate on the higher ground, livestock being in most cases protected from rain, mosquitoes and other biting flies in huge, circular thatched cattle byres (*luak*).

The cultivation of cereals, mainly sorghum, is a laborious and precarious process. Assuming that enough rain has fallen at the outset to provide drinking water for those who have returned to clear the ground, rain that follows can be poorly distributed; young crops frequently shrivel and die during short, intermittent periods of drought. Later heavy thunder storms may lay the crops and accumulated rainfall and creeping flow may cause damaging floods. Late crops, particularly on lighter soils, may not mature for lack of moisture at the end of the rains. Pests, particularly birds, and plant disease also take their toll.

Most Nilotes would prefer to live on the products of livestock alone and regard themselves as cattle-owning pastoralists par excellence. They are, however, forced to rely on other products - grain, fish, wild vegetation, and the meat of wildlife in some areas - since endemic and epidemic disease can decimate their herds, just as climate can ruin their crops (Howell, *et al.*, 1988, p.226). The traditional economy, rarely at more than subsistence level, is therefore precarious and can only be sustained by an opportunistic balance of all sources of sustenance, food security being provided by the system of reciprocal obligations of the kinship system set up by the exchange of cattle in marriage (*idem.*, p.237). For the most part these people, except in the northern parts of the region, are not self-sufficient in grain. Statistics between 1930 and 1954 show only two years when there was not a net import of sorghum from the north, although there was great variation in success and failure from year to year throughout the area (SDIT, 1955, p.137, Table 49). Figures for subsequent years are not available owing to the civil war, but there is evidence of substantial imports of grain, especially in the Bor area, during the 1970s and early 1980s.

In these circumstances, it might be thought that almost any viable form of alternative livelihood would be preferable to the present. In fact, as we shall see later, alternatives such as irrigated agriculture or mechanised crop production are not only economically dubious but the same combination of climate and soil has proved, up to now, that they are technically unsound.

Effects of the Jonglei Canal Stage I

The canal has not been completed, but detailed surveys were undertaken to determine a whole range of effects, many of which will be shown to be disadvantageous to the inhabitants of the Jonglei Area. We have seen that the *toich* or river-flooded grasslands are an essential seasonal resource during the driest months of the year. Not only is there drinking water available in the rivers, but the process of seasonal inundation itself produces species of grasses which sustain the herds from about January until April. There are no other alternatives as the grasses of the high land are exhausted or reserved for the livestock (mainly smaller stock) held by the few people who elect to remain behind, and the unburned rain-flooded grasslands have become woody and unpalatable, and produce little or no regrowth after burning. It follows that the river-flooded grasslands are crucial to the pastoral economy at this time of the year. It is, however, just these grasslands which may be reduced by the operation of the canal. Though this is not a firm figure, we have seen that the water benefit of the canal downstream will be in the region of 4.7 km^3, a substantial percentage of the average 'losses' by evaporation and transpiration, the natural production of river-flooded grasses being a function of the annual fluctuation in river discharge and thus of the annual variation in area flooded. In other words to the local inhabitants these are not losses in water at all, though we have seen that at times the waters are excessive and the cause of damaging floods, as in 1964.

Toich-land varies in extent from year to year according to inflows into the area as well as periodically varying very extensively as recorded above. The effects of the regime of the canal, operated at 25 millions m^3 per day, were calculated by the use of the model described earlier under three different projected river regimes.

Reductions in Dry Season Grazing

Table 3 Percentage reductions in areas of swamp and *toich*

(1950-61 data)		(1961-80 data)		(1905-1980 data)	
Permanent Swamp	*toich*	Permanent Swamp	*toich*	Permanent Swamp	*toich*
57.1	30.7	26.6	20.9	42.6	27.0

These figures represent reductions in swamp and river-flooded grassland or *toich* expressed as a percentage of the average areas of vegetation during the time period given. The availability of *toich* grazing grasses has varied from year to year within those periods and has also varied in different locations. For example, *toich* grazing in the vicinity of Bor, and for some miles north of it, has been in short supply (1961-80) because the river has been neither high enough to flood the surrounding grasslands nor low enough to allow access to the floodplain. Some Bor Dinka have been denied access to the Aliab valley across the Nile since it has been almost permanently under water and more Dinka than before have been forced to drive their cattle southwards into Mandari grazing grounds, resulting incidentally in clashes between the two peoples. By contrast dry season grazing has been more readily available to Nuer farther north (see Figure 4 and compare with Figure 5). Nuer on the Zeraf island, however, were hard pressed for grazing, and their cattle decimated during the peak floods of the 1960s, many people being forced to take up residence with their fellows on the mainland or to turn to fishing and hippo hunting for a livelihood.

The most adverse scenario under the canal regime would be if the river returned near to the mean discharge of the years between 1900 and 1960 (see Figure 6 and compare with Figure 3). It will be noted, however, that in these circumstances conditions would be not unlike those prevailing during that period when on the whole adequate dry season pasture was available, though there could be particularly low years when shortages would occur. It will be noted that there would be much greater reductions in permanent swamp. This would not affect pastoral interests, but a decrease in area of the order of 57% (1950-61 data) could adversely affect commercial fisheries, and those of the small communities of fishermen (*monythany*) that live in the *Sudd* proper.

Flood protection

The floods of the 1960s, reaching a peak in 1964, caused great damage to human interests; on the Zeraf island alone it was reckoned that 130,000 cattle were lost owing to exposure and lack of grazing since practically the whole area remained under water for a long period. Similar disastrous effects occurred west of the Bahr el Jebel in the vicinity of Adok. It follows that any reduction in peak flows could be protective and beneficial. The same model can be applied to give some indication of the effect of the canal on areas of flooding, the figure for a canal diversion of 25 million m^3 per day reducing the area of flooding by about 19 per cent at the peak discharge in 1964 (Howell *et al.*, 1988, p.393 *et seq.*). This is not as much as the rather exaggerated claims of those advancing the benefits of the Jonglei Canal Stage I, but it would probably be sufficient to relieve from flooding a good deal of the higher better drained land occupied by man and beast at the height of the rains and where cultivation is practised [3].

Discharge regulation

The main objective of the canal is to maximise the quantity of water within its capacity passing through it with, one would hope, the minimal environmental damage. It should, however, be possible to go further than this and when necessary reduce the flow in the canal and increase that in the natural channels in such as way as to inundate the *toich* for a sufficient depth and duration to assure dry season grazing in the following year and without materially

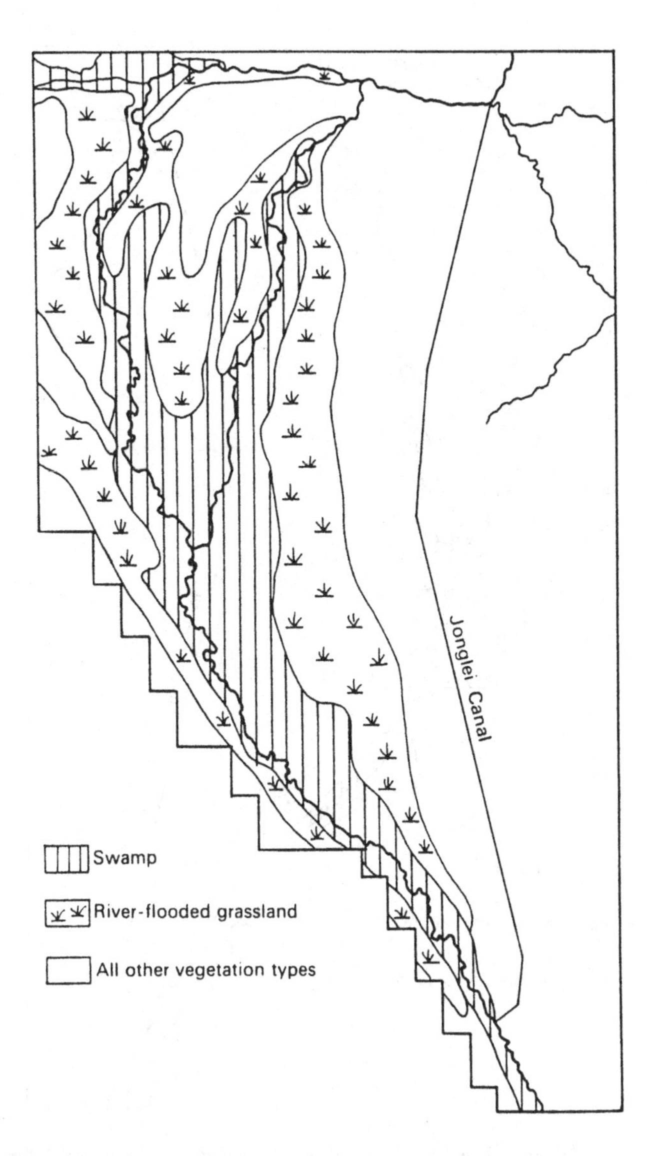

Figure 5: Expected extent of permanent and seasonal swamp if the Nile continues to discharge at 1963-80 levels. Cf Figure: 4.(Source: Howell *et al,* 1988)

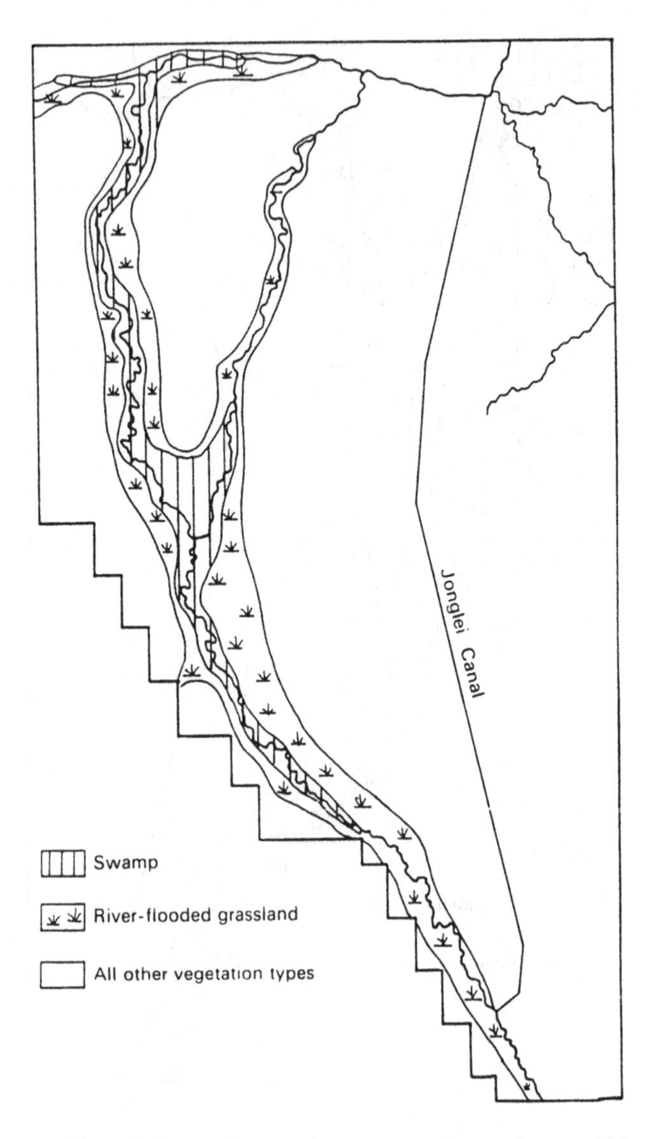

Figure 6: Expected extent of permanent and seasonal swamp if the Nile returns to 1906-60 discharges.(Source: Howell *et al*, 1988)

affecting the overall canal benefit downstream. Such manipulations of flow between canal and river channels have been examined (Sutcliffe and Parks, 1982) with at least encouraging indications. The possibilities need to be examined in greater detail and particularly in relation to the Jonglei Canal Stage II, where the effects on areas of flooding will be greater but with increased opportunities for storage and control behind the Albert dam which would be part of the scheme [4].

Effects on climate

One of the first questions posed when considering the Jonglei canal and its effects is whether the reduction in spill and hence evaporation will cause drier conditions within the area or farther north. In 1954, the Jonglei Investigation suggested that the extra convective activity from the heating of greater dry land surface would counterbalance the decline in the amount of water entering the atmosphere by evaporation and transpiration, so that the net effect would be negligible. This was subsequently challenged (Mann,O.,1977) partly based on the area of higher rainfall marked over the northern part of the Sudd on many maps. This rests on the evidence of a single gauge which was subsequently found to be unreliable (Sutcliffe and Parks, 1982). This view is therefore discredited, and in any case there is no evidence that the huge increases in swamp area from 1961 onwards affected rainfall. It therefore seems probable, according to the Mefit-Babtie Survey, that the reverse effect - the reduction in swamp - would have no impact either.

Communications

Until the outbreak of civil war in 1983 river transport was the main life-line to the south and the Jonglei Canal Stage I brings the obvious advantage of shorter river communications between Khartoum and the main urban centre of the southern Sudan at Juba, in effect reducing the length of the journey by 300 km. The canal will also bring communications, as well as water, to a particularly remote part of the Sudan, which is inaccessible during the rains and largely abandoned in the dry season. Passing points and berthing places are part of the design and will lead to the creation of small ports which are likely to develop and open up contact with the hinterland in much the same way as those along the natural channels of the river have done. There is, however, likely to be considerable disadvantage to the people of the Zeraf Island and those living west of the Bahr el Jebel, in that mainstream traffic will follow the canal and the old western landing places will be ill-served. In the past, moreover, river traffic has been a major factor in keeping these channels open. Oil prospecting is likely to restart once peace has been restored and this may mean that the companies concerned will wish to keep the channels clear, but if discharges drop to the low figures prior to 1961, they could become too shallow for commercial traffic and for the movement of fisheries barges. A minimum flow in the Bahr el Jebel may prove essential, one suggested figure being 35 million m^3 per day (Mefit-Babtie, 1983).

The design of the canal includes an all-weather road along the top of the canal bank. This seems optimistic since the surface would have to be metalled throughout and there are no suitable materials in the area; the work would be very costly to surface and maintain. If the surface is left as it is it would quickly become rutted and impassable, and probably hazardous, for long periods in the rains (see also Chapter 5).

Physical effects of the canal embankments

The canal will in many areas drive a barrier between wet season villages and dry season grazing grounds along the river channels and therefore dislocate the pastoral cycle. Many people living east of the canal will have to cross it with their livestock when regrowth from rain-flooded grasslands is exhausted and they have to move westwards to the river-flooded grasslands of the Nile. A good deal of this problem could have been resolved had the Permanent Joint Technical Commission agreed to the most easterly realignment of the canal recommended by consultants [5], but this was not to be. A compromise was reached on a revised alignment but this was not far enough to the east to alleviate all the difficulties. Three bridges were planned and designed, but finance had not been assured or contracts let before the canal had reached 260 km from the Sobat mouth when it was halted by southern rebel action. The idea of man-handled chain operated ferries was turned down for reasons that are unclear, for they are common in Africa and, indeed, in parts of the Sudan. Instead 12 motorised ferries with elaborate berthing points were planned, without regard to the crucial problem of maintenance. Concentration of large numbers of livestock round a broken down ferry could well cause difficulties of crisis proportions, given the limited pasture in the vicinity.

It is worth adding in parenthesis that the Nuer and Dinka will probably prefer to swim their cattle across the canal when the time comes. This is not new to the Bor Dinka, for example, who swim their cattle across the Nile to the Aliab Valley when conditions are suitable. The authorities would do well to consider support in the form of reinforced structures at various points along the canal to facilitate the crossing of livestock without damage to the embankments and to provide suitably designed boats more efficient than the usual 'dug-out' canoe.

Crossing the canal will present a massive logistical problem and besides, raises questions of land ownership among those who may need to cross the canal and cross each others' territory in order to do so. The scale of the problem can be seen by figures quoted by Mefit-Babtie. They estimated that 800,000 cattle crossings would be required, though this might be reduced by some 100,000 if grasslands of the river-flooded type develop from accumulations of water against the eastern side of the canal. At least 250,000 people would have to cross and with them 100,000 smaller stock - sheep and goats (Howell *et al.*, 1988, pp.424-7).

Some people will have to make four crossings; those living west of the canal alignment, who start the dry season by driving their herds eastwards to graze regrowth after burning of the rain-flooded grasslands will have to cross the canal in an easterly direction to do so. They will then have to cross again to the west to reach the riverain pastures, return to the east side at the beginning of the rains, again to benefit from regrowth after the first rain, and finally west again to return to their settlements on higher ground.

Flood water against the eastern canal embankment

Another of the many important effects which can bring benefits if properly controlled but grave damage if not will be the ponding of rain water against the east bank of the canal. Sheet flooding or 'creeping flow' coming in from the south-east is substantial, though as yet no attempt has been made to quantify the volume of water (which varies greatly from year to year). A number of *khors* or drainage channels cross the line of the canal but no plan was made to syphon the water to the other side. The Mefit-Babtie Survey suggested that something between 500 and 1000 km^2 of dry season grazing of river-flooded species might be produced by ponding of this kind. This could be a benefit, but flood protection measures would also be needed to prevent damage to dwellings and cultivations as well as the provision of causeways to facilitate communications in the rainy season [6].

A bonus will be the perennial water supply the canal will provide in an area that is for the most part waterless in the dry season. This would be potable for livestock, but would not reach WHO quality standards for humans and would require treatment. This could be by relatively simple plant, but would nonetheless be costly and require careful maintenance.

Remedial development and environmental constraints in the Jonglei Area

It is sometimes said that the earlier Jonglei Investigation ignored development possibilities. For curious bureaucratic reasons it was under instructions to avoid development issues and to concentrate on 'remedial measures' and the costs thereof (Collins, R.O., 1990, Chapter 6). This was an unimaginative approach and in practice these instructions were largely ignored. Remedies were in any case unavoidably synonymous with development projects, though economic viability was a secondary consideration because remedies were to be financed from compensation money from Egypt. A good deal of trial and experimental work was successfully conducted on irrigated crops, mainly sorghum, rice and cotton, though this was at Malakal where the land is well above the river and drainage feasible, which is not the case in the Jonglei Area farther south. Trials were also made in irrigated crop production in *toich* land when exposed during the dry season or enclosed by embankments when the river is high to allow controlled gravity irrigation, particularly of rice. Special attention was paid to the potential of rain-flooded grassland which was found to be valueless except, as we have seen, after burning at the end of the rains when there is sufficient moisture in the soil to produce regrowth and at the onset of the rains which produces the same results. By far the greatest area is rain-flooded grassland and it needs to be said again that the Jonglei Investigation found it virtually useless (except for thatching) by the middle of the dry season when it is woody and unpalatable and during the rains when it is coarse and anyway so waterlogged that livestock could not graze there[7]. The analysis performed as part of the Mefit-Babtie study confirmed these conclusions. Experimental work in bunding to impound more water, increase the depth and duration of the flooding, and hence alter the grass species, was undertaken with only limited success - but the idea was taken up by researchers later (Mefit-Babtie, 1983).

The successor team, the Southern Development Investigation Team, at the end of 1953 extended the work outside the Jonglei Area to all parts of the southern region and their report - much of which is still relevant - contains detailed information about agricultural development potential and particularly the constraints (SDIT, 1955, pp. 9-17 and Appendix IV). Its work was to cover a three-year preliminary programme but was unfortunately reduced to a few months in the field, since it was disbanded after independence.

Announcement of the impending Jonglei Canal Stage I, without prior consultation with those who lived in or represented the area, caused considerable local anxiety, eventually expressed in riots in Juba in 1974 [8] (See Chapter 5). This prompted the Sudanese Government to tie the project to the idea of local economic and social development (Alier, 1990). The National Council for Development Projects in the Jonglei Area and its Executive Organ (JEO) were created and their projected programme was to bring to the inhabitants of the area great expectations, which in the event were largely unfulfilled (Howell *et al.*,1988, Chapter 19). Plans included integrated rural development of the Kongor area funded by UNDP, fisheries development by FAO, and major agricultural schemes by the Dutch International Land Development Consultants (ILACO), around Bor.

Transformation versus improvement policies

John Garang de Mabior, now coincidentally Commander-In-Chief of the Sudan Peoples Liberation Army, in his doctoral thesis voices the aspirations of most educated Nilotes: 'transformation', by which he meant massive changes to the existing traditional, and marginal, subsistence economy in the form of large-scale agricultural schemes (Garang, J. de M. 1981) . These included the 'Jonglei Irrigation Project' (JIP), quoted variously at 100,000 to 200,000 feddans and located just north of the village of Jonglei. In fact not even the pilot scheme was begun. It was realised all too soon that except where it crosses watercourses and depressions the canal for most of its length would be below ground level. Gravity irrigation would not therefore be possible [9], and even if in the long run cheap power (perhaps gas) became available from the oil fields of the southern Sudan, the physical obstacles to large-scale farming in this part of the Flood Region were apparently insuperable. The ground slope in that area is less than 10 cm/km so that drainage would be difficult, the risk of salination possible and the impermeable soils virtually impossible to work except by hand when they became soaked and tacky. The project was therefore dropped. The only possibility seemed to be small-scale irrigation in the relatively restricted areas where the land is below the canal water level. These were seriously mooted, but no attempt was made to tap the canal water by pipes through the canal bank, which in any case the PJTC omitted to provide when the canal was under construction. Small-scale projects of this kind would be valuable but hardly modernisation in the sense of Garang's 'transformation'.

The Dutch consultants, ILACO, also turned to the possibility of mechanised crop production, mainly in the Pengko area east of Bor and one which was typical of the rain-flooded grasslands of the Flood Region, and carried out extensive experiments and trials. Yields of the crops under trial - sorghum, maize and rice- were found to be promising on a small scale but were much less so when applied to larger areas. The soils within the experimental area, affected by *gilgai* structures, were found to be not homogeneous, sandy topsoils alternating with heavy clays and truncated termite mounds. This led to "micro-topographical differences within the area which prevented proper drainage, leading to serious waterlogging problems and, in turn, poor plant development and a luxuriant growth of weeds" (ILACO, 1981; Howell *et al.*,(1988) pp. 326-7, 437-40).

The use of ridges and cambered beds produced higher yields but generated their own problems. Pests in the form of weeds, birds, etc, also proved limiting factors. Above all the soil proved too hard for mechanised cultivation at the end of the dry season and inaccessible to and unworkable by tractors in the wet. The period when tractors could be used was therefore very limited which would mean wholly uneconomic use of plant. Yields per hectare high enough to cover recurrent costs, excluding capital costs, were never reached even in the small and carefully monitored trial plots. It was concluded that there was no future for fully mechanised production of sorghum or maize. Trials in the production of rice were also undertaken with only limited success, though here further experiment is clearly needed. The conclusions by 1981 were that transformation techniques were unlikely to succeed in the Canal Zone, though mechanised agriculture was feasible north of Malakal, rather outside the Jonglei Area. As will be seen later all this is of particular significance when considering the impact of the Jonglei Canal Stage II and because similar environmental conditions prevail throughout the Flood Region, with only minor variations, the results are relevant to the other major drainage schemes.

The attention of the JEO and development aid workers then turned to improvement techniques. So far as crop production is concerned limiting factors are not only climatic. There are labour shortages because, even in times of peace, many of the male population have come to migrate to the north to earn money to meet taxation, school fees, the cost of veterinary medicines and above all to buy cattle to start their own herds or supplement existing ones depleted by disease or recent bridewealth payments. Others may be preoccupied in herding the cattle away from the homesteads and cultivation areas at the start of the rains, just the time labour inputs are needed. The pools near to the fields which in most cases are the only source of drinking water in the early dry season may not be filled so that people cannot return to their cultivation soon enough to start clearing the fields at the right moment.

Shifting cultivation, almost universal in the Flood Region, is not entirely haphazard but a systematic procedure of rotation with prolonged periods of bush fallow. It conserves fertility, prevents erosion and the multiplication of pests, weeds and plant disease. It is, in the circumstances, a stable system suited to a subsistence economy. But it is in itself a constraint on labour because much time is taken up in moving to new sites which usually involves the building of new houses and cattle-byres, and it is perhaps also a constraint on progress more generally. Shortage of well-drained land demands this rather crude rotation system. Mono-cropping might in certain conditions be possible; an intensive system based on crop rotation seems more desirable. This requires a whole range of experiment and trial recommended by the Southern Development Investigation Team which concluded that "until sufficient information is accumulated to provide a sound basis for the design of crop rotations suited to the various environments found in the southern Sudan, and until such rotations are thoroughly tested at experimental stations, it would be dangerous to advocate any drastic changes in the present farming systems" (SDIT, 1955, pp. 166-8). This caveat applies nowhere more than to the Flood Region.

There are, of course, many obvious improvements possible. A first priority is the provision of permanent water supplies in the villages and cultivation areas to allow people to go back there earlier in the rains. There are, however, difficulties. Boreholes and diesel pumps provided during this period suffered from maintenance and logistical problems in providing imported spare parts, so much so that the UNCDF financed programme was abandoned (Howell *et al.*, pp. 443-6). BADA, the Dutch funded development project in the Bor area, provided 35 boreholes served by diesel engines but had quickly concluded that hand-pumps were more reliable. The alternative was water tanks (*hafirs*) and a programme was started by the JEO with some success until halted by hostilities (Mefit-Babtie, 1983). A whole section of the Mefit-Babtie Report is devoted to water supplies.

The difficulties of crop production in most parts of the Jonglei area and the Flood Region as a whole, at any rate south of the Sobat, are in fact more complex than the brief outline given here. Some areas are better off than others, but there is almost inevitably an overall shortage necessitating the distribution of cereals within the area through trade or reciprocal obligations inherent in the social system. In many years and in many places the import of grain from the north is standard. Given these circumstances the target is inevitably food-sufficiency.

Researchers only began to turn their attention to small-scale replicated trials, the introduction of new tools and techniques - in effect the "improvement policy" - a year or two before all development effort ceased with the outbreak once more of civil war (Howell *et al.*,1988, Chapter 19).

Pasture management and livestock development

Experimental work on the use of bunds and embanking to contain sheet flooding and to utilise the vast areas of rain-flooded grassland more effectively were likewise terminated by the outbreak of hostilities, though it needs to be said that trials had not produced very promising results. During the decade of peace, progress was made in the establishment of livestock markets and, more particularly improvement of the veterinary services needed to obviate the threat of large-scale losses from disease. The Mefit-Babtie team concluded that given the existing pasture resources there was little room for expansion of the rudimentary livestock industry in the Dinka Bor and Kongor Districts through which the canal alignment runs, but greater scope for the Nuer farther north. Predictions of the potential in this respect when the canal was completed and in operation were not possible because so much depends on climatic conditions and future discharges in the Bahr el Jebel.

Conclusions

In summing up it is worth quoting the conclusions given in *The Jonglei Canal: Impact and Opportunity* (1988):

> "The final impression which emerges from this relatively brief period (c. 1976-84) is that development solutions to the particularly difficult circumstances of the environment will require many years of patient trial and investigation. Patience may be difficult too; the demand for immediate and positive action will be urgent and exacting in circumstances that require rehabilitation measures in a situation far worse than any form of drainage scheme could create.......The prospects and the challenge are there, but any new programme of development must begin where others were forced to leave off and must be preceded by a thorough examination of the experience gained in this interlude in the violent and damaging history of the area in the last 33 years." (Howell *et al.*, 1988, pp. 446-7)[10]

The violence has, alas, increased in intensity since.

We now turn to the proposals for further water conservation in the Flood Region, the effects of which could be far more damaging than those of the Jonglei Canal Stage I but in which the constraints and obstacles to remedial development are similar and just as daunting.

Jonglei Canal Stage II

Except that it is to have an additional carrying capacity of 25 millions m^3/day, thus doubling the throughput and reducing the flow in the natural channels proportionately, engineering plans for the second stage of the Jonglei canal have not been published. It can be assumed that it will roughly parallel the alignment of the canal stage I, though experience gained in the construction of the first 260 km of the latter may favour a more easterly line. The distance between the two canals will also be important; wide separation would exacerbate the problems of seasonal movements of people and livestock.

Stage II will involve storage in Lake Albert, which, indeed, was the case with nearly all previous canal plans other than the current Stage I. Little is known about the river regime envisaged under these arrangements, though it seems likely the dam and storage therein will be operated so as to maximise savings in evaporation in relation to the Lake Nasser reservoir behind the Aswan High Dam, as well, perhaps, as to meet periodical shortages in the Blue Nile. This would require the agreement of Uganda and the international perspective would be extended by the fact that part of the Lake Albert shore as well as the delta of the Semliki river are in Zaire. The net evaporation in Lake Albert is, however, relatively high compared with Lake Victoria. To balance any sustained shortages in the Blue Nile might require either higher levels than 25 m in Lake Albert or, alternatively, to draw on storage in Lake Victoria.

It will be noted in Chapter 4 that the conclusion of the Uganda Government in 1957 was that storage in Lake Albert up to a limit of 25 m on the Butiaba gauge and with a dam at Mutir was acceptable, given adequate compensation for all losses and the cost of relocating the 20,000 Ugandans, who would be displaced by flooding, on land in the vicinity. As part of the Equatorial Nile Project, not then overtaken by the Aswan High Dam, this was regarded as a legitimate bargaining factor in obtaining recognition of Uganda's irrigation needs, at that time restricted by the Nile Waters Agreement of 1929, as well as a higher level of discharge through the turbines at Owen Falls constrained by the Owen Falls agreements of the 1950s. This would apply to any other hydro-electric sites which might have been developed in the interim, e.g at Bujagali, Ayago, or Murchison. It is not known what attitude would be adopted by the Uganda Government of today. The Nile Waters Agreement of 1929 has been repudiated by Tanzania and Kenya, and though no specific official statement on this count appears to have issued from Uganda, it must be assumed that she will follow suit if it is in her interests to do so. That being so, storage in Lake Albert will not be required as a bargaining counter and her agreement to it will rest on other incentives, presumably mainly financial, if indeed she agrees at all.

It has been rumoured that the Permanent Joint Technical Commission (PJTC) of Egypt and the Sudan have in mind a return to an earlier proposal under which the Albert Dam would be many miles downstream at Nimule and with a maximum rise in level in the lake to 35 m on the Butiaba gauge. The effects of this would be to flood huge areas of low-lying land between the mouth of Lake Albert and Nimule unless protected by very costly and perhaps unreliable embankment, and much greater land losses round the perimeter of the lake. Evaporation from a much larger area would also be greater. Up-to-date population figures are not available, and the size of the human problem is not known. The scheme would result in the flooding of large parts of the Murchison Falls National Park, including much river-edge grassland which is part of the grazing cycle of the wild life of the area. This was one of Uganda's largest tourist attractions and of great economic benefit to the country and, it is hoped, will become so again.

So far as the Sudan and the Jonglei area in particular are concerned, it is not possible to calculate areas of reduced flooding and hence losses of dry season pasture because all will depend on the operational regime of the dam and the two canals. It is likely that the dam will be used to store and discharge the mean, in which case variations within and between years will not arise, unless discharges are raised to meet shortfalls in the Blue Nile from time to time. It is, however, probably a conservative estimate to suggest that losses in river-flooded grasslands would not be less than twice the area estimated for Stage I. If so the outlook for the livestock population and animal industry in the Jonglei area would be bleak and these plans likely to be vigorously resisted. Fisheries would likewise be much affected and the effects on wild life migrations very damaging.

Moreover, except for possible flood protection advantages, Stage II would bring no additional benefits to the area. The benefits of water supplies to the canal zone and a shorter navigation route will have already been supplied by Jonglei Stage I. Disbenefits, apart from losses in essential pasture resources, in the form of e.g. canal crossings, would be greatly multiplied. Some people and their cattle might have to make 8 crossings in the year.

The difficulties of providing satisfactory alternative livelihood have been outlined above in connection with Jonglei I. Controlled spillage, however, to reproduce natural conditions and provide adequate river-flooded grassland might be easier - an effective but comparatively uneconomical use of water to meet basic pastoral needs.

The South-Western Area and the Bahr el Ghazal Basin

This scheme is often referred to as the Bahr el Ghazal Drainage Scheme but in fact covers rivers to the south that have no connection with it. The area is nonetheless a single hydrological unit. With the exception of the hilly country in the south, where drainage lines run more directly towards the Nile, the whole area is inclined from the Nile-Congo Divide in a north-easterly direction towards the outlet of the Bahr el Ghazal at Lake No. The Nile Congo Divide is a plateau at 800-1000 m altitude, with a few isolated peaks up to 1700 m, and a rainfall considerably in excess of that which occurs over the Flood Region. The rivers drain across an ironstone peneplain, which in turn is overlaid at a lower level by clay plains north-east of a line Aweil-Tonj-Rumbek-Yirol-Juba. East and north of this line the rivers fan out in deltas and swamps, in some cases periodically connected with the Bahr el Jebel in the southern part and the Bahr el Ghazal towards the north. The whole of the plateau and peneplain is covered by woodlands except where shallow soils, seasonal waterlogging or grass burning inhibits tree growth, while the clay plains are characterised by open grassland.

The topography and ecological features of the clay plains are much the same as the Bahr el Jebel floodplain in the east: varying configurations of higher, sandier and better drained ridges of land, huge expanses of 'intermediate' or rain-flooded grassland on heavy cracking clays, and river-flooded grasslands or *toich* related to the numerous small rivers, particularly towards the east. Apart from the Bahr el Ghazal itself, which contributes very little to the Nile, these rivers never reach the Nile in defined channels but end in permanent swamps, in some cases connected with those of the Bahr el Jebel, the dividing line between them and water spilling westwards from the latter river being very indeterminate and varying from year to year according to Bahr el Jebel flows. Earlier estimates of this westwards spill have been as much as 6 km^3 (Butcher 1938), but there is very little evidence of this order of magnitude. The Southern Development Investigation Team (1954) classified the rivers in the South-Eastern Area and Bahr el Ghazal basin as: rivers connected with the Bahr el Jebel swamps (notably the Gwir, Tapari and Lau) with an estimated average flow totalling 1.58 km^3; rivers connected with the eastern Ghazal swamps (Naam, Gel and Tonj) with an average combined flow of 2.3 km^3; and those connected with the western Ghazal swamps (the Jur, Lol, Pongo and Bahr el Arab and rivers east of it) with a combined average flow of 9.17 km^3 - a total of 11.47 km^3 for the Bahr el Ghazal basin . The first three rivers mentioned above are not part of the Bahr el Ghazal system, but are mentioned because they may be included in the overall drainage scheme for the area (see Figure 7). Chan and Eagleson (1980), taking as a basis records from Vol IV of *The Nile Basin* and Supplements up to No 8, which include data to 1967, estimated a rather higher total figure of 12.7 km^3 for the Bahr el Ghazal inflow, of which only an average of 0.3 km^3 (as measured at Khor Doleib on that river) reaches the Nile, half of the recorded average discharge of 0.6 km^3 at Lake No on the Bahr el Jebel being attributable to spill from the latter river above the junction of the two rivers.

The main point is that up to 97% of the inflow into the basin is lost by evapo-transpiration, but producing permanent swamp, which is virtually useless, and large tracts of river-flooded grasses (*toich*) which, as we have seen, are of vital importance to the pastoral sector of the local economy. Sutcliffe and Parks have now brought these calculations a stage further and used recent measured tributary flows to bring estimates of Bahr el Ghazal inflows up to date (11.32 km^3) and also to estimate areas of flooding from the water balance. Their analysis is summarised in Chapter 13. Figures show that the inflows and the flooded areas have if anything decreased in recent years in contrast to the massive expansion of flooding in the floodplain of the Bahr el Jebel (see pp.8-9). There appears to have been some correlation between the flows of the Bahr el Ghazal tributaries and those of the Blue Nile and its tributaries which have also decreased. This is, perhaps, an extension of the Sahelian rainfall phenomenon. There is also evidence from the hydrological study that flooded areas of the Bahr el Ghazal basin vary seasonally more than those of the Bahr el Jebel. Comparing the hydrology of the Bahr el Jebel System with that of the Bahr el Ghazal basin, the Southern Development Investigation Team commented as follows:

> "It is significant to ask why, when half the water pouring into the Bahr el Jebel swamps is collected at their tail, only a twentieth of that which enters the Ghazal swamps emerges, in spite of the fact that there is more direct rainfall on the latter. There appear to be four main reasons:
>
> (i) The Bahr el Jebel enters the swamps in a single channel. Six major channels debouch radially into the Ghazal swamps, the sill length for spill being correspondingly increased.
>
> (ii) High ground, which might limit floodplains, is less in the Bahr el Ghazal system.
>
> (iii) Major slopes on the lower Ghazal are less than those in the lower reaches of the Bahr el Jebel.
>
> (iv) Whereas four-fifths of the Bahr el Jebel is clear, perennial water from Lake Albert and only one-fifth is silty, torrent water flowing during and immediately after the rains, the whole of the Bahr el Ghazal is of the latter type. This results in much greater silting of river beds where slopes slacken with consequent diminution of channel capacities.
>
> It is interesting to note that the River Jur, the one river in the Bahr el Ghazal area which traverses the swamps in a continuous channel, has the shortest length to traverse (and hence greatest slope) and also the largest annual flood and the greatest clear, low season flow of all the Ghazal tributaries" (SDIT, 1955, p.50).

The large-scale drainage scheme to tap these waters is included in the Egyptian Nile Water Master Plan, though precise engineering details have not been

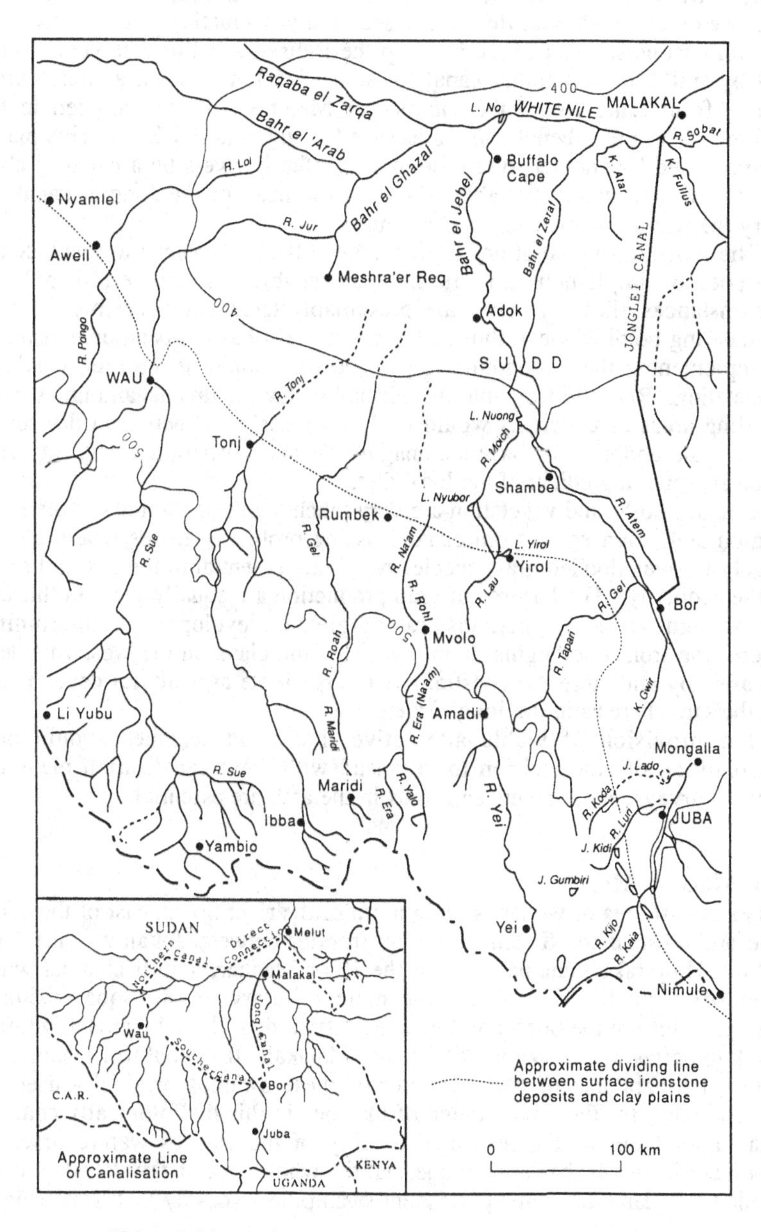

Figure 7. Rivers of the South-Western Area (Source: JIT, 1954)

released. Broadly the plan is that a north-bound canal would harness the Jur, Pongo, Bahr el Arab and the Bahr el Ghazal to a point near Lake No, a distance of about 425 km. Because of seasonally fluctuating levels of discharge in both channels, water might back up the Bahr el Jebel or vice versa and be lost by spill. Hence another canal (of some 225 km), known as the "Direct Line", from Lake No to the vicinity of Melut would be included in the scheme. The gross benefit in water saved is given as 5.1 km^3. This plan is accompanied by another to tap the more southerly rivers by a canal of about 300 km which would probably link with the head of the Jonglei canal and carry the water downstream by that route.

These plans, and variations of them, are sufficiently vague to preclude any prospect of implementation in the foreseeable future even if political circumstances allowed. They are presumably technically feasible from the engineering point of view, though the cost:benefit has not been demonstrated. If implemented their economic impact on the people of the area would be devastating. Reduced flow into the defined valley floodplains and unrestricted flooding areas of the rivers would lead to a massive reduction in dry season grazing and undoubtedly have a damaging effect on fisheries, a very important albeit seasonal ingredient of the local diet.

Climate, soils, and vegetation are very much the same throughout the Flood Region and, apart again from canal crossing problems, losses would also be largely in river-flooded grass species which are essential to the pastoral sector of the economy. The hazards of crop production are equally great in this area and though there are, perhaps, rather greater development opportunities where the ironstone begins to emerge from the clays on the western side of the area, by and large the constraints on large-scale agricultural development are the same here as in the Jonglei area.

The provision of viable alternative livelihood schemes should be a precondition of any decision to go ahead with these canalisation plans and further surveys, experiments and trials in the area are essential.

The Machar Marshes

This is a vast area of wetlands lying north of the river Sobat, east of the White Nile and west of the Sudan-Ethiopian frontier between Jokau and the Khor Yabus. They take their name from the Khor Machar, a spill channel which takes off from the river Baro, the main tributary of the Sobat rising in Ethiopia, close to the border and extends eastwards and northwards to within a few kilometres of the White Nile below Malakal. It is doubtful whether the different inhabitants of the perimeter of these swamps originally used the name Machar for the whole range of marshes in this hydrologically complex area rather than for the restricted locality of the various watercourses, of which the Khor Machar is only one, but the whole area of rain-flooded, river-flooded grassland and some permanent swamp now goes by that name (Figure 8).

The area is relatively unknown. There is no perennial navigable channel like the Bahr el Jebel through the Sudd or seasonal river route as in the case of the Bahr el Ghazal/Jur system. It has been little explored or surveyed and the

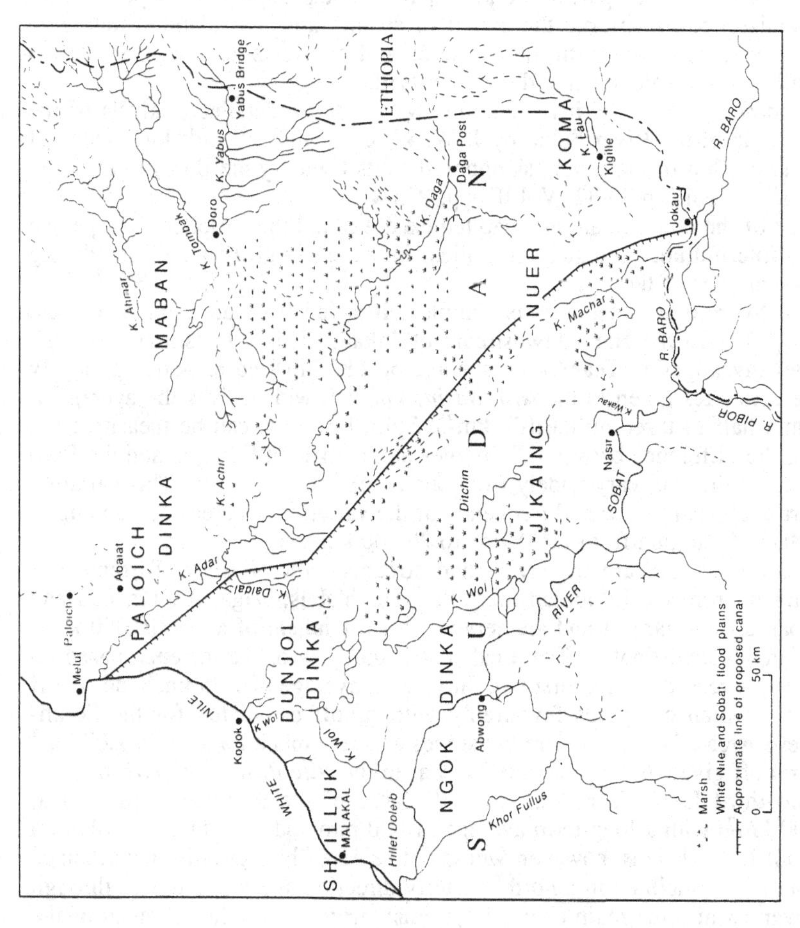

Figure 8: The Machar Marches.
(Source JIT, 1954, Fig K15)

only comprehensive account derives from the Jonglei Investigation (1946-54), though it would not have been affected by any Jonglei canal. But the Team considered the area under-utilised and therefore a possible alternative source of dry season grazing for those farther south who would suffer losses in this respect. The area was therefore closely observed, mapped using American Air Force photographs which had become available in 1945 and a number of astronomical fixes was made to control the maps in the making. Surveys were undertaken of at least part of the principal channels, supported by periodic air reconnaissance to observe the extent of swamp and marshland which from year to year is never static but expands and contracts according to annual rainfall over the catchment and the area itself.

To these surveys and investigations was added a detailed analysis of the flood cycle of the River Sobat by J. W. Wright, which extends knowledge of the relationship of that river (as opposed to its tributary the Baro) to the area. (JIT, 1955, Vol.I, p.30-32; Vol.III, pp. 973-4)

Most of the marshes are uninhabited except round the periphery, but people make intermittent excursions into them in search of grazing and fish during the dry months of the year.

The hydrology of the area is complex. It is fed from three main sources. First spill from the Baro downstream of Jokau. This was estimated by the Jonglei Investigation Team at an average of 2.82 km^3, corresponding closely to the estimates given in the *Nile Basin, Vol VII,* which gives the average at two and half to three milliards" (km^3). Spill, however, can be measured by taking the difference between discharges at Gambela in Ethiopia and the Baro mouth and this can vary widely from one to six km^3 annually. This variation in part explains the size and frequency of the variations in areas of flooding in the Machar, ranging from 6000 km^2 to 20,000 km^2.

Secondly, there are the rivers and streams - the 'Eastern Torrents' - coming in from the Ethiopian foothills. The Yabus, Daga, Ahmar, Lau and Tombak are the largest and are served by a catchment of about 10,000 km^2. The Jonglei Investigation Team calculated the flow of the torrents, based on three years records of the first two only, at an average of 1.75 km^3 and runoff of 15%, which compares favourably with runoff calculated for the South-Western Area of 4 - 7% where the slopes are very much flatter. (cf. 2.07 km^3 with runoff given as 14 % for the Machar in the *Nile Basin, Vol VIII*, p. 271).

The third factor is rainfall over the area of the marshes, estimated at 20,000 km^2, with a long-term average of 800 mm and an annual contribution of about 15 km^3. This, however, would not be saved by a canalisation scheme.

Drainage is inclined in a north-westerly direction and water passes through the swamps in three main channel systems: from the northern branch of the Khor Machar to the Khor Adar and then to the White Nile near Melut; from water from the more northerly of torrents, notably the Daga, which also joins the Khor Adar; and from a southerly branch of the Khor Machar which eventually links with the Khor Wol, running roughly parallel to the Sobat and exits into the White Nile just south of Kodok (JIT, 1954, Vol.I, p.29; Vol. III, pp. 971-984, 913-69).

In fact it is only in years of exceptionally heavy rainfall that there is any appreciable discharge from these marshes through the Khor Adar and Khor

Wol into the Nile, together rarely exceeding 0.3 km^3, though there have been deduced discharges up to 1.8 km^3 in very wet years such as 1946-7. Various estimates have been given of total average inflow: Baro spill 2.8 km^3 (JIT, 1954 and Hurst 1966) and 3.54 km^3 (el Hemry and Eagleson, 1980); eastern torrents 1.75 km^3 (JIT, 1954, pp. 973-4), and 1.4 km^3 (Hurst 1966), and 2.0 km^3 plus 1.6 km^3 inferred runoff from the plains (el Hemry and Eagleson, 1980), the latter giving the average outflow into the White Nile of 0.12 km^3. This gives a total average inflow varying between 4.2 km^3 and 7.1 km^3, indicating that firm estimates have not yet been possible and await a more efficient and wide-ranging network of gauges. The area has suffered greatly from grave civil disturbance during both civil wars and records have lapsed.

Spill from the Sobat through channels that leave the river west of the Baro junction appears to be relatively small (JIT, Vol.III., p.913 *et seq*). The Khor Wakau is the only one to have been gauged; on a rising flood, water travels from the Sobat along this channel into the Machar (c.0.15 km^3), but as the river levels drop, more (0.4 km^3) returns to the Sobat than enters it. This may also apply to channels from the Sobat to the Khor Wol downstream.

The bulk of the Khor Machar and adjacent marshes are claimed by the Eastern Nuer. To the west and north-west is Dinka territory, and in the north are Maban, joined periodically for dry season pasture by Rufa el Hoi Arabs. Nuer migrate into the Machar in the dry season, especially in years when the river-flooded grasslands bordering the Sobat are less productive owing to inadequate flooding. However, because of the large and unpredictable fluctuations in the area of the marshes, and particularly of *toich* or river-flooded grassland, the area is rather less regularly utilised by Nuer for dry season pasture than other similar resources elsewhere, but as an area of potential expansion for livestock as well as fisheries it is economically important. The Dinka on the western side of the Machar Marshes are often hesitant to migrate into the area in the dry season because water supplies can dry out between them and the river, making retreat difficult and, by contrast in some, years the hinterland is permanently under water and inaccessible to livestock. While the marshland certainly varies in extent, its economic potential could be greatly increased and exploited by the provision of *hafirs*, boreholes or other forms of water supply in the dry season.

Though these may vary widely from year to year, the amounts of water lost by evapo-transpiration is equal to pretty well the total inflow plus rainfall. Plans have therefore been drawn up to reduce losses by canalisation along a line which runs from Jokau to Melut, a distance of approximately 300 km, with possibly some embankment of the Baro. It is estimated that this would give a benefit of 4.4 km^3, a figure which appears optimistic.

Drainage of the swamps would undoubtedly affect local interests and much of the area would turn to grasslands of the rain-flooded type which, as we have seen, have only limited seasonal grazing value and grow on land which is no use for mechanised crop production.

As for remedial development to meet losses, the area is more promising than the Jonglei or South-Western area. Despite lower annual rainfall, crop production among the Dinka of the White Nile is more reliable and successful than other parts of the Flood Region. There are also areas among the Eastern

Nuer capable of producing good cereal crops, but one of the principal handicaps is river flooding, and a reduction in peak flows in the Sobat system might be beneficial. These statements are conjectural and it is clear that the project must be preceded by surveys, experimental work and field trials.

It should be noted that an alternative to the Machar canal scheme would be a dam close to Gambela, which would impound 25 km^3. Discharges would be regulated to prevent spill and as far as local interests are concerned the effects would be very similar to those of canalisation though the locations could be different. Hurst (*Nile Basin, Vol III)* appears to have favoured this dam, noting that in dry years there might be little or no yield from a canal.

Conclusions

Data, except for the Bahr el Jebel, are tentative and for the Machar widely differing and probably unreliable. It would, however, seem from such figures as have been given that the average 'loss' by evapo-transpiration from the Flood Region as a whole is between 33 and 37 km^3. The estimated total benefit in water saved for transmission downstream is given as 18.9 km^3, though we believe the estimates for the Machar, and possibly for the Bahr el Ghazal, to be optimistic. However, they give some guide to the measure of the problem which means an overall reduction of perhaps nearly half the water naturally available. This is an enormous off-take and the environmental impact is bound to be massive; from the point of view of the local economy it would probably lead to an unsupportable reduction in vital natural resources.

We have seen that one form of remedial measure would be to control discharges as between canal and natural channels in such a way as to inundate sufficient areas of river-flooded grassland to meet grazing needs when exposed in the dry season. In the case of the Jonglei Stage I, while the operation will depend on the run of the river, this would be possible given the headworks across canal and river. The Jonglei Canal Stage II would be controlled from the Lake Albert dam and it should be possible to simulate natural conditions by calculated releases to modify reductions in river-flooded grassland. The nature of the diversionary structures for the other schemes is not clear, but with so many sources of inflow from the many smaller rivers discharge regulation might be less easy.

Inflow into the Bahr el Ghazal basin, as will be seen from Chapter 13 varies very considerably from year to year. This seems to be also the case with the Machar marshes. Moreover the extent of the reduction by drainage in vital dry season grazing in the floodplains would be much greater in years of low rainfall, thus aggravating what already occurs naturally - years of adequate pasture in the dry months in some years alternating with years of lesser growth and sometimes shortage.

We have also seen the natural constraints to alternative development of the economy which stem from a whole range of hostile factors in the environment. Even consideration of these drainage schemes must therefore be preceded by close investigation - as was the case with the now defunct Equatorial Nile Project and the now halted Jonglei Canal Stage I. The effects must be determined and measured. Simultaneously, experimental work and trials must be initiated once more with a time factor sufficient to ensure that the technical feasibility and economic viability of alternative livelihood schemes are determined well ahead of decisions to proceed with the drainage projects.

References

Alier, Abel (1990). *Southern Sudan*: Too Many Agreements Dishonoured, p.200 *et.seq*. Exeter, Ithaca Press.

Chan, Siu-on and Eagleson, P.S. (1980). *Water Balance studies in the Bahr el Ghazal Swamp*. Department of Civil Engineering, Massachusetts Institute of Technology, Report No. 261.

Collins, R.O. (1990). *The Waters of the Nile: hydropolitics and the Jonglei Canal 1900-1988*. Oxford, Clarendon Press.

Euroconsult (1981). *Jonglei Environmental Aspects*. Arnhem, the Netherlands.

Garang, John de Mabior. Doctoral Thesis (1981). *Identifying, selecting, and implementing Rural Development in the Jonglei Projects Area, Southern Sudan*. Iowa State University.

El Hemry and Eagleson, P.S. (1980). *Water Balance Studies in the Bahr el Ghazal Swamp*. Department of Engineering, Massachusetts Institute of Technology, Report No. 260.

Howell, P., Lock, M., and Cobb, S. (1988). *The Jonglei Canal: Impact and Opportunity*. Cambridge University Press.

Hopper, D. The Development of Agriculture in Developing Countries. In *Scientific America*, **235** (1976), pp. 197-205.

Hurst, H.E., Black, R.P., and Simaika, Y.M. (1946). The Nile Basin Vol. **VII.** *The Future Conservation of the Nile*. Cairo, Ministry of Public Works.

ILACO (1981). *Pengko Plain Development Study, Vol. 1. Evaluations and Conclusions*. Arnhem, the Netherlands.

Jonglei Executive Organ (1979). *Comparative Socio-Economic Benefits of the Eastern Alignment and Direct Jonglei Canal Line*. Khartoum.

Jonglei Investigation Team - JIT - (1954). *The Equatorial Nile Project and its Effects in the Anglo-Egyptian Sudan*. Khartoum, Sudan Government.

Mann, O. (1977). *The Jonglei Canal. Environmental and Social Aspects*. Nairobi, Environmental Liason Centre.

Mefit-Babtie (1983). *Development Studies in the Jonglei Area. Technical Assistance Reports for Range and Swamp Ecology Surveys*. Glascow, Khartoum, Rome.

Mefit-Babtie (1983). *Range Ecology Survey: Livestock Investigations and Water Supply*. Glascow, Khartoum, Rome.

Southern Development Investigation Team - SDIT - (1955). *Natural Resources and Development Potential in the Southern Provinces of the Sudan*. Khartoum, Sudan Government.

Sutcliffe, J.V. and Parks, Y.P. (1982). *A Hydrological estimate of the effects of the Jonglei Canal on areas of flooding*. Wallingford, U.K. Institute of Hydrology.

Sutcliffe, J.V. and Parks, Y.P. (1987). Hydrological Modelling of the Sudd and Jonglei Canal. *Hydrological Sciences Journal*, **32**, pp. 143-59.

Notes

1. e.g. Dr John Smith, Director of Agriculture in the 1950s. Collins R.O. (1990), p.228
2. The Mefit-Babtie survey found that the end of the dry season on the *toich* sees the lowest milk-yields and lowest level of condition in cattle during the year. The Jonglei Investigation Team considered the lowest ebb in animal condition to be towards the end of the rains when livestock are restricted to 'highland', areas which are limited in extent and hence carrying-capacity by the surrounding floods. The difference may well be due to the narrowness of the *toich* belt in the Mefit-Babtie study area, and the poor river flood in the years of the main study.
3. Investigation into this aspect of the canal's effects has been inadequate. Also the possibly local backwater effects up the Zeraf river, lower Sobat, and Bahr el Jebel when the canal is in full discharge and the natural channels at high levels need to be considered.
4. See also JIT (1954). "Revised Operation of the Project", pp. 545, 705, 818, 821-1076.
5. A more easterly alignment was put forward by the Dutch engineering consultants to the project, Euroconsult. A more detailed socio-economic survey was carried out by Dutch Consultants (Hoek, B. van Der and Zanen S.) on behalf of the Jonglei Executive Organ (JEO), the body set up by the Sudanese Government to oversee development projects designed to alleviate local disadvantages of the canal and to develop the area. Their report vigorously argues for a much more easterly alignment and suggests that this would reduce rather than increase the costs of the project since less would be needed for the kind of remedial infrastructure mentioned above. (JEO 1979).
6. See Collins, R. O. (1990), Chapter 6. Very heavy flooding against the abandoned canal banks has been reported in recent years, particularly in 1988. In some cases the inhabitants have themselves breached the banks to allow flood water to pass (Johnson, D.H., Personal communication).
7. See Howell, *et al.* (1988), Figure 7.9 *Nutrition available in* Hyparrhenia *grassland.*
8. The effects of this canal were not known at the time and unease stemmed from educated classes who had read the earlier report of the Jonglei Investigation Team but had not realised that the two proposed canal schemes were very different in size, operation and effects. See Collins, R.O. (1990), p.318 and Alier (1990).
9. Researchers in the 1950s had stressed the value of irrigation to supplement rainfall, pointing out that lower evaporation rates in the south made the use of water more economical than in the Gezira. They also stipulated that the functions of navigation and irrigation should be performed by separate canals, a canal for the latter to run parallel and above ground level, with a capacity of 5 million m^3/day and cross regulators to be sited to serve irrigation units of 100,000 feddans. It must be pointed out, however, that

this was intended to be for forage crops alternating with rice, though even in these circumstances drainage difficulties were accepted as the main constraint. See also SDIT, p. 234.

10. It is not yet possible to gauge the devastating effects of the last eight years of civil war. Huge numbers of people have died and will die from starvation and disease fostered by malnutrition as well as the direct effects of hostilities. Insecurity and the dislocation of war have added to the climatic and other environmental risks involved in crop production. And a high proportion of the population is living as refugees in the north or until recently in Ethiopia.

13
The water balance of the Bahr el Ghazal swamps

J. SUTCLIFFE and Y. PARKS

Introduction
The Bahr el Ghazal swamps are unusual in that they are fed by a number of seasonal tributaries whose flows are almost entirely lost within the basin. The swamps have not been studied in recent years since the analysis by Chan and Eagleson (1980) and more flow records are now available. A water balance model has been developed for the Sudd (Sutcliffe & Parks,1987) and subsequently applied in a comparison of other African wetlands (Sutcliffe & Parks,1989). The application of similar analysis to the Bahr el Ghazal swamps serves to compare their regime with that of the Sudd or Bahr el Jebel swamps.

Previous studies
Previous accounts of the Bahr el Ghazal basin are included in *The Nile Basin, Vols I & V* (Hurst & Phillips,1931,1938), JIT (1954), SDIT (1955), and Chan & Eagleson (1980). Brief descriptions of these studies are followed by discussion of the main topics relevant to this study.

The Nile Basin, Vol I, contains descriptions of the topography of the basin and individual rivers, illustrated by a number of photographs. The standard pattern of the Bahr el Ghazal tributaries is that of rapid runoff from an elevated perimeter along the Nile-Congo divide with good drainage and some rapids, through a zone where the river meanders between alluvial banks in a defined and widening valley into a zone of unrestricted flooding over clay plains (SDIT,1955, see also Chapter 12).

A preliminary water balance of the Bahr el Ghazal swamps was attempted in *The Nile Basin, Vol V*, where measurements of the Jur at Wau were supplemented by estimates for other tributaries based on rainfall and percentage runoff. These were compared with estimated evaporation from the swamps in the lower Bahr el Ghazal basin, whose extent was estimated from survey maps.

At the same time, Butcher (1938), in attempting to estimate the water balance of the Bahr el Jebel swamps, had deduced from comparisons at different latitudes that there was a spill of about 6 km^3/year from the Bahr el Jebel to the west.

Following the preliminary reconnaissance of Hurst, river flow stations had been established about 1941-1942 at a number of sites along the main road from Shambe to Wau and Nyamlel, which roughly coincides with the boundary of the ironstone plateau and thus between the zone of runoff generation and that of spill and losses by evaporation.

Summaries of estimated or measured flows along this line of gauges were included in JIT (1954), where the evidence for spill from the Bahr el Jebel to the west is also discussed. Estimates of all the main rivers were also included in SDIT (1955), but these estimates have been revised in subsequent studies. In particular, Chan & Eagleson (1980) used records published in *The Nile Basin, Vol. IV, Supplements 3-8* as a preliminary input to a water balance study of the Bahr el Ghazal basin. River flow measurements have been continued in recent years, and revised statistics differ somewhat from earlier estimates.

The picture which emerges from these studies is of complex rivers collecting runoff from a relatively impermeable plateau which then meander through floodplains where spill spreads over limited areas and converge in a complex swamp from which only limited flow reaches the Nile. The runoff is highly seasonal and generated in the upper reaches where average rainfall of about 1200-1400 mm occurs between March and October; the runoff is concentrated between June and November and averages some 60 to 100 mm over the upper basins. The river spill floods local and more general floodplains in a seasonal pattern which gives rise to seasonally-flooded grassland and more permanent papyrus swamp. The objective of the present study is to use the present hydrological records to estimate the extent of this seasonal and permanent flooding.

The topics which require evidence from previous investigations may be summarised briefly as the extent of the permanent and seasonal swamps, the annual and seasonal variations of river inflows, and whether spill from the Bahr el Jebel to the Bahr el Ghazal swamps is significant.

Area of Bahr el Ghazal swamps

The only direct estimate of flooded areas in the Bahr el Ghazal basin appears to be that given in *The Nile Basin, Vol V*, p 185. Areas of swamp were taken from the Sudan 1/250,000 maps, and included those marked as swamp or reported swamp on these maps. It was noted that the tendency of survey work had been to lessen the areas considered to be swamp, though large areas were roughly indicated as being swampy in the rains. The areas of swamp fed from the Lol, Jur, Tonj, Meridi and Bahr el Arab were given as 7700 km^2 definite swamp and 6800 km^2 reported swamp, with an additional 2100 km^2 fed from the Naam and Lau, the latter of which is considered to be part of the Bahr el Jebel swamps.

River inflows

River flows have been measured regularly at a number of sites along the main road from Juba to Wau and Nyamlel since about 1942, though some measurements were taken earlier. The early records have been published in *The Nile Basin, Vol IV and Supplements;* the records were interrupted about 1961, but have been resumed since 1971 at several major sites by the Sudan authorities and recent records are printed in Sudan yearbooks.

Summaries of mean flows at all sites estimated at different dates are compared in Table 1. It will be seen that at those sites where recent records are available, mean flow estimates have in general decreased. It appears that the flows of the Bahr el Ghazal tributaries are correlated with those of the Blue Nile and such tributaries as the Rahad and Dinder, rather than the outflows from the East African lakes which supply the main flows of the Bahr el Jebel. Thus the Bahr el Ghazal tributary inflows have decreased in recent years, in contrast to those of the Bahr el Jebel.

Table 1 Estimated flows of Bahr el Ghazal tributaries (km^3)

River	SDIT(1955)	MIT (1980)	Present study
Bahr el Jebel tributaries			
Gwir at mouth	0.12		
Tapari at mouth	0.44		
Lau at Yirol	1.02	(2.060)	
Eastern Ghazal swamps			
Naam at Mvolo	0.64	0.476	
Gel at RB	0.55	0.520	
Tonj at Tonj	1.11	1.600	1.363
Western Ghazal swamps			
Jur at Wau	4.52	5.220	4.496
Pongo }	4.23	0.575	
Lol at Nyamlel }		3.900	3.243
Bahr el Arab	0.32	0.300	
Raqaba el Zarqa	0.10	0.100	

The seasonal distributions of the Bahr el Ghazal tributaries may be compared from the monthly normals published in *The Nile Basin, Vol IV, Supplements 7 & 8,* as summarised in Table 2.

In their water balance study, Chan & Eagleson (1980) first used essentially these same river flows. However, they then increased these measured flows after a preliminary water balance and conceptual modelling of the various tributary basins. They compared the measured flows with the estimated evaporation from the areas of swamps given by *The Nile Basin, Vol.V*, taking the full area estimated as 16,600 km^2, and adding evaporation from other grassland areas. They noted that there was no detailed topographic description of the area below the road linking the gauging stations, which they termed the Central Swamplands (84,950 km^2). Nevertheless they assumed that over half was inundated either permanently or seasonally. Evaporation from these areas was estimated as 2.2 m/year from the permanent papyrus swamp (16,600 km^2), 120 mm/month for 12 months from the flooded grassland (28,500 km^2) and for 7 months from the unflooded grassland (39,900 km^2). A preliminary water balance of the whole area left a shortfall of some 15 km^3, even after allowing for 6 km^3 spill from the Bahr el Jebel, and this was thought to be

Table 2 Normal discharges of the Bahr el Ghazal Basin (million m^3)

	Jan	Feb	Mar	Apr	May	Jun	Jul	Aug	Sep	Oct	Nov	Dec	Year
R Lol at Nyamlel (1944-62)	26.8	8.4	3.0	1.8	20.5	128	306	750	1180	1070	342	65.8	3900
R Pongo d/s Rd Bridge (1944-60)	9.7	3.8	0.2	0.0	3.0	21.9	42.4	86.1	147	154	93.1	13.7	575
R Pongo at Rd Bridge (1944-60)	1.6	0.0	0.0	0.0	3.7	23.4	41.1	87.0	157	168	98.8	15.9	597
R Geti thro' Rd Bridge (1944-60)						0.6	8.3	21.6	31.2	25.6	7.5	0.7	95.5
R Jur at Wau (1942-61)	44.3	11.0	0.0	30.3	130	248	436	803	1310	1380	646	180	5220
R Tonj d/s Road Bridge (1944-60)	22.4	13.0	10.7	10.1	36.7	75.1	133	198	306	292	135	25.7	1260
R Tonj thro' Road Bridge (1944-60)	34.7	19.4	16.6	15.6	49.2	99.1	165	241	390	363	169	36.3	1600
R Gell d/s Road Bridge (1942-59)					6.8	43.9	42.8	67.8	104	76.0	42.8	1.0	385
R Gell thro' Rd Bridge (1942-60)					8.6	49.7	72.7	115	161	96.2	19.4	0.6	523
R Wokko thro' Rd Bridge (1942-60)						16.1	20.3	39.3	33.7	16.3	4.6	0.0	130
R Naam at Mvolo (1942-52)					22.7	77.1	55.1	122	107	69.8	20.6	2.0	476
R Yei d/s Mundri Bridge(1944-60)	8.6	0.0	0.0	20.6	139	124	235	519	445	331	142	33.6	2000
R Yei thro' Rd Bridge(1944-60)	13.2	0.2	0.2	15.7	131	144	240	517	467	336	154	46.6	2060
Bahr el Ghazal d/s Khor Doleib (1937-64)	22.6	31.5	41.9	41.7	29.4	22.6	24.6	28.5	17.9	18.2	13.7	12.5	305

Derived from *The Nile Basin, Vol IV, Supplement 8*, up to 1967 explained by unmeasured runoff or groundwater flow from the various tributaries. This estimate was increased to some 20 km^3 unmeasured flow after comparing the simulated annual catchment yield based on the Eagleson conceptual model with the total measured inflow.

Although the flows of the Bahr el Ghazal tributaries are fairly low when expressed as percentages of rainfall, it seems to us unlikely that flows through the gauging network have been underestimated by two-thirds as a result of spill upstream and flow down the river valleys or underground flow. We have therefore used the measured flows and taken advantage that flow measurements have continued on the main rivers to revise and bring up to date the earlier estimates.

Spill from Bahr el Jebel to Bahr el Ghazal

Chan & Eagleson (1980) also included an estimated spill of 6 km^3/year from the Bahr el Jebel to the west to the Bahr el Ghazal swamps north of Shambe. They quote the authority of JIT (1954) for this estimate, but this as noted later, derives ultimately from the study of "latitude flows" by Butcher (1938).

Previous discussion of this problem is to be found in a number of sources. Hurst & Phillips (1931), after air reconnaissance, observe that "the swamps of the Bahr el Jebel and Bahr el Ghazal north of Hillet Nuer are connected by depressions which are for the most part full of papyrus". They also note that the swampy plain between Shambe and Meshra el Rek may be dry in the dry season or after a succession of low floods, but when the water level on the Jebel is high it is probably almost continuous swamp. Newhouse (1928, p.12) discusses the geography of the west side of the swamps below Shambe. He describes channels which take off and rejoin the main river, and in particular draws attention to Gage's Channel which starts at Hillet Nuer and, though completely overgrown by papyrus, has been traced to Khor Doleib joining the Bahr el Ghazal; but he comments that it is of no importance at present, the discharge being trifling. Butcher (1938) carried out water balances between a series of latitudes where discharges had been measured, and deduced that only some 40% of the losses could be explained by evaporation, and that the most likely explanation was flow towards the Bahr el Ghazal, though no trace of a passage had been found. However, his estimate of evaporation from papyrus was based on tank experiments where evaporation was greatly under-estimated because of stagnant conditions which affected the growth of the papyrus; his estimate of evaporation was 1533 mm/year less 912 mm rainfall. On the other hand, calculations based on Penman estimates of open water evaporation, which are likely to compare with evaporation from flooded papyrus in optimal conditions (Penman,1963), amount to some 2150 mm/year, which doubles the net loss. Hurst & Phillips (1938) also examine the possibility that water flows from the Bahr el Jebel to the Bahr el Ghazal swamps and conclude that this explanation of the losses on the Jebel is untenable.

The Jonglei Investigation Team (1954) paid some attention to connections between the Bahr el Jebel and Bahr el Ghazal and found, as had others, that there were physical channels but that these were overgrown and quoted flows are small. Nevertheless, it is suggested that an estimated "6.0 milliards of Bahr el Jebel spill..... flow westwards between Lake Nuong and Buffalo Cape." This estimate is probably based on Butcher (1938), who was unable to account for the losses from the Bahr el Jebel swamps without this large spill, which is topographically feasible but for which there is little direct evidence. Once a more realistic estimate of evaporation rates is included in a water balance study of the Bahr el Jebel swamps, the necessity for this spill disappears (Howell *et al*, 1988, p. 103).

Recent hydrological data

The main tributaries of the Bahr el Ghazal swamps are the Jur, the Tonj and the Lol. Recent discharge measurements exist for the Jur at Wau, the Tonj at Tonj and the Lol at Nyamlel; these sites are above the areas of main spill from these rivers into the swamps where, as noted, almost all the inflows are lost by evaporation. Estimates exist for the other tributaries but these three gauging stations account for over 80 per cent of the estimated total inflow. Thus a reasonable extension and long-term estimate of the inflows may be obtained from these three records; the records of the minor tributaries are too intermittent to use directly.

Flows are available at the three sites between 1942 and 1986, but each of the records is intermittent, with a gap at all sites between 1963 and 1969. There are problems in gauging these rivers because some spill occurs at high flows, the flows are highly seasonal with negligible or zero flows during the dry season, and access has not been easy in recent years. Thus the flows of the Jur at Wau record zero flow for several months in the dry season in some years, and have gaps in the record in several dry seasons. Inspection of the records has shown that gauge levels and flows are often not recorded when flows are low, but nevertheless the bulk of the flows occur during the wet season and recession months of June to November. Examples of other problems are found in the case of the Tonj at Tonj, where two gauges were maintained which were recorded as "through road bridge" and "downstream road bridge" with different flow records. The former, larger, flows have been used in this study.

To deduce the total inflow from the three records, these were first completed as far as possible; during the recession period monthly flows were estimated for short gaps by comparison of adjoining records. During other months, which were usually in the dry season, average flows for that month were inserted for missing data provided that the measured flows registered a high percentage of the year's total; where this was not the case the year was treated as missing. By this means 36 years of data are available for the Jur, 26 years for the Tonj and 34 years for the Lol. Monthly and annual averages are summarised in Table 3. The seasonal distribution of flows are similar at the three sites (Figure 1a-c), and the annual flow series are related (Figure 2a-c).

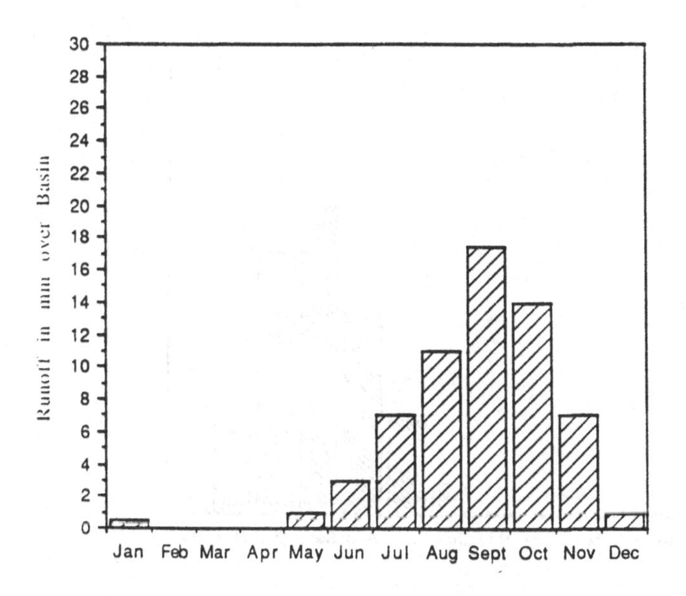

Figure 1,a Seasonal flow of the Tonj at Tonj (mm over basin)

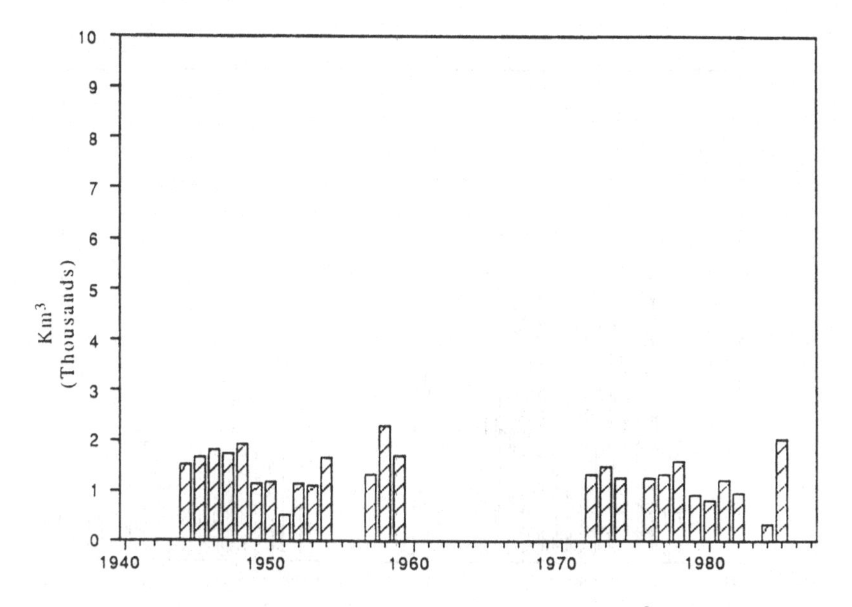

Figure 2,a. Annual flow of the Tonj at Tonj (km^3)

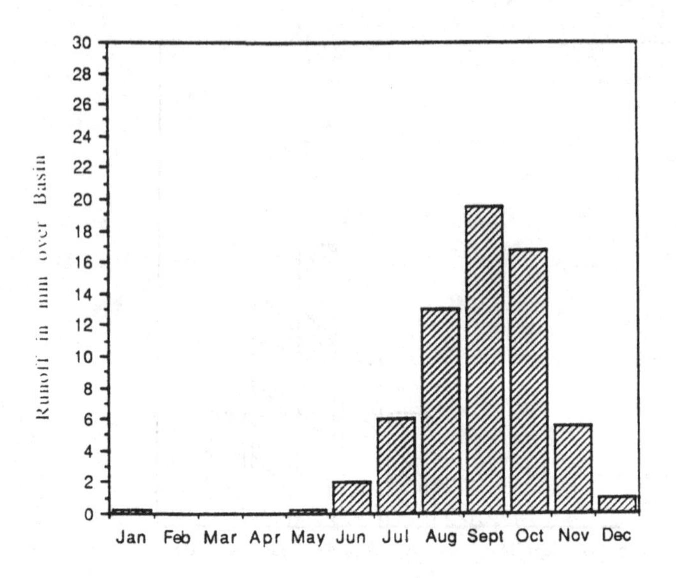

Figure 1,b. Seasonal flow of the Lol at Nyamel (mm over basin)

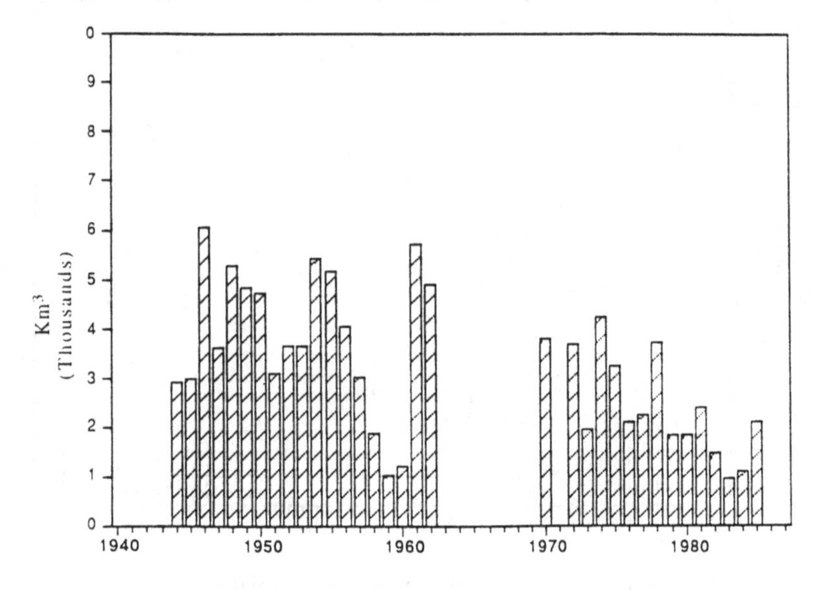

Figure 2,b. Annual flow of the Lol at Nyamel (km^3)

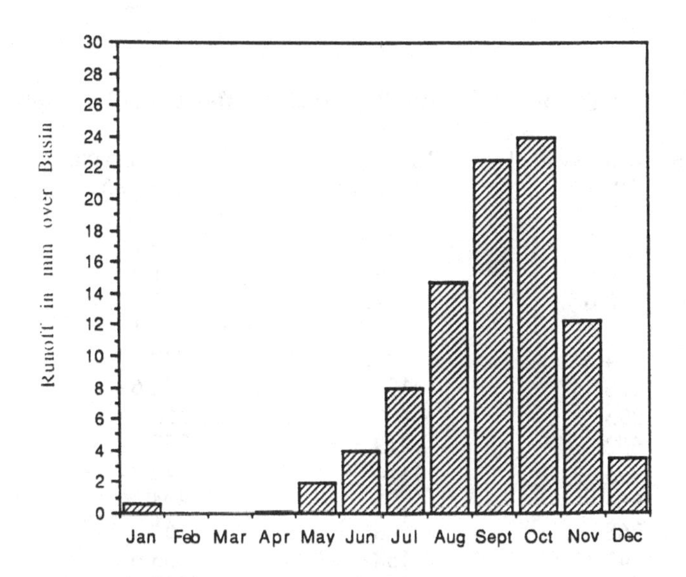

Figure 1,c. Seasonal flow of the Jur at Wau (mm over basin)

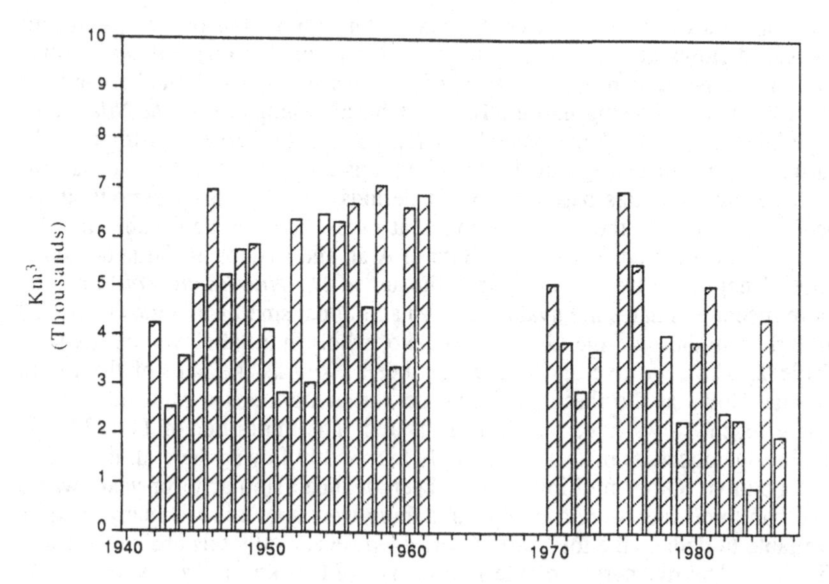

Figure 2,c. Annual flow of the Jur at Wau (km^3)

Table 3 Present estimates of monthly average flows (million m^3)

	R.Lol at Nyamel (1944-85)	R.Jur at Wau (1942-86)	R.Tonj at Tonj (1944-85)
Jan	10.2	23.6	4.8
Feb	2.0	4.4	1.3
Mar	0.6	2.0	1.4
Apr	1.6	5.6	1.1
May	14.6	85.8	30.9
Jun	106	195	74.6
Jul	293	400	158
Aug	652	719	250
Sep	995	1114	342
Oct	851	1181	304
Nov	268	608	161
Dec	49.1	158	33.6
Year	3243	4496	1363

The total inflow series was completed as far as possible by comparing the measured flows at these three sites with the estimated mean flows at other tributaries. Normal monthly and annual flows may be obtained for all the main tributaries flowing into the Bahr el Ghazal swamps from *The Nile Basin, Supplement 8,* and these have been reproduced in Table 2. Although the earlier normal discharges at the three stations are somewhat higher than the more recent estimates based on longer periods including drier periods, it has been assumed that the earlier normals at other sites are acceptable for this study. The estimated flows of the Bahr el Arab and Raqaba el Zarqa quoted in Table 1 appear to be derived from estimates in *The Nile Basin*, *Vol.V*, and no more recent estimates are available. Two additional small tributaries neglected in earlier studies are the Geti and Wokko, whose normal flows are given in Table 2. The total average inflow has been taken as the sum of the recent average flows at the three sites and the earlier normal flows at other sites, where recent flows have not been measured. The multiplier of 11.323/9.102 has been deduced as in Table 4 to apply to monthly flows where all three sites are operative. Other multipliers have been deduced similarly for years where less than three sites were available, and these were applied to as many years as available to deduce the total inflow series whose annual totals are illustrated in Figure 3. The discharges of the Lau or Yei (2.060 km^3), Tapari (est. 0.440 km^3) and Gwir (est. 0.120 km^3) have not been included in the water balance of the Bahr el Ghazal swamps as they flow towards the Bahr el Jebel, although it is unlikely that they contribute significantly to the flow of the main river.

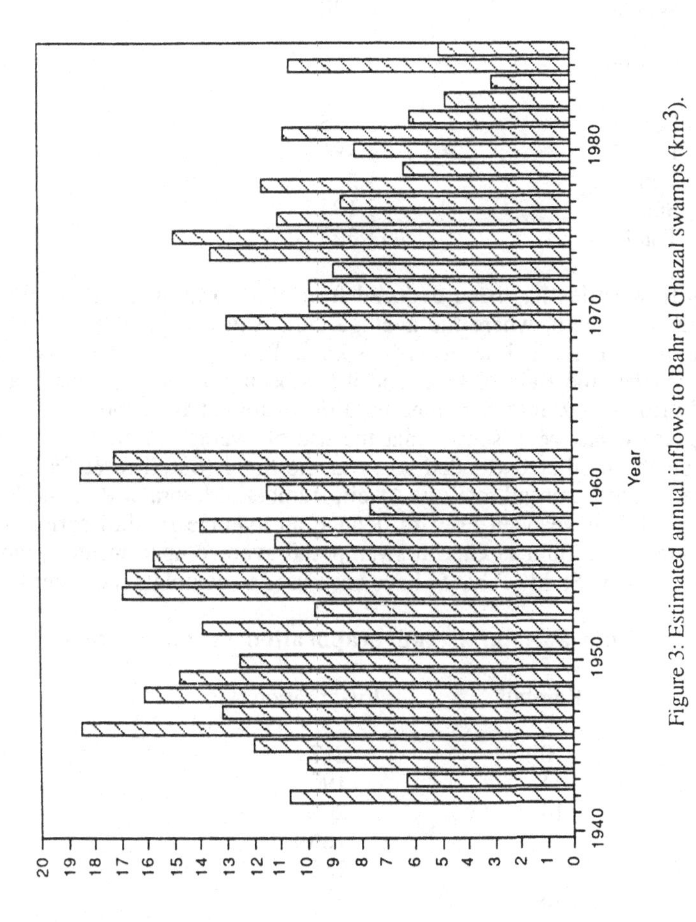

Figure 3: Estimated annual inflows to Bahr el Ghazal swamps (km^3).

Table 4 Average annual flows used to deduce multiplier

Tributary	Average annual flow (km^3)
Bahr el Arab	0.300
Raqaba el Zarqa	0.100
Pongo	0.597
Geti	0.095
Gel	0.523
Wokko	0.130
Naam	0.476
Sub-total	2.221
Lol	3.243
Jur	4.496
Tonj	1.363
Sub-total	9.102
Total	11.323

The outflow series has been deduced from the flows of the Bahr el Ghazal below Khor Doleib, where the average annual discharge is only 0.312 km^3 compared with the inflow of 11.323 km^3. Thus less than 3 per cent of the inflow reaches the Bahr el Jebel and it has been reasonable to insert average monthly flows in years where measured flows are not available.

Because it has been shown that the use of average rainfall has had little effect on the predicted flooded areas in the Sudd (Sutcliffe & Parks, 1989) average monthly rainfall totals at Aweil, Rumbek, Meshra el Req and Shambe (Shahin,1985) have been used to deduce an average rainfall series for this modelling of the Bahr el Ghazal swamp and average open water evaporation estimates for the Sudd (Table 5) have been used to complete the water balance.

Table 5 Rainfall and evaporation estimates (mm)

	Rainfall	Evaporation
Jan	0	217
Feb	4	190
Mar	14	202
Apr	49	186
May	110	183
Jun	143	159
Jul	175	140
Aug	184	140
Sep	141	150
Oct	69	177
Nov	10	189
Dec	1	217
Total	900	2150

The water balance model

A simple water balance model (Sutcliffe & Parks,1987,1989) was developed to use the fact that inflow and rainfall equal evaporation, storage change and outflow for any wetland area, and that evaporation can be estimated as approximating to open water evaporation for flooded wetland. The model also allows for infiltration into newly flooded ground, and uses an empirical observation based on surveys around the Sudd that flooded volume is linearly related to flooded area, or in other words that the average depth of flooding remains constant as flooding spreads. The model does not include a water balance for those areas where rainfall recharges soil moisture storage in grassland which is not inundated from the river, on the basis that these areas do not contribute significantly to the water balance of the inundated areas; the model may, however, for this reason underestimate areas where rainfed grassland is available during the dry season.

This model has been applied to the hydrological data described previously to give a monthly series (Figure 4) of estimated areas of flooding for those years between 1942 and 1986 when records are available. This series may be compared with the estimate of areas flooded by the Bahr el Ghazal tributaries given in *The Nile Basin, Vol.V* (1938); it is not clear whether the estimate of 14,400 km^2 refers to the average, minimum or maximum extent of the Bahr el Ghazal swamps. The results of this study give estimates of flooded areas which vary between about 4000 and 17,000 km^2, before the lower inflows of the most recent years.

Since this assessment of tributary flows and swamp modelling was carried out, we have discovered that Harris of MSSL (1991) is developing a technique for remote sensing of wetland areas using thermal infrared satellite data based on the temperature contrast between dry and inundated land. He has provided us with a typical AVHRR band 4 image which shows the outline of the wetland areas of the Sudd region (see Figure 5). It further shows the extent of the inundated floodplains of the Bahr el Ghazal tributaries, and their locations can clearly be distinguished by comparison with a map of the river system.

On the date of the image which is 30 December 1986, the areas flooded from each tributary are quite separate and relatively limited in area rather than extensive continuous swamps; it is interesting to note their correspondence with the distribution of population recorded in 1955 (SDIT, 1955). Indeed, areas flooded from each of the tributaries may be compared with estimated mean flows of the individual tributaries and the two correspond well (Figure 6) although 1986 flows were well below average. Because the date of the image corresponds with the end date of the wetland modelling, a comparison is possible. The total flooded area of the Bahr el Ghazal swamps on this date may be estimated as 4600 km^2 ; subjective interpretation of cloud and wetland outline on the image has been necessary, so that a range of estimates of 4000-5000 km^2 is realistic. The model prediction on 1 January 1987 depends on extrapolation of tributary recession flows, but the flooded areas are estimated by the water balance model as 5000 km^2. This comparison suggests that the estimated inflows and flooded areas are reasonably realistic. It is worth noting that the areas flooded from the Bahr el Ghazal tributaries were only about a third of those of the Bahr el Jebel swamps on this date.

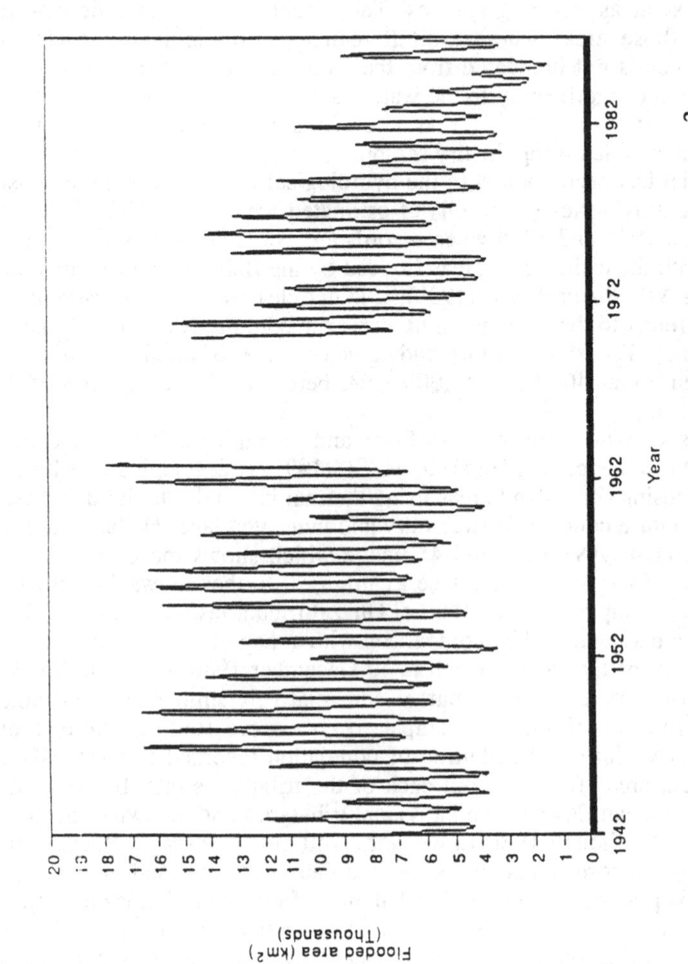

Figure 4: Estimated areas of river flooding in Bahr el Ghazal swamps (km^2).

Figure 5: Band 4 thermal infrared satellite image of the Sudd region of 30 December 1986.
(Reproduced by kind permission of the Mullard Space Science Laboratory, University College , London)

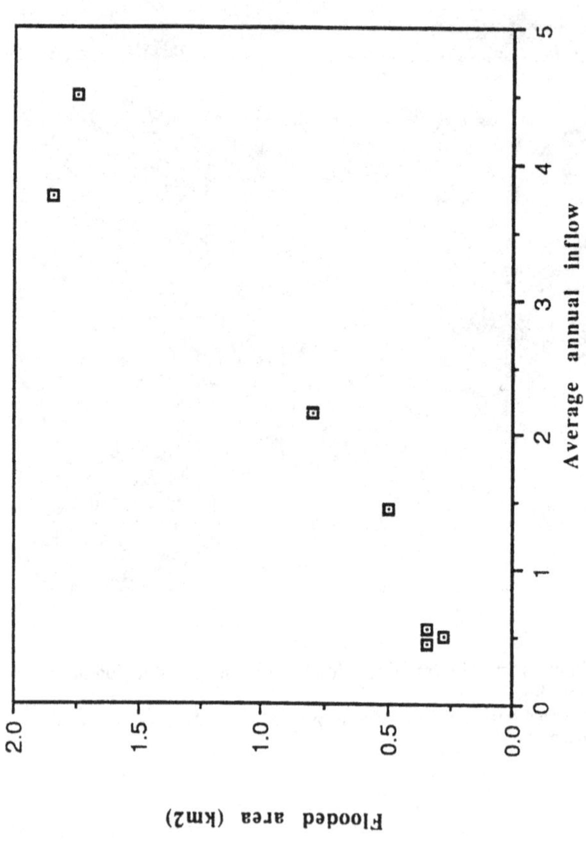

Figure 6: Comparison of flooded areas (km^2) on different tributaries. (from satellite imagery of 30 December 1986 compared with estimated average annual flows in km^3)

Further refinement of the model would be made possible by study of satellite imagery over a period of years. The range of flooded areas can be increased or decreased by altering the parameter which corresponds to the mean depth of flooding, which has been taken as one metre; this would have little effect on the average areas of flooding, which depend on total inflow and the rate of evaporation. Field investigation might show that the evaporation rate from the relatively stagnant papyrus of the lower basin is lower than assumed.

This study nevertheless provides an estimate of the seasonal cycle of flooding and uncovering of the swamps which may be compared with the behaviour of the Sudd or Bahr el Jebel swamps. It also reveals that the extent of flooding has in general decreased in recent years in distinct contrast to the marked increase in the area of the Sudd since 1961. This contrast is due to the different behaviour of the Bahr el Ghazal tributaries, where flows have decreased in line with the Blue Nile and its tributaries in the Sahel drought of recent years, while the flows of the Bahr el Jebel have been high since the rise in Lake Victoria in the years 1961-1964.

Thus, although this study of the Bahr el Ghazal regime is preliminary in the sense that further examination of the tributary flows would be desirable, and that satellite imagery could be used to calibrate the model, it has brought out important contrasts between the behaviour of the Bahr el Ghazal and Bahr el Jebel swamps.

References

Butcher, A.D. (1938). *Sadd Hydraulics.* Cairo, Government Press.

Chan Siu-On, & Eagleson, P.S. (1980). *Water Balance Studies of the Bahr el Ghazal Swamps.* Dept of Civil Eng, Massachusetts Institute of Technology, Report No.261.

Harris, A.R. (1991). *Remote Sensing of Wetland areas using infrared satellite data.* (forthcoming).

Howell, P., Lock, M. and Cobb, S. (1988). *The Jonglei Canal: Impact and opportunity.* Cambridge University Press.

Hurst, H.E. & Phillips, P.(1931). *General Description of the Basin, Meteorology, Topography of the White Nile Basin. The Nile Basin, Volume I.* Cairo, Government Press.

Hurst, H.E. & Phillips, P. (1933). *Ten-day Mean and Monthly Mean Discharges of the Nile and its Tributaries. The Nile Basin, Volume IV & Supplements.* Cairo, Government Press.

Hurst, H. E. & Phillips, P. (1938). *The Hydrology of the Lake Plateau and Bahr el Jebel. The Nile Basin, Volume V.* Cairo, Government Press.

Jonglei Investigation Team [JIT] (1954). *The Equatorial Nile Project and its Effects in the Anglo-Egyptian Sudan. Report of the Jonglei Investigation Team*, Khartoum, Sudan Government.

Newhouse, F. (1928). *The Problem of the Upper Nile.* Cairo, Government Press.

Newhouse, F. (1939). *The Training of the Upper Nile.* London, Pitman.

Penman, H.L. (1963). *Vegetation and Hydrology,* Commonwealth Agricultural Bureaux, Farnham Royal.

Shahin, M.M.A. (1985). Hydrology of the Nile Basin, *Developments in Water*

Science, **21**, Amsterdam, Elsevier.
Southern Development Investigation Team [SDIT] (1955). *Natural Resources and Development Potential in the Southern Provinces of the Sudan.* Khartoum, Sudan Government.
Sutcliffe, J .V. & Parks, Y.P. (1987). Hydrological modelling of the Sudd and Jonglei Canal, *Hydrol. Sci. J.*, **32**, 143-159.
Sutcliffe, J.V. & Parks, Y.P. (1989). Comparative water balances of selected African wetlands, *Hydrol. Sci. J.*, **34**, pp. 49-62.

Part IV

Economic, international and legal issues

14

Evolving water demands and national development options

J.A. ALLAN

Introduction

The nine states of the Nile Basin though connected by the river are not moved to relate intensely to each other. The river itself is of major current strategic economic importance to only one country, Egypt, although it will become increasingly significant to one other, the Sudan, in the near future as the utilization of Nile water reaches the maximum level sustainable by existing river flows as currently managed. The other measure of interaction and interdependence, trade, indicates that the countries of the Nile Basin have little reason to communicate. The dominant trading partners of all nine countries, for exports as well as imports, are from the industrialized economies of Europe, North America and the Far East. Nor are there any uniting elements in the field of international relations; there are no strategic groupings based on shared anxiety or mutual interest with respect to factors particular to the Nile and its catchment. The major international forum at the continental level, the Organization of African Unity (OAU), embraces all African countries and not just those with parts of their territories in the Nile Basin. It has a potential mediating role but is unlikely to be the body which steers basin-wide relations. The reason for this is the imperative that future development of Nile water will require substantial investment at levels which will be extremely difficult to mobilize from indigenous Nile Basin resources.

All countries of the basin have significant and sometimes dominating agricultural sectors which use, or potentially use, Nile water. The significance of the renewable Nile water is partly that it is an element in the suite of economic resources essential if the increasing demands for food are to be met. And the dominating feature of this challenge is the universal problem of increasing population. Nile water is also significant in other sectors of national economies in that hydro-power is being, or could be generated, in all the Nile Basin countries and power *is* already an exportable commodity.

Meanwhile the process of economic development is associated with the tendency for agricultural sectors to form a progressively smaller proportion of a national economy as industry and services expand with general economic development. It is also common for the water element of the renewable natural resources supporting agricultural production to be used more intensively, and therefore in general more effectively, since such intensive use enhances economic returns to water. However, it must be emphasized that while agriculture is the largest user of water in all countries of the Nile Basin, the agricultural sectors will play progressively smaller comparative roles in their economies in future. This crucial but little regarded factor is just beginning to become recognized to be significant in Egypt but it will eventually be the concern of other governments in the basin.

Demographic issues and food supply

All countries of the basin have rapidly rising populations and some have problems, periodically extreme, in feeding burgeoning numbers. The demographic element of the food supply problem facing the Nile countries will continue to undermine the attempts of all governments to meet food demands. For a time during the 1970s Egypt's rate of population growth dropped to under 3 per cent per year, but it has risen again and elsewhere in the basin the rates of population growth are well above three per cent. The possibility of achieving a significant reduction in the rates of population growth in the next two decades is not high and the consequent economic burden will haunt the political leaderships of the Nile Basin countries. Demographic trends will continue to be matched by increasing levels of food demand and in political terms reductions in the levels of population growth would be the single most beneficial development in terms of addressing problems of feeding populations, as well as in enabling sustainable and economic water use policies. Reductions in the levels of population growth would also be the single most beneficial development in the political arena through reducing current and future internal national, and international, tension over Nile water availability.

Egypt continues to feed its population but only by regularly importing a substantial proportion of its food staples. The Sudan has sufficient natural resources to feed its numbers but as recent history testifies natural and political events in parts of the vast country, mainly beyond the reach of Nile water, have devastated food production systems so that emergency food has had to be imported on numerous occasions during the 1970s and the 1980s. The countries which currently use significant volumes of Nile water for agriculture, Egypt and the Sudan, use just over 50 billion cubic metres per year, or about 60 per cent of the total flow of the lower Nile system representing a volume close to the maximum utilizable for agricultural production. The other 40 per cent of water is only available to a minor extent for economic use. About 20 per cent necessarily flows to the Mediterranean Sea to flush out the heavily salinized water at the end of the system (See Chapter 8), and a quantity variously estimated between ten and 15 per cent is lost by evaporation and seepage at Lake Nasser. The remaining water, only about about 7-8 per cent of the total flow, is currently used for urban and industrial use. Because the perceived economic returns from urban and industrial uses will always be judged to be higher than those from agriculture, a number of Nile Basin countries are likely to follow the example of Egypt, the largest Nile water user, and reduce allocations to agriculture in the decades ahead.

The demand for water by non-agricultural sectors of the national economies of the states of the Nile Basin will increase steadily if not dramatically. But since the overall water demand is already close to the maximum level sustainable, the options for agriculture in the lower Nile Basin are few and as far as the downstream countries are concerned they will have to be concerned with increasing water-use efficiency rather than with expanding agricultural water-use. Egypt has been following this strategy for many years while at the same time searching for upstream co-operation to increase the volume of water reaching Egypt.

The conflicting trends of increasing food demand and of reducing water availability for agriculture exacerbate the tension attending the evaluation of downstream water supplies. The first trend, increasing food demand, is fuelled by rising populations but is mainly driven by changing patterns of consumption enabled by increasing standards of living. The second trend, declining water availability, is the one which brings international repercussions in that in the past Nile water has been adequate to meet existing and emerging demands for water by the agricultural sectors of the Nile Basin countries. This is no longer the case.

The agreement reached between Egypt and the Sudan in 1959 was intended to secure Egypt's water for the foreseeable future and at least into the twenty-first century. In the agreement there was provision both for the security of the flow to Egypt through monitoring arrangements by Egyptian engineers on the Blue Nile and elsewhere, as well as the construction of jointly funded structures such as the Jonglei Canal. (Nile Waters Agreement, 1959, third provision) Meanwhile Egypt has already adjusted to deficiencies in water availability for agriculture (See Chapter 8). It has strictly limited its current and future water allocations to the 55.5 billion cubic metres recognized by the Sudan and it has during the past three years actually cut the supply of water to some crops and areas. It has also substituted capital and international assistance for water by arranging to import food. And it is predicted that Egypt will mainly meet its increasing food requirements from imports rather than from existing or new Nile water. The Sudan on the other hand will be able to increase agricultural production in so far as it will use its remaining 'share' of Nile waters according to its 1959 agreement with Egypt. The Sudan has adequate renewable natural resources to meet its current and future food demands but it will require dramatic changes in its economic and social infrastructures, as well as in its national security, if it is to fulfil its agricultural potential and thereby feed its population.

The extent to which Nile water is essential to securing the future food supplies of the other seven states varies considerably both in scale and imminence. Ethiopia is the country which could use from the Blue Nile alone between six billion cubic metres (US Department of the Interior 1964) and 12 billion cubic metres or km^3 (Abate, 1991) and while it is currently contemplating using the waters of the river known in the Sudan as the Sobat, the scale of Ethiopia's use of water will be modest and is unlikely to be significant for over twenty years. The impact of water development in the other states of the basin is negligible, mainly because their withdrawals would take place at points in the catchment south of the Sudd of Sudan. From these marshes the depletion of Nile water by evaporation and evapotranspiration is the greatest single 'consumer' of water(See Chapter 12 p.) after Egyptian use for agriculture. Losses estimated at nineteen billion cubic metres per year far exceed estimated losses at Lake Nasser (ten billion cubic metres per year) as well as the current total annual water use of the Sudan. Any reductions of flow to the Sudd which might be caused by abstraction in the upper riparian states in the White Nile catchment are therefore less significant than those caused on the Blue Nile.

In brief the problem of food supply for the countries of the Nile Basin, while being related to that of Nile water supply, is by no means determined by the availability of water. Egypt has already had to allocate other resources generated in other parts of the economy, or use international political credits mainly with the United States, to provide a secure supply of food and this is the way that the problem will continue to be solved for the foreseeable future. It will be students of political economy rather than those of environmental science or market place economics who will be able to explain postures and policies. At the same time it is essential for those attempting to provide explanations in terms of the political economy of internal and external Nile Basin relationships to recognize that the political leaders of the Nile countries and their advisers have a very sophisticated knowledge of the latest scientific views on climate change, the probable developments in the rate of water consumptive use in the agricultural sector and the costs of developing irrigated land. All of this makes the national water policy and international relations of the Nile countries on occasions impenetrable as well as compellingly interesting.

An economic factor which will have a major impact on the rate at which food production based on Nile water will go ahead is the sheer cost of implementing land reclamation. It is impossible to generalize about the true 'economic' cost of reclaiming a hectare of irrigated land throughout the basin as climate and other environmental factors vary so much. Also the perceptions and expectations of rural communities vary immensely and this affects their propensity to become involved in and even more importantly to succeed in the very difficult enterprise of reclaiming land (Allan 1983, p.247.). If farmers are to be successfully induced to live in what are regarded as relatively remote reclamation schemes elaborate social amenities must be provided, involving investment in transport, education and health services. The consequence of omitting these services has been that schemes have been abandoned or run ineffectively. The sheer cost of providing the adequate agricultural and social infrastructures requires the investment of well over US$25,000 per hectare on farm investment, suggesting levels of investment of US$25 mn per 1000 hectares. Since countries are talking of reclaiming tracts of more than 500,000 to one million hectares, investments of US$12 bn to US$25 bn are implied. These sums would only cover local minor water distribution systems and on-farm investments. In addition there will be substantial civil works to construct and the major engineering infrastructure for regional water distribution. No country of the region can from its own resources service such investment flows.

The consequences are numerous in that the problem has its internal and external dimensions. Internally the pressure on scarce investment resources leads to conflict between spending agencies of governments. Choices have to be made between investment in agricultural expansion and other productive sectors as well as between such investment and investment in measures to improve the social infrastructure. Externally the problem is expressed in the relationships of the Nile countries with international agencies and in the myriad of multi-lateral and bilateral government relationships in the world of international assistance. All countries of the Nile Basin receive international aid and are substantial debtor economies owing large sums to or are deeply involved with the World Bank and the International Monetary Fund and as a result of the unwise policies of the 1970s also to the international departments of banks throughout the Western world depending on the extent to which such debts have been cancelled. Future agricultural investment on the scale envisaged to meet the food gap requires a commitment to the reclamation of land and is a policy for which the officials of international banks have shown very little enthusiasm for over a decade, and it is very unlikely that capital will be mobilized internationally on the scale required to bring into production the amount of productive irrigated land needed to service the growing food requirements of the Nile Basin countries. This was the position before the events of late 1989 in Eastern Europe which significantly altered the options available for the allocation of Western capital investment and internatonal assistance, whether such allocations were straightforwardly economic or influenced by humanitarian considerations. These new allocations have tended to reduce the share of assistance going to Nile Basin countries.

Security

The option to invest is not just affected by the availability of indigenous and/or international investment resources, nor by the presence of national commitment nor even by the ability to mobilize international assistance. In two of the largest countries of the region, both with a potential substantial share of Nile water, Ethiopia and the Sudan, political circumstances have been very disturbed for more than a decade. In these countries enterprises requiring undisrupted politics and a stable economy cannot flourish, and new initiatives are hazardous and usually fail, especially when they are rural based and require a sustained commitment from the local farming community. The insecurity which afflicts major parts of the Nile Basin has economic as well as political implications.

The economic basis of the development and further development of Nile water

As all the countries of the Nile Basin are short of investment resources it is relevant to review the irrigation development potential and prospects of the major potential users of water. The issue can be addressed in terms of food needs, water availability and the resources needed to fund the development of such agricultural potential. The following data are based on a very rapid assessment of the current extent of irrigated farming in the Nile Basin countries supported by Nile water. Only in the case of Egypt are the data based on published plans and government policy. It is to be expected that the aggregated intents of the individual countries will be shown to exceed available resources. This is the nub of the problem to which this and other chapters relate.

The economic context

The countries of the Nile Basin are all deficit economies facing difficult challenges with respect to economic development. Only Zaire and Uganda have had positive trade balances in the last decade. The largest and most populous countries, Egypt, the Sudan, Ethiopia and Tanzania, have negative trade balances approaching, or in excess of, 50 per cent. Meanwhile only Egypt has an economy of a sufficient size to finance major capital works such as are essential preliminaries to land reclamation and afterwards the additional expenditure needed to provide the engineering and social infrastructure to sustain a stable and self-sufficient rural community. Even so the land reclamation costs alone would be beyond the competence of the Egyptian economy, requiring an amount equivalent to over half the annual GNP of the late 1980s. Because the Egyptian economy does not have the capacity to meet this level of investment Egypt has found it very difficult to escape dependence on international agencies and other bilateral sponsors to fund its land reclamation programme with all the attendant frustrations. Economies enjoying annual GNPs of more than ten times that of Egypt find it difficult to mobilize the levels of investment required to extend Egypt's irrigated area. The United Kingdom has found it difficult to mobilize the both the political will and the much more modest number of billions of dollars required to build the Channel Tunnel and related infrastructure out of its private and public resources which in gross terms total over US$600 per year. By contrast Egypt has mobilized the political will many times over; it only lacks the means to implement this political will.

The other countries will be quite unable to finance civil works and agricultural infrastructure investments based on Nile waters without the support of international financial investment. It is well understood by the governments of those countries which have not made much use to date of Nile water that to develop it with international money will bring constraints. Outside agencies take a broader view than national bodies and the experience of Turkey in its control of the Euphrates over the past decade is relevant in this regard. This experience shows how the relationship of an ambitious upstream riparian can evolve with its major donors. Turkey has now completed two of the major structures on the Euphrates, those at Keban and and Karakaya. These dams were built with the assistance of World Bank investment and with World Bank influence on the way the work was specified and scheduled. The Turkish authorities found this so irksome that they took the financially difficult decision to fund the third and largest of all the structures, the Attaturk Dam, from national resources. Turkey's ability to take this decision could not be matched by any of the Nile upstream states, but Ethiopia at least has indicated its awareness of the unconstructive tension which attends the involvement of international agencies in both river basin monitoring and planning as well as in the design of water management systems.

The major relevant economic indicators shows very clearly how the upstream riparians on the White Nile have relatively minor investment options but even these will be impossible to fund internally in the short term (within the next two decades) and the role of international funding bodies will be as important on the East African plateau as they will be in Ethiopia, certainly in the short term and probably also in the medium term.

The political context

It has been indicated above that the Nile Basin countries have few economic links other than the background concern over the regular arrival of water across the border from their upstream neighbours. There was a time during Britain's very brief imperial interlude in both the populous and heavily irrigated downstream Egypt as well as in the hydraulically virtually undeveloped East African plateau territories that engineers and administrators from the imperial power foresaw dangers for Egypt if the hydraulic potential of the upper Blue and White Nile and their tributaries were to be developed. In the unique position in which they found themselves the British concocted a series of agreements which constrained the actions of their own colonial authorities in the East African territories to take no measures which would reduce the flow of water to the north. Attempts were also made to gain the acceptance of the colonial government of Ethiopia of the same principle (See Chapter 17).

Not since the 1950s when most of the Nile Basin countries gained independence has it been possible to contemplate a co-ordinated approach to the management of Nile waters. The arrangement of the Permanent Joint Technical Commission between Egypt and the Sudan after the 1959 Nile Waters Agreement was a more apparent than real example of the co-ordination of water planning and use. The reality of the arrangement was only tested in the 1980s when flows were reduced and the use by the Sudan began to move to within 20 per cent of its 1959 agreed allocation and beyond the volume of water being conveyed during the low flows of the Nile in the mid-1980s. (PJTC 1989)

The 1988 flood removed the accumulating pressure which had been evidenced by the falling volume of water stored in Lake Nasser and by the anxiety of both Egyptian and Sudanese officials and politicians faced with the painful adjustments required to reduce the use of water in agriculture and to cope with the imminent termination of the generation capacity at Aswan. Hydropolitics became palpable and not the relaxed relations of the 1960s and 1970s. Nor can they ever again be relaxed with respect to water for the demographic, food needs and economic reasons discussed above.

Are there any other factors which will affect or determine the relationships of the Nile Basin states? Are there existing links or discontinuities which would affect inter-state relationship and are there natural or historical factors which would encourage deals to be made between the governments of the post-independence Nile Basin states? The answer to these questions would seem to be no. The Islamic dimension is one which might seem to encourage an affinity between Egypt and a majority in the Sudan. However, Egypt is inspired by both secular and Islamic traditions and as a consequence it is difficult to identify any symmetry betweeen the two governments as the emphasis changes from year to year. But it is clear that despite the Government of the Sudan's generally fundamentalist position it is national interest which informs the approach of the Sudanese officials and engineers when discussing water planning and allocation and not cultural sympathy.

The separateness of Ethiopian history, attitudes and expectations from those of the states downstream of the rivers which rise in the Ethiopian highlands extends over many millennia and the events of the past decades have done nothing to change the position. Ethiopia's officials and people see no precedents which would guide them to recognize obligations to downstream states. They are pleased that their water development schemes would not necessarily impair the flow of water to the north, and they see no reason to impede the development of water to secure the position of the Sudan and Egypt. Similar attitudes obtain in the East African highlands, but here the consequences of water resource development are less likely to have a significant impact on downstream states.

Have the downstream states any significant leverage on the upstream states which would make them amenable to recognizing the interests of the Sudan and Egypt? Such leverage could be economic, military, political, or surrogate. Clearly neither the Sudan nor Egypt have any economic leverage as they are themselves in economic difficulties. The Sudan is involved to the point of exhaustion with a civil war and could not for the foreseeable future muster a force to defend its hydraulic interests. And while Egypt has considerable military strength, and this strength is no longer tied down confronting Israel, it has not the capacity to embark on campaigns in remote parts of the Nile Basin especially as the returns from such a military investment would be far outweighed by the sheer economic cost, not to speak of the immeasurable political costs of such a venture.

Nor have the Sudan and Egypt any political leverage over the upstream riparians. There are no alliances whether under the umbrella of the OAU or the UN which will with any certainty bring outcomes which they would regard as equitable. The most likely sources of influence are surrogate. The international community has influence because it can determine the flows of economic assistance on which all the countries of the Basin are dependent, in some cases for periodic emergency relief, and in all cases for development spending. The consequent relationships are not welcomed by the Nile Basin governments but are seen as an inescapable part of the future economic and political scenario. The reservations are more explicitly expressed by some governments than others. The Ethiopian Government is probably the least biddable but even it has in the recent past shown a capacity to become more approachable.

Such approachableness is stimulated by the protracted and economically and militarily debilitating civil war centred on the northern provinces of the country. Without external military support the struggle could prove politically extremely embarrassing and while there may be a temptation for countries such as Egypt to exploit the predicament of the Government in Addis Ababa any short term gains would be outweighed by the longer term costs of such patronage which would in the end be interpreted as interference in the internal affairs of Ethiopia. Similarly support for the opposition groups in the south of the Sudan would compound the confusion in the Sudan, and while the confused politics in Ethiopia are to the advantage of the users of water in Egypt and the northern Sudan in that while the civil wars continue, major hydraulic civil works are impossible to implement, the disruptions in the southern Sudan delay the development of the Sudan's hydraulic potential. But the disrupted politics have many consequences and evaluating their positive and negative features are not easy. It could as easily be argued that the major consequence for Egypt of the protracted conflict in the south of the Sudan is the non-completion of the Jonglei projects which would have brought to the agricultural sectors of both Egypt and northern Sudan two or more billion cubic metres of water per year. Such quantities, though small, have disproportionate significance because they come on top of hard pressed existing water budgets.

Politics also affect the mobilization of basin-wide meteorological and hydrological monitoring and the early warning systems which they support. The impetus for such initiatives comes from the downstream states. The upstream riparians do not have the same problems as the downstream users and do not see the urgency to set up such projects. At the same time they do not see the over-arching scientific institutions as politically benign, but rather as an insidious reflection of the interests of the downstream users supported by principled international civil servants more aware of trans-national than national interests. This suspicion is reinforced by awareness that some of the major international players and sponsors, USAID and the World Bank, have Egypt as a deeply dependent client. And it is difficult to envisage such sponsors doing anything to worsen the positions of their deeply indebted clients. Viewed from upstream Ethiopia it is difficult to discern objectivity in the approach of the international agencies and the major bilateral lenders and donors. Thus the urgency with which USAID and the UN agencies invest in the creation of basin-wide monitoring and water planning institutions proves the bias of the alliance of the downstream states and the outside agencies and puts them in conflict with the priorities of the upstream riparians.

The time factor

The differing approach of upstream and downstream states to Nile water development is very clearly epitomized by the urgency with which the respective governments approach water planning and use. The downstream states are deeply concerned and give the Nile issue a great deal of attention in their Ministries of Foreign Affairs and in the relevant ministries concerned with water management and land development. By contrast the governments of upstream states accord meetings about international rivers a much lower priority and do not see the problems as urgent because they do not have the resources to develop water and even if they did the engineering lead times are such that the economic impacts of the implementation of major water management schemes would not be evident for another decade or more. Meanwhile the mobilization of capital to initiate these major works must remain for the foreseeable future in the world of fantasy.

It will take some years after the achievement of domestic accord before Ethiopia will have the political will as well as the means to give priority to water planning and development. The other upstream riparians do not have to end civil wars to begin a consideration of water development, but they do lack the will to address the issue with urgency because of the absence of the means with which to make the relatively massive investments.

Conclusion

Nile waters appear to have a convenient unity. And if politics would allow the overall resource to be considered as a whole then a number of economically rational and environmentally sensible decisions could be made which would maximize the returns to the limited water resource of this international river. The different riparians variously esteem the resource and because of their differing immediate and potential dependence on the Nile flow they are articulating their needs in very different ways and with equally different levels of urgency. The international community is not regarded as benign, whether for reasons of imperial taint or the biases deriving from the entanglement of hopelessly deep client-patron relationships. The possibility of a set of co-operative institutions evolving in the short term from the current unpromising circumstances is remote, but provided the climatic factor does not prove to be a progressively negative feature of the hydrology of the Nile Basin then it is possible that Ethiopia and Egypt will be able to adjust their water management policies so that there is no disenhancement of the flow of the Ethiopian tributaries into the Sudan. If Ethiopia is to be persuaded to take an international view of its Nile waters it behoves Egypt to continue to signal that it expects to budget its water allocations from the lake in the way that it did in the 1980s when releases were reduced in 1987 and 1988 in line with the level of storage in Lake Nasser. If, however, the flow of the river is affected by a significant and continuing reduction similar to that experienced in the 1980s when it was between 10 and 20 per cent below the long term average, then the likelihood of amicable adjustment will be remote and the necessary sacrifices will be internally unacceptable. In these circumstances the pressure on the United States, the only agent which has had the resources to enable Egypt to overcome the even greater economic problems which it has faced since the late 1960s, will intensify enormously to provide the resources which will substitute for the absent water.

References

Abate, Z. (1991). Planned national water policy: a proposed case for Ethiopia. Paper presented at World Bank International Workshop on Comprehensive Water Resources Management Policy,Washington D.C.

Allan, J.A., (1983). Natural resources as national fantasies, *Geoforum*, 4:3, pp 243-247

Nile Waters Agreement, (1959). *Agreement between the Republic of Sudan and the United Arab Republic on the full utilization of the Waters of the*

Nile, Ministries of Foreign Affairs of Egypt and Sudan, Cairo and Khartoum.
PJTC (Permanent Joint Technical Commission), (1989). Data on water use by Sudan and Egypt quoted by Chesworth, Chapter 3.
US Department of the Interior (1964). *Land and water resources of the Blue Nile Basin: Ethiopia*, Washington DC

15

Nile Basin Water Management Strategies

J.A. ALLAN

Introduction and background

The Nile Basin is one of the most studied natural resource systems in the world. The hydrological studies and basin-wide reviews completed in the past were produced under different historical and regional circumstances. Some of the most comprehensive by Egyptian and British engineers (Egyptian Government (Macdonald) 1921, McGregor 1945, Morrice and Allan 1958, and Hurst 1933, Hurst et al 1946, Hurst 1952) were prepared when a single power, Britain had significant influence over five of the Nile Basin states and as a result a more comprehensive and integrated approach was taken in these reports and publications than has been possible since.

With the change of government in Egypt in 1952 and the cataclysm in international and regional relations following Suez in 1956 a new approach to resource security was adopted by Egypt which involved the building of a major water storage structure, the Aswan Dam at its southern border. This was completed with Soviet support and Egypt continued its Soviet orientation until 1973 when Sadat turned to the United States for practical economic and strategic support. The building of the Aswan Dam was not just a feature of water management, however, it was also very significant in terms of future political relations in the Basin.

At first sight the structure may seem to be of importance only to Egypt and Sudan, the main users of Nile water in the mid-1950s and still by far the major users. However, its existence has ongoing political and possible legal significance and far from being a permanent feature, it will have to be seen as a negotiable asset if in future the best use is to be made of Nile flows.

The creation of the Lake at Aswan was only possible because it followed detailed negotiations between Egypt and Sudan. These negotiations were based on assumptions concerning the average flow of the river, as well as legal precedent concerning the division of that average flow and of any future water that might be gained from drainage schemes in Southern Sudan. Ethiopia declined to particiapte in the negotiations and has at no point recognized the Nile Waters Agreement of 1959. Principles of prior use were proposed by Egypt and accepted by Sudan which led to the agreement that the existing estimated average flow should be shared with three out of four parts going to

Egypt and one part out of four going to Sudan. Future water resulting from equally shared investment would, however, be shared equally. Egypt had to move from the idea of prior use to equal shares for additional water in order to gain the agreement of Sudan and in addition Egypt had to agree to finance the dislocation costs of the Nubian farmers displaced by Lake Nasser/Nubia which flooded many scores of kilometres of the Nile valley at Sudan's northern border.

The Aswan Dam has been a remarkable engineering and economic success in protecting Egypt from flood and drought. It was well worth the modest price paid by Egypt to gain Sudan's agreement to the project. In one regard, however, the dam will be of questionable effectiveness in future in that while the evaporation losses were bearable so long as there was sufficient flow in the system to support Egyptian agricultural and power demands, now that Sudan is using close to its one quarter share of the flow and Lake Nasser's/Lake Nubia's storage is no longer augmented by the unused share of Sudan, both Sudan and Egypt want to find additional water. Further, the 1980s were years of serious drought in the Sahel and Ethiopia and the flow of the Blue Nile, previously the provider of 86 per cent of the annual flow of the river beyond Khartoum, has been 20 per cent below the long term average (51.7 cubic kilometres per year) since the early 1970s (43.4 cubic kilometres per year 1973-86).

Because the political situation has been so conflictual in the south of Sudan throughout the 1980s the water management structures associated with the Jonglei Canal have been vetoed by Southern interests. Almost three quarters of the Canal has been constructed but it is unlikely to be competed in the forseeable future. The important principle highlighted by the situation in Southern Sudan is that the interests of those living in tracts with a water surplus should not be underestimated, or worse, ignored.

Political strife has also affected Ethiopia both in terms of security and its economic development. Ethiopia has been unable to develop its huge power potential and to irrigate from its Nile tributaries in its remote north western region. The development of the necessary civil and irrigation works will require a decade or more of effort and investment. In order to bring them about Ethiopia's economic situation and its economic and political relations externally will have to be transformed and if the developments are to take place with the approval of the downstream states, with the downstream riparians feeling that they have achieved some reciprocal benefit, then it will be necessary for the states involved to devise a framework for evaluating regional water budgets and the benefits and disbenefits of upstream development in both economic and resource security terms.

Climate scenarios

Current scientific opinion is pessimistic on the likelihood of the long-term relevance of the assumptions made in the mid-1950s concerning the average annual flow of the Nile. These assumptions are of political significance because Egyptian engineers used them when planning the capacity of the High Dam. The assumptions appeared to have been under estimates up to the end of the 1970s at which point engineers were concerned that an overflow structure was needed to enable the release of seasonal surplus water. The 1970s did, however, prove that the storage at Aswan was crucial and did protect Egypt from drought more than once in the 1970s. In the 1980s, however, the deficiency of the Ethiopian tributaries together with the reduction of White Nile flows after their unusually high levels of the 1960s and 1970s exposed Egypt to shortages of water unprecedented since the completion of the High Dam.

Climatologists have not yet been able to explain the processes which lead to variation in rainfall and temperature as they affect the Nile Basin countries. (Hulme 1990, p71, Said 1990, p5) The consensus appears to be that the levels of rainfall enjoyed during the first half of the twentieth century were high compared with those of previous centuries and it was unlikely that the annual average Nile flow of 84 cubic kilometres would prevail in future. Future water planning, national and international, should be carried out on the basis of lower water flows than in the past.

In the light of this probably worsening climate scenario the following consequences can be deduced:

- It would seem appropriate for Egypt to anticipate flows of 55.5 cubic kilometres for the immediate future but Egypt should anticipate reduced flows from the end of the century unless the schemes in the southern Sudan can be completed and others constructed to yield new water for the Sudan and Egypt. The reduction in available water will result partly from the greater utilization of water by the Sudan as it takes up its share agreed with Egypt in 1959 and partly from the likely continued reduced flow in the Ethiopian tributaries.
- **The Sudan** will be able to utilize more water partly because it has not yet developed the volume agreed with Egypt. Meanwhile because of the problems of bringing the southern Sudan drainage schemes into effect and because of the anticipated reduced flows in the Ethiopian tributaries it is unlikely that the Sudan will have access to the 18.5 cubic kilometres which it has been planning to use since the late 1960s.
- **Ethiopia** will not be affected by the anticipated reduced annual flow in the streams which rise in the Ethiopian highlands. Ethiopia small volumes of water currently either for power generation or irrigation and when it does in due course the volume used will only be a small proportion of the water which it currently exports. It could be that there will be no reduction in annual average flow through the construction of storage systems because of the benefits deriving from the ability to store the seasonal surpluses of high rainfall years in locations where rates of evaporation are low. It is important to note that with respect to Nile resources Ethiopia will not be affected by any reduction in flow through declining average annual rainfall. Ethiopia has, therefore, a very different set of priorities from its downstream neighbours. It should be noted that in national terms the Nile resource could be seen as potentially compensating resources for those tracts of the country adversely affected by the deepening droughts experienced

for example in the 1980s.

A worsening climate scenario with respect to the part of the catchment which lies in **the East African highlands** is not anticipated to be so significant as that affecting the Ethiopian highlands. The impact of climate change with respect to the White Nile tributaries is expected to be of less significance than those deriving from the probable lower rainfall levels in Ethiopia because:

- the amplitude of any variation in flow from the East African Highlands to Sudan are reduced by more than 50 per cent by the transmission and evaporative losses from the Sudd swamps.

At the same time the fact that there has been no correlation between climate events in Ethiopia and those in the East African Highlands, (Evans 1990, p. 39, Hulme 1990, p. 79) means that the impact of White Nile flows cannot be calculated as potentially compensating for the deficiencies in Ethiopian tributary flows, even if in the 1960s and the 1970s this is precisely what they did provide.

The problem with the influence of future climate is that it appears to be a progressively deteriorating factor rather than a progressively enhancing influence. All systems whether ecological or economic are disproportionately affected by such constraints. They aggravate already difficult development circumstances in that even more initiative, imagination and ultimately tangible resources than those currently deployed have to be mobilized to achieve beneficial combinations of natural and investment resources. Every cubic kilometre of water which does not arrive through the system as a free and legitimate entitlement, or at least as a free good, has to be gained by negotiating other economic or strategic assets. Where the legitimacy of an entitlement is in dispute then even existing allocations come into question and prices may have to be paid for a resource which has previously been apparently secure according to precedent.

Water management scenarios

The current scenario

The current management of Nile waters is one based on the views articulated by British imperial interests in the first half of the twentieth century which were basin-wide and based on the assumption that Egypt's interests were paramount. These ideas have been augmented and given partial substance by Egypt's policy since its Revolution to protect its supply of water not by taking an integrated basin-wide approach with storage and transmission works optimally sited, but through total control at Aswan to maximize autonomous control and therefore national security even if evaporation and seepage losses had to be endured. These losses have averaged the anticipated ten cubic kilometres per year but have ranged from over 13 cubic kilometres in the late 1970s, when the lake was full, to about 6 kilometres in the late 1980s when live storage in Lake Naser/Lake Nubia fell to little more than 10 cubic kilometres. (Evans 1990, p. 36 and personal communication) The international agreement which determined many features of the current scenario was the 1959 Nile Waters Agreement. (United Nations, 1962)

The current scenario is seen as a reasonable basis for future discussions by Egypt provided that the Aswan storage remains a feature of Nile management. All the other states regard Aswan storage as useful only to Egypt; the Sudan is the only state which feels any obligation to regard the structure as a permanent feature of Nile management as a result of the 1959 Nile Waters Agreement and the existence of the Permanent Joint Technical Commission (PJTC).

With respect to the middle reaches of the river and its tributaries the current scenario includes major political problems with a civil war in the southern Sudan and political instability in Ethiopia. These circumstances have effectively prevented any water resource development in these areas and in the case of the southern Sudan have demonstrated how political interests can effectively veto the construction of water management structures.

In the upstream countries of East Africa the current scenario is one of relative inactivity. Any plans for water resource development are modest and would not affect the volumes of water leaving the lakes of the highlands by more than an annual average of 2 or 3 cubic kilometres per year. As indicated above this diminution of flow would be drastically reduced in its effects through the consequences of evaporation in the swamps of the Sudd.

Alternative Ethiopian tributaries scenario

One of the two most significant scenarios with respect to effects on the flow of the river would be the development of major works on the Ethiopian tributaries for the generation of power. It has been shown that Ethiopia has a huge potential hydro-power resource. (US Department of the Interior, 1964) The potential for irrigation is less clear and the tracts suitable for development are not accessible and would have to bear significant development as well as operational overheads as a result of location, but it seems clear that Ethiopia could utilize 6 cubic kilometres of water in irrigating newly developed land. The management regimes for optimizing power and irrigation use of water are rarely in accord. The former requires a generally steady pattern of flow, while agricultural use requires that water is released when there is little or no rain and this last gives the releases a very seasonal emphasis.

The benefits of storing water in the high and cool sites in the Ethiopian uplands would enhance the water available for both Ethiopia and its downstream neighbours. Evaporation losses amount to little over one metre depth per year in Ethiopia compared with over three metres depth at Aswan - the highest actual and potential rate of evaporation in the world. The US study (1964) calculated that the storage works required to generate power would enhance the average annual export of water by Ethiopia. No detailed information is in the public domain, however, which models the regime which would result from water management for irrigated agriculture. Irrigated tracts in the high temperature tracts of north-western Ethiopia would require applications of water of at least 12,500 cubic metres per hectare and if Ethiopia were to contemplate the development of 500,000 hectares (Egypt currently cultivates almost three million hectares and irrigates almost all of them for two crops per year), Ethiopia would require 6.25 cubic kilometres of water. After passing through the soil profile some of this water would return as degraded water to the Nile via natural and engineered drainage systems. However, it is clear that the dimunition in flow would not be fully compensated by Ethiopian storage. It would only be possible to compensate the states downstream from Ethiopia by arranging to save a proportion of Aswan water in Ethiopian storage works. Even the experience of the past twelve years has shown that evaporation losses at Aswan can be halved if the volume of water stored there is reduced.

It is clear that there would have to be a significant shift in Egyptian expectations if the scenario suggested here were to be recognized as a secure alternative to the existing system. Such a shift could be brought about if Egypt were to anticipate reductions in flow imposed unilaterally by Ethiopia in meeting its irrigation water requirements and without any recognition of Egypt's needs during a period of filling new storage structures. Far better would be a preliminary agreement concerning the management of the system to maximize the benefits and minimize the disbenefits to all parties, including the Sudan, as the result of the development of Nile waters in Ethiopia.

For the foreseeable future potential changes in the volume of flow from Ethiopia would be relatively small, in the order of two to four cubic kilometres per year, if Ethiopia were to mobilize investment in water management. Such volumes would be significant, however, in that Egypt is already short of water and is finding it difficult to engineer new water to irrigate the tracts which its proposes to reclaim from that which arrives each year at Aswan. A reduction of even one cubic kilometre would be significant to Egypt.

Alternative Southern Sudan scenario

The Sudd and its surrounding grazing areas provide livelihoods for a number of tribes which make up the state of the Sudan. Activist elements of these peoples have indicated their resentment at not being consulted by those who devised and initiated the Jonglei Project and have prevented the completion of the project. The scheme remains the only major potential source of new water to extend irrigated agriculture in the Sudan and Egypt. The PJTC devised the Jonglei I Project to enhance the flow to Khartoum and beyond by four cubic kilometres annually and by a further four cubic kilometres through a Jonglei II phase.(See Chapter 12)

The Jonglei I Scheme will be completed and another phase initiated only if the peoples of the southern Sudan feel that they are being compensated for the water which would be conveyed to the north. A new scenario for the southern Sudan would involve political gestures by the governments of the Sudan and Egypt which would recognize the value of the water through some tangible economic gesture by Egypt and the Sudan which would lead to economic and social development in the South.

Alternative East African Highlands scenario

Emphasis given to the needs of the East Afriican countries would introduce yet another scenario. As water master plans are not available for these states it is not possible to deduce their priorities. Uganda is not short of water, rather it needs to drain some tracts in order to make them economically useful. The transportation of water to Uganda's dry areas is not regarded as economically feasible as the lifts involved would be very expensive. Kenya could utilize a cubic kilometre of water per year and Tanzania as large a volume in due course.

An East African scenario would, therefore, involve the reduction in White Nile flow from the East African highlands by up to five cubic kilometres per year but such a reduction is unlikely to have an impact until beyond the end of the century. Meanwhile the significance of such a reduction in flow would depend on the implemenation of the evaporation reducing drainage schemes in Southern Sudan.

The interests of downstream Egypt and the Sudan in Nile water development in East Africa are similar to those which they have in the Ethiopian tributaries. The maintenance of the annual average flow should be their main concern and the identification of reciprocal political and particularly economic initiatives and arrangements which would induce the upstream states to effect their water development in such a way that the diminution of flow will be minimized through judicious water storage and considerate release of seasonal and average flow.

Political reciprocity

All the above alternatives to the existing scenario require a new an imaginative approach by downstream states to securing their existing share of Nile water or at least to a substantial proportion of it. An integrated approach is required which will bring about studies of the environment as well as of appropriate institutional, political and legislative arrangements which will enable mutually agreed water management policies.

Part of the process will involve the identification of reciprocal economic arrangements which would encourage those with surplus water to share that water with countries which have already well developed irrigation and power generation systems. The notion of the equitable sharing of the benefits of water needs to be established and while this will take into account the prior use of water, it will also give weight to the use of water by the agricultural sectors in countries which have not previously been much involved in irrigated agriculture.

The sort of reciprocal economic gesture which might induce the governments which were not party to the 1959 Nile Waters Agreement to enter treaty arrangements promising the downstream states a secure share of water would involve the investment by the downstream states in the storage and transmission structures upstream and especially in those elements of them which might improve the efficiency of the storage and transmission systems in terms of water conservation. Such proposals will be resisted by downstream states as long as they feel that the recognition by the international community is secure. However, as this is position is unlikely to endure, it is probable that the notion of political reciprocity will enter the debates in the very near future.

References

Egyptian Government, (1921). *Nile Control,* Vol 1, Edited by Sir Murdoch MacDonald, Government Press, Cairo.

Evans, T.E. (1990) History of Nile Flows, in Howell, P.P. and Allan, J.A., *The Nile, Resource Evaluation, Resource Management, Hydropolitics and Legal Issues,* SOAS, University of London.

Hulme, M. (1990). Global Climate Change and the Nile Basin, in Howell, P.P. and Allan, J.A., *The Nile, Resource Evaluation, Resource Management, Hydropolitics and Legal Issues,* SOAS, University of London.

Hurst, H.E., Black, R.P. and Simaika, Y.I. (1946). *The Future Conservaion of the Nile,* Vol VII, *The Nile Basin,* Cairo.

Hurst, H.E. (1959). *The Nile,* Constable, London.

McGregor, R.M. (1945) 'The Upper Nile Irrigation Projects', W N Allan Papers, University of Durham

Morrice, H.A.W. and Allan, Wm.N. (1958). *Report on the Nile Valley Plan,* 2 vols, Khartoum

Said, R. (1990). *Nile Discharges Correlated to Climatic Events,* Paper at a conference in London on The Nile, SOAS, University of London.

United Nations, (1962). *UN Document ST/LEG/SER.B/12,* United Nations, New York. The Nile Waters Agreement was signed in Cairo on 8 November 1959; the Permanent Joint Technical Commission was set up by agreement in Cairo on 17 Januaru 1960. The texts have been recorded in the UN Legislative Series.

US Department of the Interior, (1964). *The Land and Water Resources of the Nile Basin: Ethiopia.* 17 vols., Washington D.C.

16

History of The Nile and Lake Victoria Basins through treaties

O. OKIDI

Introduction

This chapter is concerned with the history of the treaties for the consumptive utilization of one of the world's largest fresh water basins. Lake Victoria is the second largest fresh water lake in the world after Lake Superior (assuming the waters of the latter can still be considered fresh). The Nile is the longest river in the world. Both basins together are bordered in different degrees by nine states, namely: Kenya, Tanzania, Uganda, Ethiopia, the Sudan, Egypt, Zaire, Burundi and Rwanda. The status of the last two is unique in that they are brought into the basin by virtue of the Kagera River, which drains into Lake Victoria. The entire basin area has been estimated at 2.9 million square kilometres, which represents approximately one tenth of the continent.[1]

Despite these superlatives, the water of the basins or drainage system, is a scarce resource in more than one sense. First, the Nile is a source of livelihood for the desert states of Egypt and the Sudan. A review of the treaties on the consumptive utilization of the Nile and Lake Victoria will show how Egypt has strenuously sought to ensure security of the water flowing down the Nile.

Second, the water is not evenly distributed upstream either over the year or geographically. For instance, Ethiopia contributes approximately 85 per cent of the volume of water which flows annually past Khartoum. Yet most of the Ethiopian heavy rain is confined to a few months of the year and falls only over a part of the country leaving Ethiopia a country of perennial droughts and famine. Similarly, in Kenya, another substantial contributor of water through the six major rivers flowing into Lake Victoria, two-thirds of the entire territory is classified as arid or semi-arid. Thus, in both Ethiopia and Kenya, the long-term agricultural strategy must entail irrigation and often inter-basin water transfers for major irrigation works. This long-term strategy must be considered for Tanzania too. In the Kagera basin, a source of approximately 25 per cent of the Lake Victoria waters, major irrigation programmes are being planned under the aegis of the Kagera Basin Organization discussed below.

In view of this broad array of issues and the possible claims on the water resources it might be expected that a correspondingly broad range of agreements would exist among these independent African states on the waters of the drainage system. In actual fact, no agreements have been signed on the consumptive utilization of the waters since 1960. The years 1959 and 1960 saw one agreement and one protocol between Egypt and the Sudan for the utilization of the Nile waters.[2] Before 1959, there were about a dozen agreements focusing on the Nile and concerning almost exclusively Egyptian interests. Meanwhile, the range of issues on uses to which the water is put have increased and there may well be several other plans under consideration for still wider ranges of water usage. Given the scarcity of water resources as suggested above, there is an evident need for appraisal of the record and clear safeguards against international conflicts. As the pressures mount it must be clear, very quickly, that the pre-1960 agreements are, at best inadequate and, at worst irrelevant or even contrary to the present exigencies of development. These are the factors which necessitate the review of the existing treaties.

For the African countries it will be clear that the solution to the perennial problems of widespread famine and general development lies in the comprehensive planning, management and utilization of natural resources, principally water.[3] These may be realized only within a mutually understood and non-conflictual international environment.

Geographical and technical perspectives

The total surface area of Lake Victoria[4] (also described in chapter 7) is approximately 68,000 Km^2 of which the Kenyan portion is about 10 per cent, Ugandan 40 per cent and Tanzania 50 per cent[5]. Surface water contributed by rivers comes almost entirely from Kenya and Tanzania, the main Kenyan rivers being the Kuja, Awach or Kibuon, Mirin, Nyando, Yala, Nzoia and Sio, and from Tanzania the Mara which crosses into Kenya. On the south-western side is the Kagera[6], significant because it drains the territories of Rwanda and Burundi, extending the Nile Basin farther in that direction and important because of development plans for its utilization and the subject of special international arrangements which are described later[7].

The only drainage outlet is at what is now the Owen Falls dam, commissioned in 1954[8] to provide storage in Lake Victoria for Egypt and Hydro-electric power for Uganda.[9] Water from this source[10], net of evaporation losses in the Sudd region (See Chapter 4,), contributes on average between 20% an 25% of total Nile flow to Egypt which is dependent on Nile waters for survival. The contribution of the East African catchment is therefore small compared with that of Ethiopia, but is relatively steady and does not feature the marked seasonal and annual fluctuations in rivers coming from the latter region.[11]

Below Jinja the Nile flows through Lake Kyoga to Lake Albert where it is supplemented by flows from the Semliki River deriving from Zaire, which thus also has an interest in the river and its usage. The course of the Nile below this point and its salient hydrological features are described in detail in other chapters, but for the purpose of international and legal policy perspectives there are certain features that need emphasis. In particular there is the imbalance, mentioned above, between the annual contribution of the Blue Nile from Ethiopia and its marked seasonal variation and the steadier much smaller contribution of the White Nile deriving from East Africa.

The average volume of water each riparian contributes to the Nile Basin might be taken into account in deciding how much water a country might properly divert for national use. In the case of East Africa, Lake Victoria's contribution is easily determined from discharge records at the Owen Falls Dam, but for the purpose of policy the exact proportion of the annual outflow of each country separately needs to be established. This line of analysis should use percentage of volume rather than absolute quantity because when an upper riparian diverts water flowing through its territory from an international basin the fear of deprivation or injury expressed by a lower riparian is clearest when expressed in proportions.

Agreements on the Nile and Lake Victoria Waters

We shall limit ourselves to arguments dealing with consumptive uses only, omitting those purely on navigational uses, as well as demarcation of boundaries and spheres of influence.[12]

Certainly one of the foremost considerations of the treaties on the Nile waters is that Egypt, as a desert state and the lowest riparian of the Nile, would be a party to each of the treaties, especially those dealing with consumptive use of the waters, while all the upper basin states would be involved in the different stages. There are about ten agreements dealing with consumptive use of the waters of the Nile and Lake Victoria. Prior to World War 1, the treaties show Great Britain, for Egypt, as the contracting state. The United Kingdom, then the administering colonial power over the Sudan, signed an agreement with Italy (1891)[13], Ethiopia (1902),[14] the Independent State of Congo (1906),[15] and with Italy and France (1906).[16] There is further agreement with Italy, signed by Britain, in 1925.[17] Since then, Britain and Egypt signed all agreements on the Nile waters beginning with the 1929 agreement dealing with Egyptian rights generally vis-à-vis those of Sudan [18], and ending with the agreements for construction and maintenance of the Owen Falls Dam achieved by Exchange of Notes between 1949 and 1953.[19]

The year 1953 is historically significant as the time when Egypt was proclaimed a Republic, and Nasser emerged as the real power, bringing about a change in relations with Britain even though the *de facto* break did not come until the Suez crisis of 1956. The Sudan also became independent in 1956. It was after that time that the fourth and final set of agreements was signed in 1959 between Egypt and the Sudan on the utilization of the Nile waters,[20] and followed by a protocol establishing a Joint Technical Commission in 1960 (see also Chapters 4 and 5).[21]

Pre-World War I agreements

Italy and the United Kingdom (UK) signed a Protocol for the demarcation of their respective spheres of influence in eastern Africa at Rome on April 15th, 1891. Of interest is a provision in Article III which stipulated that 'The Government of Italy undertakes not to construct on the Atbara any irrigation or other works which might effectively modify its flow into the Nile'.[22] The agreement, by its very nature, ceased effect with the end of Italian and British colonial rule in the region.

Ethiopia and the UK signed a treaty at Addis Ababa on May 15th, 1902, regarding the frontiers between the Anglo-Egyptian Sudan, Ethiopia, and Eritrea. Article III provided:

> His Majesty the Emperor Menelek II, King of Ethiopia, engages himself towards the Government of His Britannic Majesty not to construct or allow to be constructed, any work across the Blue Nile, Lake Tana, or the Sobat which would arrest the flow of their waters into the Nile except in agreement with His Britannic Majesty's Government and the Government of the Sudan.[23]

The view of the present government in Ethiopia towards agreements signed by the imperial government is not clear, but it could be expected that their binding force cannot be taken for granted. Dante Caponera once observed that Ethiopia questioned the validity of the agreements for the following reasons:

> 1. The agreements.... between Ethiopia and the UK have never been ratified. Customary rights which might appear from the behaviour between lower riparians and Ethiopia would not be binding on the latter country if a purely positivistic approach toward interpretation of the sources of international law would be upheld.
>
> 2. Ethiopia's 'natural rights' in a certain share of the waters in its own territory are undeniable and unquestioned. However, no treaty has ever mentioned them. This fact would be sufficient for invalidating the binding force of those agreements, which have no counterpart in favour of Ethiopia. In Roman law such a pact would be null and void; it is likewise in international law. This is explainable by the international political conditions of Ethiopia in 1902.
>
> 3. The agreements were signed between Ethiopia and the UK (for Egypt and the Sudan). Since the latter question the validity of their own water agreements, Ethiopia, which had not one single benefit from them, had even greater reason for the claiming of their unfairness and invalidity. The research for new agreements by Egypt and Sudan demonstrates the non-viability of these agreements.
>
> 4. The UK in 1935 recognized the annexation of the Ethiopian Empire by Italy... UK's recognition of annexation is an act which invalidated all previous agreements between the two governments. Ethiopia has never asked for renewal of the Nile agreement after such recognition.[24]

The points listed here are important because they underscore the fact that Ethiopia did not, in the 1950s, recognize the treaty as binding. Whether the arguments are persuasive is a different matter. For example, there is nothing in international law which prevents any state from entering into a treaty which benefits only one of the parties. An extension of this point would perhaps include treaties which extend rights to third parties.[25]. On the other hand, the argument about British recognition of the Ethiopian connection might be the more forceful, although the legal consequences of war are not entirely clear-cut. It should be noted that since the 1902 treaty there has not been any agreement between the lower riparians, the Sudan and Egypt, and Ethiopia. Egypt and the UK or the Sudan have signed other agreements since 1929, but in no instance was Ethiopia a party, even though more than 80 per cent of the Nile waters reaching Egypt originate in Ethiopia.

The UK and the Independent State of Congo signed an agreement in London on May 9th, 1908, to redefine their spheres of influence in Central Africa.[26] Article III provided:

> 'The government of the Independent State of Congo undertakes not to construct, or to allow to be constructed, any work on or near the Semliki or Isango Rivers, which would diminish the volume of water entering Lake Albert, except in agreement with the Sudanese Government.'

Again, we can assume that this agreement ceased with the end of the colonial era; it has significance only as an indicator of how far back the interests of the Sudan and Egypt in Nile Basin waters have been protected. Great Britain, France, and Italy signed one set consisting of a tripartite Agreement and Declaration in London on December 13th, 1906.[27] This agreement and declaration came after Italy had failed to establish control over Ethiopia, and was a reconfirmation of the terms of the Protocol of April 1901, and the Agreement of May 1902. In the tripartite agreement at the insistence of Great Britain, Article IV provided:

> 'In the event of the status quo being disturbed, France, Great Britain and Italy shall make every effort to preserve the integrity of Ethiopia. In any case, they shall concert together on the basis of the agreements enumerated (herein) in order to safeguard:
>
> (a) The interest of Great Britain and Egypt in the Nile Basin, more especially as regards the regulation of the waters of that river and its tributaries (due consideration being paid to the local interests...)' [28]

Post World War I agreements

The above principles were reiterated in the 1925 agreement between Great Britain and Italy, but neither agreement has validity beyond the colonial era.

In an agreement by Exchange of Notes, in December 1925, at Rome,[29] the imperialist powers were to agree on how they, as well as the Sudan and Egypt, would use their influence to benefit from the Ethiopian highlands. The gist of the pre-negotiation agreement is captured in the following paragraph of a Note dated December 14 from Britain:

> 'In the event of His Majesty's Government with the valued assistance of the Italian Government, obtaining from the Abyssinian Government the desired concession on Lakes Tsana, they are also prepared to recognize an exclusive economic influence in the West of Abyssinia and in the whole of the territory to be crossed by the above-mentioned railway. They would further promise to support with the Abyssinian Government all Italian requests for economic concessions in the above zone. But such recognition and undertaking are subject to the proviso that the Italian Government on their side, recognizing the prior hydraulic rights of Egypt and Sudan, will engage not to construct on the headwaters of the Blue or White Niles or their

tributaries or effluents any work which might sensibly modify their flow into the river....'[30]

In a note dated December 20th, 1925,[31] Italy accepted the foregoing stipulation as an accurate outline of what the two countries had agreed upon as their common position in the anticipated negotiations with Ethiopia. It is obvious that the 1925 Agreement could not have been intended to be binding on Ethiopia. Simply to list it with other instruments on the Nile without pointing out its proper background and substance [32] might give the impression that the agreement had a legal effect on Ethiopia: it did not.

The first post-war agreement on the Nile waters was arranged in 1929 by Egypt and the UK (acting for the Sudan and the Eastern African dependencies), based on two Commission studies initiated by Egypt which formed a background for the agreements.[33] For the purposes of this analysis, it is sufficient to begin in 1920, when the Egyptian Minister of Public Works issued a report on the scheme for control and use of the Nile waters.[34] That report, which suggested five dams and a reservoir on the Nile and elaborated in later proposals (e.g.*The Nile Basin* Vol. VII (1946) *Future Conservation of the Nile*) which were finally abandoned by Egypt in favour of a single high dam at Aswan.The Egyptian government appointed a Nile Projects Commission that same year, to give its opinion on the projects 'with a view to further the regulation of the annual supply to the benefit of Egypt and the Sudan' and to report on the propriety of the manner in which the increased supply of available water would be allocated at each stage of development for Egypt and the Sudan.[35] These terms of reference indicate that Egypt was concerned about the interests of the Sudan, but did not seek any way of cooperating with Ethiopia or the Central African states within the upper basin, including the area around Lake Victoria.

The Commission's report stated that Egypt's rights were limited to a supply of water sufficient to irrigate an area equal to the largest area which had been irrigated in any single year since the Aswan Dam in its present form was completed, and that Egypt has an established claim to receive this water at the particular seasons when it is required.'[36] They added further that the largest area which Egypt might thus claim would be five million feddans, which were under cultivation in 1916-17.[37] There was no agreement within the Egyptian government regarding the merit of the report, and the resolution was left tied to the political future of the Sudan. However, when the British Governor-General of the Sudan was assassinated in Cairo in 1924, the British government in the Sudan threatened to increase irrigation uses of water in that country.[38] As a result, Egypt sought a fresh study for which the new Nile Waters Commission was set up in January 1925.[39] The Commission consisted of a Dutch engineer as an independent chairman, one British and one Egyptian member. Their recommendations provided the basis of the 1929 Nile Waters Agreement and were, in fact, annexed to that agreement.[40]

The 1929 Nile Waters Agreement was achieved by an Exchange of Notes between Mohammed Mahmoud Pasha, President of the Egyptian Council of Ministers, and Lord Lloyd, British High Commissioner in Cairo, on May 7th, 1929, and came into force the same day.[41] The Egyptian government pointed out that, while conceding and entering into an agreement with Britain on the utilization of Nile waters before political settlement was reached on the future of the Sudan, Egypt reserved the right to renegotiate the issue at the time of consideration of the future of the Sudan.[42] In the first paragraph Egypt made it clear, as a matter of principle, that the 1929 agreement was to be temporary, and its terms viewed as conditional on future political developments. This point is restated emphatically in the last paragraph of the Pasha's Note where he wrote:

> ' The present agreement can in no way be considered as affecting the control of the river which is reserved for free discussion between the two Governments in the negotiations on the question of the Sudan! [43]

The statement is important as it is the only point in the agreement which indicates the duration that the agreement was to remain in force.

The Pasha admitted, secondly, that:

> 'It is realized that the development of the Sudan requires a quantity of water greater than that which has so far been utilized by the Sudan. As your Excellency is aware, the Egyptian Government has always been anxious to encourage such development and will therefore continue that policy and be willing to agree with His Majesty's Government upon such an increase of this quantity as does not infringe Egypt's natural and historical rights in the waters of the Nile and its requirements of agricultural extension.' [44]

It is possible to read emphasis into the reference to Egypt's 'natural and historic rights' ; this writer believes the significance of the paragraph is that Egypt recognized Sudanese rights to develop, and to use the Nile waters for that purpose. That is a significant departure from the position taken before the 1925 Commission, which had been rejected as a negation of the right of the Sudan to exist as a viable State. To the extent that Egypt accepted the right of the Sudan to an increasing quantity of water for its development, Egypt also had accepted that the rights to use varying quantities of water would depend on the needs of the moment of negotiation. This interpretation seems to be supported by the fact that when the 1920 Commission was faced with the question of how much water Egypt was entitled to, it simply suggested that Egypt must claim the quantity of water necessary to irrigate the five million feddans under cultivation in 1916-17.[45] There was no 'natural' figure discernible in history.

The principle of prior appropriation which one commentator has suggested[46] as an ideal interpretation of historic rights is not really helpful. Prior appropriation would refer only to the precise quantity that had been appropriated, and no more. Changing circumstances would be negotiated for separately and according to what was equitable and reasonable at the time. Similarly, if for any reason additional quantities of water were available, i.e by draining the Sudd in the southern Sudan, then the division of that new quantity would be negotiated separately.

Egypt did not object to use of the Nile waters for construction of control works and subsequent irrigation in the Sudan, but did insist on prior consultation and explicit agreement on what such construction would entail. Thus, the Pasha added in paragraph 4 (ii) of his Note to Lloyd that:

> 'Save with the previous agreement of the Egyptian Government no irrigation or power works or measures are to be constructed or taken on the River Nile and its branches, or on the lakes from which it flows, so far as these are in the Sudan or in countries under British administration, which would, in such a manner as to entail any prejudice to the interests of Egypt either reduce the quantity of water arriving in Egypt, or modify the date of its arrival, or lower its level.'

It seems clear that the two countries, Egypt and the Sudan, would have to agree before the Sudan could abstract the water of the Nile to an extent that would change the quantity of the water flowing to Egypt. Sub-paragraph 4 (iii) of the Pasha's note stated that Egypt would carry out a complete study of the hydrology of the Nile in the Sudan, and that the Sudan should provide all necessary facilities and access. In this regard the Sudan permitted Egypt to construct and maintain, in Sudanese territory, any structure it might need for study of the hydrology of the river. In the event of any dispute arising on the interpretation and application of the agreement, the parties would in good faith seek a mutually acceptable solution. If that failed, the matter would be referred to 'an independent body with a view to arbitration', as stated in paragraph 4 (vi). The response from Lord Lloyd [47] confirmed the accuracy of the Pasha's letter as a reflection of the agreement they had reached, and assured that the agreement was directed toward regulation or irrigation arrangements of the Nile and had no bearing on the status quo in the Sudan.

In summary, Egypt enjoyed overwhelming rights in the utilization of the Nile waters; the quantity of water to which Egypt was entitled was not specified; and the agreement did not have a specific duration.

What is the current status of the 1929 agreement vis-à-vis the former British dependencies referred to in paragraph 4 (ii) of the Pasha's note? Because the occasion did not arise, the agreement was never invoked or applied in Kenya and Tanzania to restrain any irrigation or other consumptive uses of water (but see Chapter 5). In Uganda one could cite the Owen Falls Dam as the type of installation envisaged in 1929. With regard to East Africa the newly independent Tanganyika government took the view that an inherited agreement that purported to bind Tanganyika for all time to secure consent of the Egyptian government before it undertook irrigation, power works, or similar measures on Lake Victoria or its catchment area, was clearly incompatible with Tanganyika's status as an independent sovereign state.[48] On July 4th, 1962, its government addressed identical Notes to the governments of Britain, Egypt and the Sudan outlining the policy of Tanganyika on the use of the waters of the Nile, and the Note was also sent to the governments of Kenya and Uganda. That Note, consistent with the Nyerere Doctrine on States succession to treaties, read in full:

> 'The Government of Tanganyika, conscious of the vital importance of Lake Victoria and its catchment area to the future needs and interests of the people of Tanganyika, has given the most serious consideration to the situation that arises from the emergence of Tanganyika as an independent sovereign State in relation to the provision of the Nile Waters Agreement on the use of the

> present arrangements whereby technical experts from the United Arab Republic, the Sudan and the Three East African countries of Tanganyika, Kenya and Uganda meet at intervals to discuss common technical problems connected with the use of the waters of the Nile.'[49]

Tanzania maintained further that, since the 1929 Agreement applied to territories under British administration, the treaty lapsed, in relation to Tanganyika, on Independence Day. This became known as 'the Nyerere Doctrine' (see also Chapter 5).

On November 21st, 1963, Egypt, in a Note replying to Tanganyika, simply submitted that 'pending further agreement, the 1929 Nile Waters Agreement... remains valid and applicable.' [50]

They added that they were in favour of the continuation of the unofficial talks between technical experts from Egypt and the Sudan on the one hand, and Tanganyika, Kenya and Uganda on the other.[51] The Note was sent to the Sudan, which made no reply to either communication.

Tanganyika's Ministry of Foreign Affairs held the view that the 1929 Nile Waters Agreement was neither a real nor a dispositive agreement and, therefore, had no legal effect on an independent Tanganyika.[52]

Kenya did not respond to the Note from Tanganyika or the response of Egypt, which was understandable as the British government had not yet left Kenya. They could have found it convenient to remain silent and leave it to an independent Kenya to sort matters out. Kenya did, upon independence, adopt a position similar to the Nyerere Doctrine of succession to treaties, submitting that the Government of Kenya was willing to grant two years grace period in which the treaties would apply on the basis of reciprocity, or be modified by mutual consent.[53]. But those treaties which were not so modified or negotiated within the two years and 'which cannot be regarded as surviving according to the rules of customary international law will be regarded as having terminated.' '[54] This would indicate that the treaty ceased to have effect with respect to Kenya as from December 12th, 1965.

The same fact applying to Kenya would apply to Uganda, particularly the position relative to succession to treaties as expressed in the Independence Declaration on Treaties wherein Uganda adopted the Nyerere Doctrine.[55]

The position of the Sudan would have a bearing on that of Kenya, Tanzania and Uganda, since the Sudan was more directly involved in the treaty. At the time of Sudanese independence in 1956, the Sudan, according to Badr, declared that it 'was not bound to take over an Agreement to which it was not a party and which was, anyway, considered unfair'.[56] They stated outright that the 1929 Agreement was obsolete; and prepared to negotiate a new one.[57] There is no reason why the three East African countries only remotely referred to in the treaty should be expected to have remained bound either.

Finally as pointed out earlier, Egypt considered the 1929 Agreement temporary pending determination of the political future of the Sudan. If it was temporary for Egypt and the Sudan, there is no reason why it should have longer life for Kenya, Tanzania, or Uganda.

The Owen Falls Dam Agreements

Attempts by Great Britain to secure, on behalf of Egypt and the Sudan, an agreement with upper riparians, especially Ethiopia, to construct major storage has been described above. A focus on Ethiopia was probably due to the fact that over 80 per cent of the Nile waters reaching Egypt originate in that country. However, the upper reaches of the White Nile were not entirely ignored. In 1946 the Ministry of Public Works drew up a comprehensive plan where the main components were, a dam or dams at the Great Lakes of Equatorial Africa and construction of the Jonglei Canal in the Sudan(see also Chapter 5). The Lake Tana Reservoir; and a dam at Merowe near the fourth cataract on the Nile were also included.[58]

It was necessary to find a suitable site for construction of what H. E. Hurst, Controller of the Physical Department of the Egyptian Ministry of Public Works, called 'Century Storage' of water.[59] In the first proposal, the dam in the Great Lakes was to be constructed at the outlet from Lake Albert with only a small dam on Lake Victoria. But for Lake Albert (5,300 square kilometres) to store the required capacity of 155 billion cubic metres of water, would flood a considerable area around it, most of which lay in the territory of Uganda and the then Belgian Congo. The governments in Uganda and the Belgian Congo objected very strongly because the flooding would displace the population and cause a loss of valuable land under cultivation in a large area along the Albert Nile.

So Egypt advanced an alternative proposal for a dam at the outlet of Lake Victoria. The advantages of this site over Lake Albert were considered to be enormous: more water would be stored than in the original plan, since Lake Victoria has a total area of 68,000 square kilometres. It was estimated that the average depth of the lake was 40 metres with a maximum of 70 metres. [60] Britain, the administering power over the three states around Lake Victoria, was not opposed to the level of the lake rising by a maximum of 1.3 metres, or about four feet above the then recorded maximum, within a range of 3 metres.[61] The consequence of this rise was recounted by Hurst:

> 'The raising of the level of Lake Victoria will necessitate some changes in the lakeside ports, and will cause the removal of a certain number of huts and embanking of a few cultivated areas, for which compensation will be paid.'[62]

Uganda was to benefit from the dam in that it would produce 200 metres of head capable of producing hydro-electric power up to 15,000 kilowatts. [63]

With this background in mind, we will consider the agreements leading to the construction of the dam. Negotiated by Britain, acting for Uganda, and by Egypt through an Exchange of Notes between the two governments, it was carried out in three forms: first, an agreement regarding the construction of the dam,[64] pure and simple; second, an agreement on the granting of a contract for construction of the dam; and third, an agreement on financial arrangements for construction and maintenance of the dam.

The first of the three agreements is the core of the formal treaty. The first Note, written on May 30th, 1949, was from the British Ambassador in Cairo to the Egyptian Minister for Foreign Affairs.[65] It reflected completed negotiations, and that the agreements were in accordance with the spirit of the Nile Waters Agreement of 1929. The purpose was twofold: to control the flow of the waters of the Nile and to produce hydro-electric power for Uganda. It stated further that even though the Uganda Electricity Board would invite tenders and place contracts for the construction, specifications for the work had been prepared in full consultation and with approval of both Egyptian and Uganda authorities.[66] The flow, which is a total of what goes through the turbines and what is allowed through the sluices, was to be supervised by Egyptian engineers resident at Jinja. Paragraph 4 of the British Note stipulated:

> 'The two governments have also agreed that though the construction of the dam will be the responsibility of the Uganda Electricity Board, the interests of Egypt will, during the period of construction, be represented at the site by the Egyptian resident engineer of suitable rank and his staff stationed there for the purpose by the Royal Egyptian Government, to whom all facilities will be given for the accomplishment of their duties. Furthermore, the two governments have agreed that although the dam when constructed will be administered and maintained by the Uganda Electricity Board, the latter will regulate the discharges to be passed through the dam on the instructions of the Egyptian resident engineer to be stationed with his staff at the dam by the Royal Egyptian Government for this purpose .... '[67]

Informal sources indicate there is still an Egyptian resident engineer at the Owen Falls Dam, so it would appear that the agreement continues in force according to these terms. The British Note provided that the Uganda Electricity Board could take any action it considered desirable before or after construction of the dam, provided such measures were taken only after consultation and agreement with the Egyptian government. Any dispute which could not be resolved by negotiation or conciliation would be referred to arbitration.[68]

The reply from the Egyptian Minister for Foreign Affairs [69] dated May 31st, 1949, confirmed the formal agreement and it came into force that day. The formal agreement provided for the Uganda authorities to grant the contract for construction of the dam, with the approval of the Egyptian Government, and that constituted the second agreement.[70]

The final round of the Owen Falls Agreement concerned financial arrangements for the construction. The first Note, dated July 16th, 1952, was from the Egyptian Minister for Foreign Affairs to the British Chargé d'Affaires in Cairo.[71] Laying emphasis on the value of Lake Victoria as storage of water for Egypt, the carefully worded Note read:

> 'The Royal Egyptian Government
> (i) Will bear that part of the cost of the dam at Owen Falls which is necessitated by the raising of the level of Lake Victoria by the use of Lake Victoria for storage of water.[72] '

The ordinary meaning of this provision suggests that the engineers who designed the dam anticipated that as a result of construction, the level of Lake Victoria would rise because the very nature of the storage function of the dam would cause backwater effect. The agreement took care of the effect of the rising level of the lake. Egypt undertook to compensate those around Lake Victoria who might be affected by the change in water level in the lake. The second paragraph of the Note said that the Royal Egyptian Government:

> (ii) Will bear the cost of compensation in respect of interests affected by the implementation of the scheme or, in the alternative, the cost of creating conditions which shall afford equivalent facilities and amenities to those at present enjoyed by the organizations and persons affected, and the cost of works of reinstatement as are necessary to ensure a continuance of the conditions obtaining before the scheme comes into operation, such costs to be calculated in accordance with the arrangements agreed between our two Governments. [73]

The Note suggested further that the flow of water through the dam would be controlled for purposes other than hydro-electric power generation, noting that on occasions the flow control could be detrimental to electricity supply to Uganda. The Egyptian government agreed 'to pay to the Uganda Electricity Board the sum of £980,000 as compensation for the consequential loss of hydro-electric power, such payment to be made on the date when power for commercial sale is first generated at the Owen Falls Dam.'[74] Egypt went further and stipulated the conditions resulting from the rising level of the lake as its responsibility. Thus, the Egyptian government agreed that for purposes of calculation of compensation under the provision of sub-paragraph (ii), all flooding around Lake Victoria within the agreed range of three metres shall be deemed due to the implementation of the scheme.

In his response of January 5th, 1953,[75] the British Ambassador concurred in the obligation undertaken by Egypt, and the Owen Falls Dam was commissioned in 1954. The regime worked well if it provided Uganda with the hydro-electric power it needed and if the storage functions continued to Egypt's satisfaction. The agreement may be assumed to be binding upon Uganda whatever the change of government, so long as Uganda continues to enjoy the power supply, provided that there was no new agreement and neither party renounced this agreement.

Egypt assumed further obligations vis-à-vis the other two riparians of the lake, Kenya and Tanzania. In the event of any physical or environmental change suffered resulting from rising levels of the Lake, Egypt would pay compensation.[76] The binding force of that obligation seems to remain, even though Kenya and Tanzania have secured their independence. That Kenya and Tanzania after their independence may not have acceded to the Owen Falls Agreement is not of any legal consequence as regards the obligation Egypt undertook toward them. It seems, therefore, that under the Owen Falls Agreement, Egypt and Uganda might be under obligation to compensate Kenya and Tanzania if the latter states suffer environmental or physical injuries caused by operation of the Dam. The law of treaties requires, further, that should Egypt and Uganda decide to modify or revoke the stipulations relating to the third party rights, they are under obligation to seek the concurrence of Kenya and Tanzania.

During the negotiations for the agreement on the Owen Falls Dam, the

Egyptian government must have seen a need for research, observation, and recording of meteorological and hydrological data from the basin of the East African lakes, including Lake Victoria. This was the subject of another agreement, done by Exchange of Notes between the Egyptian Ministry of Foreign Affairs and the British Ambassador in Cairo (for Uganda), before the Owen Falls Dam agreement was completed.[77]

The substance of this agreement was contained in the Egyptian Note to the British Ambassador on January 19, 1950 [78] and it indicated the degree of cooperation which the Ugandan authorities had promised to Egypt because the data would help Egypt determine the amount of water it could receive from these upper reaches of the Nile. The Ugandan authorities agreed to establish data collection posts, marked on an enclosed map, which would not be varied without prior consultation. Further, the resident Egyptian engineer at Owen Falls Dam and his assistants would have access to all the posts situated in Uganda. The intention seemed to be that they would carry out periodic inspections of the posts 'to assure.... the posts are being satisfactorily maintained and the observations regularly collected.' [79]. Egypt would contribute toward the expenses incurred in maintenance of the posts, within a certain monetary limit. [80]. The project was a long range one; the British reply dated February 28, 1950, [81] confirmed Uganda's undertakings as outlined, and the agreement entered into force on March lst, 1950. This agreement provides Egypt with hydrological and meteorological data of the East African Lakes region and thus allows them effective long-term planning.

The 1959 Agreement for full utilization of the Nile Waters

This agreement, which ushered in a new era in the history of the Nile basin, was signed by Egypt and the Sudan at Cairo on November 8th, 1959.[82]

The preamble stated that the 1929 Agreement had 'only regulated a partial use of the natural river and did not cover the future conditions of a fully controlled river supply.' [83] To utilize the Nile waters for the benefit of the two republics required the implementation of projects for full control of the river, an increase of its water supply, and the planning of new working arrangements 'on lines different from those followed under the present conditions.' [84]

To refer to 'full utilization' and 'full control of the river' when there were only two states involved in the agreement rather than all of the basin states, especially the upper one, seems anomalous. There is no evidence that Ethiopia, which contributes so much of the gross annual flow at Khartoum, or the East African States, were invited to any of the negotiations (see also Chapter 4). Needless to say, the two parties to the agreement were both simply recipients and users, dependent on water from central Africa and Ethiopia. They needed the cooperation of those upper basin states if their goal was to be assured. In declaring that the new agreement was not only more comprehensive, but also different in spirit from preceding ones, especially the 1929 agreement, they were beginning nearly with *tabula rasa* as far as the utilization and control of Nile waters was concerned and with regard to treaties between the two states.

The point of departure between the two parties was their 'acquired rights' stemming from the Nile Waters Agreement of 1929. This assumed a total mean flow of 84 km^3 , of which 48 km^3 had been allocated to Egypt and only 4 km^3 to the Sudan, [85] the balance of 32 km^3 being uncontrolled and running to waste in the sea. The Sudan was pressing for a greatly increased quota and favoured the Nile Valley Plan, a refinement of Hurst's Century storage plans prepared by H.A.W. Morrice and W.N. Allan in 1958 (see Chapter 5). Egypt on the other hand was by then committed to the Aswan High Dam Project which would, at any rate for the forseeable future, supercede plans for integrated control throughout the Nile valley. The Sudan agreed to this in return for a greatly increased share - the allocations being Egypt 55.5 km^3 and the Sudan 18.5 km^3, it being assumed that approximately 10 km^3 would be lost by evaporation and seepage behind the High Dam. Badr [86] looked at the relative figures for Egypt and the Sudan and concluded that:

> A state is at liberty to accept less than is due to it, should it so decide, for considerations of policy of which it is the judge. But the exercise of such a liberty in an international treaty... makes it inadvisable to draw legal conclusions from such an instrument or to consider it a precedent in international law.

Thus, he opined that there was really no historical or legal basis for the proportions set aside for Egypt and the Sudan in this agreement.[87]

The control works under the Agreement were outlined in Section II of the document.[88] Perhaps the most important was the provision for the construction of the Sadd el 'Aali, or the High Dam, at Aswan, to store water for Egypt and to prevent the flow of excess volumes of water to the sea. At the same time, the dam would cause back water flooding of the territory of the Sudan, particularly of the town of Wadi Halfa. Under paragraph 6 of Section II, Egypt agreed to pay £E15 million to the Sudan as full compensation for damages to Sudanese property that might be caused by the storage of water at the Sadd el 'Aali Reservoir. Details of such compensation were outlined in Annex II to the Agreement. The Sudan also undertook to transfer its population whose property was to be affected by the storage effect of Aswan from Halfa and surrounding areas prior to July 1953.

The Agreement provided that the Sudan would construct the Roseires Reservoir on the Blue Nile and any other works deemed necessary to enable the Sudan to exploit its share of the water. This was a major concession to the Sudan because, during negotiations leading to the 1929 agreement, Egypt had strongly opposed such works in the Sudan. At that time Egypt had been concerned over possible Sudanese intentions because in 1924 Britain had threatened to increase irrigation consumption of water in the Sudan.[89] The political atmosphere in 1959 was different. It is noteworthy, however, that even though the two states could agree on construction of the Roseires Reservoir on the Blue Nile, they failed to involve Ethiopia as a party to the treaty in order to assure themselves of the volumes of water from Ethiopia.

The same disregard extended to the states of the Upper Nile Basin; this is seen clearly is Section III of the Agreement which emphasized the loss of water through evaporation in the Sudd in the Sudan.[90] The Sudan government agreed to increase the supply of water flowing down the Nile, and to drain the swamps.

Central to this pair of commitments was the Jonglei Canal Project which would at that stage run from the village of Jonglei in the south to the Sobat Mouth in the north. The two countries agreed to share the cost of the construction as well as the water released from the swamp (see Chapters 5 and 12).

Anticipated projects for the use of the Nile waters under the Agreement were to be backed by a system of technical cooperation between the two parties. Thus, the parties agreed in Section IV to constitute a Permanent Joint Technical Commission, composed of an equal number of members for each republic, to be responsible for the supervision of all working arrangements in the Agreement; carrying out necessary hydrological studies to facilitate adequate policies; and preparation of work implemented in territories outside the Sudan by agreement with their concerned authorities.[91]

Paragraph (i) of Section V commits the parties to a common front in any negotiations with third states. It reads:

> '...in case any question connected with the Nile waters needs negotiations with the governments of any riparian territories outside the Republic of the Sudan and the United Arab Republic the two republics shall agree beforehand on a united view in accordance with the investigations of the problem by the Commission. This unified view shall then form the basis of instructions to be followed by the Commission in the negotiations with the governments concerned.'

At the time of this agreement, there was a nine year old agreement between Egypt and Britain (for the Sudan) for the hydrological study of the basins of the central African lakes. Therefore, in terms of basic hydrological data on the Nile and Lake Victoria basins, the two states were ahead of the other basin states. The advantages, in the event of any negotiations anticipated in this agreement, would be significant for Egypt and the Sudan relative to the upper basin states.

The protocol concerning the Permanent Joint Technical Commission

Section IV (3) of the 1959 Agreement required the parties to form a Technical Commission to fulfil the functions already analyzed above. Four members were appointed to each party. That purpose was met by a Protocol signed by the two states in Cairo in January, 1960 [92] which was to be an integral part of the 1959 Agreement. There was a stipulation in the Protocol that should there be a need to alter any aspect of it, then that would be done by Exchange of Letters between the two parties. [93]

Agreement for the Hydrometeorological Survey of Lakes Victoria, Kyoga and Albert (Mobuto Sese Seko)

A plan of operation for hydrometeorological surveys of the above area was signed by five countries: Egypt, Kenya, the Sudan, Tanzania and Uganda, as well as the United Nations Development Programs (UNDP) and the World Meteorological Organization (WMO), and declared operational from 17th August, 1967.[94] Its purpose was to evaluate the water balance of the Lake Victoria catchment in regard to control and regulation of the lake level as well as the flow of the water down the Nile. Funding for the project was to come from the UNDP, while WMO was the executive agency.

As background preceding the 1967 Agreement, Egypt and Britain had signed an Agreement for cooperation in meteorological and hydrological surveys of the Lake Victoria catchment by an Exchange of Notes in 1950.[95] Following that, Kenya, Tanganyika, and Uganda set up an East African Nile Waters Coordinating Committee to establish and maintain 'a common East African case and point of view on the Nile waters'[96] (see Chapter 4). Theoretically, the Committee was to consist of three Ministers concerned with water resources in the three East African states, but in fact the Ministers never met as a Committee. Instead, the participants were technical and administrative officers. On a few occasions, members of the committee, and members of the Permanent Joint Technical Committee of the Nile (of Egypt and the Sudan) held consultative meetings to discuss such matters as control of discharge at Owen Falls Dam, the future storage of waters in Lakes Victoria and Albert, and irrigation requirements of the East Africa countries in the lake drainage area (see Chapter 4)[97]. By 1960, the Coordinating Committee had, after preliminary discussions, endorsed the need for a survey of the hydrometeorology of the catchment area of Lake Victoria[98]. In 1961 the three East African governments requested the UN Expanded Programme of Technical Assistance (EPTA) for aid to conduct a preliminary hydrometeorological survey of that catchment[99]. In response, a team of three consultants from WMO and FAO carried out a preliminary survey in early 1962, and submitted a report to the three governments in 1963.[100]

A discussion of that report convinced the three governments that the survey should be extended to include Lakes Kyoga and Albert catchments, and that they should include Egypt and the Sudan as participants. A review of the proposal by the UN Special Fund in 1965 approved the project and Egypt and the Sudan were invited as participants in the hydrometeorological survey.[101] At a meeting in Nairobi in August 1965, the representatives of the nine countries formulated a project proposal and submitted it to the Special Fund. It later was adopted by the UNDP for funding.

That is the background of the 1967 Agreement. As the project progressed, the five participants had consultations with Rwanda and Burundi to extend the project area to cover the Lake Victoria catchment in those countries as well.[102]

The Agreement for the Establishment of the Organization for the Management and the Development of the Kagera River Basin (The Rusumo Treaty)[103]

The Kagera basin drains four states, namely: Burundi, Rwanda, Tanzania and Uganda. But only the first three of them signed the agreement at Rusumo on 24th August,1977. Uganda was to accede to the treaty in 1981.

The Background

The agreement has its origins in the diplomatic intitiatives of the Presidents of Rwanda and Tanzania, when they exchanged a visit at the border village of Rusumo in 1976 to discuss matters of mutual interest. During their discussions the two agreed, *inter alia*, to cooperate in the construction of a bridge across the

Kagera River at Rusumo to facilitate transport and trade between their countries, and to initiate technical studies towards harnessing hydro-electric power at Rusumo Falls on the Kagera. But it was, in fact, the hydro-electric power project that necessitated the immediate involvement of Burundi, as an upper riparian and Uganda as the lower riparian of the river. The long-term security of the power would depend on an assured flow of the river, while a dam for the project would be downstream with possible effects on the lower riparian.

Consequently, the four countries agreed immediately to request UNDP funding for the planning and development of the Kagera basin and its waters. In July 1969 the UNDP sent a fact-finding mission to consult with the four governments, with the eventual concurrence that a UNDP-sponsored project be established to coordinate orderly regional planning; that a technical committee composed of representatives of the four governments be established for this purpose; and that such a project be coordinated with the on-going projects in the region, particularly, the Hydromet and the mineral research projects in Burundi and Rwanda.

Uganda opted to be an observer at these activities, even though she agreed, in principle, with the concept of regional and basin-wide planning. Burundi, Rwanda and Tanzania for their part, established a Technical Committee which submitted a joint request for project funding to the UNDP in July 1970. The request was approved by the UNDP Governing Council in January 1971, and the Kagera basin development studies were inaugurated in June 1971, with a project headquarters established at Bukoba, Tanzania, in August 1971. Active fieldwork commenced in September 1971 with the collection and analysis of the existing data, identification of gaps in the data, recommendations for essential additional studies and the preparation of the second phase. These activities, comprising Phase I, were completed in June 1973.

Phase II was required to prepare an Indicative Basin Plan based largely on data available to the three states, taking into account national priorities and, bearing in mind, the need for the harmonious development of the basin. The specific items covered in the study contracts were aerial photography, tourism, hydropower potential, fisheries, and institutional arrangements. The report was submitted at the end of 1976.

During that period there were two related developments agreed upon by the Technical Committee. First, the Committee decided in June 1976 that the project headquarters be transferred from Bukoba to Kigali. That was effected in November 1976. Second, it was decided to commission a specific study on the Rusumo hydro-electric project and a protocol was signed at Kigali, on 22nd October 1976, with the Belgian Government. The latter, in turn awarded a contract to a consortium of Tractionnel/Electorobel consultants requesting that they study the hydropower project and the implications of the dam for irrigated agriculture, settlement, environment, fisheries and tourism.

The tentative results of the studies and the range of possible activities were clear enough to warrant the signing of the treaty. Burundi, Rwanda and Tanzania Heads of State signed the Agreement to establish the Organization for the Management and Development of the Kagera River Basin (commonly known as the Kagera Basin Organization or the KBO) at Rusumo on the 24th August 1984.

The Agreement[104]
The Organization was established by Article I of the Agreement, with the territorial jurisdiction to cover the entire catchment of the Kagera river. Even though the parties were only three, the founding states anticipated the future participation of Uganda. Thus, they reserved Article 19 exclusively to the provision that: 'The present Agreement is open to accession by Uganda'

Of course one would have expected that; because of its status as the lowest riparian, Uganda would require participation in the treaty in order to control the security of the water, especially in view of possible irrigation programmes. But the situation is possibly explainable, in part, by the changes and chaos which followed the 1971 military coup d'etat against President Milton Obote. Thereafter, President Nyerere made no secret of the fact that he wanted no dealings with Idi Amin as President of Uganda.

As explained earlier, Uganda acceded to the Agreement on May 19th, 1981. The instrument of accession was in the form of an agreement between the three original contracting states and Uganda, and signed by all four at Bujumbra. Article 3 of that instrument notes that the amendments to the KBO agreement were mutually accepted by the original parties and by Uganda.[105]

The application of the agreement, *ratione materiae,* was covered in Article 2 and this derived largely from the specific studies in the Indicative Development Plan. The article states that 'The objectives of the Organization is to deal with *all questions* relevant to the activities to be carried out in the Kagera Basin, (emphasis added), notably:

a. Water and hydropower resources development.
b. The furnishing of water and water-related activities for mining and industrial operations; potable water supplies for other needs.
c. Agricultural and livestock development; forestry, and land reclamation.
d. Mineral exploration and exploitation.
e. Disease and pest control.
f. Transport and communication.
g. Trade.
h. Tourism.
i. Wildlife conservation and development.
j. Fisheries and agricultural development.
k. Industrial development, including fertilizer production, exploration and exploitation of peat
l. Environment protection.

In the Final Reports produced by the KBO/UNDP studies the projects were articulated in four key sectors, namely: (1) water related projects, including hydropower production, irrigated agriculture, rainfed agriculture, forestry, livestock and fishing; (2) transport and communication; (3) industries; (4) training and manpower development.

Within the context of water resource scarcity one would be conscious of the main consumptive uses of water resources, namely irrigated agriculture and industries. It is to be noted, for instance, that Phase II studies had proposed a combination ranging from 90,000 hectares of irrigation plus improved rainfed agriculture to 200,000 hectares, irrigation of new land to be supplemented by irrigation/rainfed operation of small perimeters of privately farmed land. Land to be opened for agriculture in the region was expected to expand to 500,000 hectares with variable proportions for irrigation. But no estimate had yet been made for the planned water consumption by the industries. Ultimately, however, what such possible irrigation expansion would mean for the water storage in Lake Victoria is beyond the scope of this chapter.

The organs of the KBO are outlined in Chapter II of the agreements (Art. 4-11). As distinct from most other basin organizations, the KBO agreement does not specifically provide for the assembly of heads of state, even though there is an annual meeting. Article 4 provides that the principal organs of the organization are: the Commission and the Secretariat headed by the executive secretary. The Commission is composed of one representative from each of the contracting states and it is the main policy-making organ with the mandate to determine projects, to solicit funds and to control and manage the budget. The Secretariat is the permanent bureau and the executive arm of the organization.

The KBO has met with difficulties in mobilizing resources and implementing its wide array of projects. With the growing population and the pressure to increase consumable goods as well as to dissipate the weight of international economic problems, the KBO states may eventually mount their water consuming projects, as envisaged in the Rusumo Treaty and the studies.

Recapitulation

In retrospect, not until World War I were important agreements reached on utilization of the Nile. Even then, the 1925 Agreement was an unusual one, certainly void in relation to Ethiopia both then and now. The first full-scale agreement on the Nile came in 1929. Again, the background of that treaty was so riddled with political complications that it could be no more than a temporary agreement, even though it assured Egypt that its water needs would be met. A more stable treaty was signed in 1959, between Egypt and the Sudan, which remains in force between the two parties.

Remaining in force as well is the Owen Falls Agreement signed between Egypt and Britain (on behalf of Uganda). The obligation seems to have fallen on Uganda by virtue of its continued use of hydro-electric power from the dam, and because it has not renounced the treaty responsible for the generation of that power. Egypt is interested in the storage value of the dam and Lake Victoria. Because of that continued force of the treaty, it appears that Kenya and Tanzania retain the third state rights extended to them in the event of injuries resulting from the rising level of the lake. Under the treaties examined here, Ethiopia, Kenya and Tanzania are not under any obligation regarding the use of the waters flowing to Lake Victoria and the Nile Basin. At least there has been no agreement on the utilization of the waters of Lake Victoria directly involving all the riparian states. Tanzania clearly rejected the 1929 Agreement and

Kenya's position is similarly clear. Moreover, we have not seen a treaty imposing any obligations on Zaire, Rwanda or Burundi, although they are Kagera Basin states. However, all of these may be subject to limited obligations under general international law to negotiate with the lower riparian states for an equitable share of the water, the exact modalities being subject to fresh negotiation.

Post 1959 Agreement developments

Since the signing in 1959 of the last agreement on the Nile Basin with any relevance to the Victoria catchment area, and the supporting Protocol in 1960, several policy actions have been taken and implemented by the basin states. Some have involved consumptive uses of the waters to an extent that would affect the hydrological and meteorological regime of the Lake Victoria and Nile drainage basins. Such policy measures may necessitate consideration of a legal regime beyond that analyzed above. Some of those policy statements and measures are outlined briefly here, with no order or priority, and only by way of illustration of what more could occur in the basins.

Egypt

It may be assumed that the regular rise in population in Egypt would raise the country's needs for irrigation for food production beyond what was needed in 1959. Egypt has doubtless agreed with the Sudan on quantities of water for their respective uses. There are, however, two considerations that might dramatically increase Egypt's need for water.

Egypt has an ambitious land reclamation policy. The development of irrigated farming in Sinai is a particularly prominent project and in December 1975, Egypt announced that it would open pipelines to carry water across the Suez Canal to the Sinai desert for irrigation[106]. The project was supposed to commence with irrigation of some 5000 feddans, to be increased later to provide 100,000 refugee families from the Gaza Strip with livelihood[107].

Additionally, Egypt has commissioned studies of the possibility of piping 'the Nile waters to Jerusalem for Jewish, Christian and Moslem pilgrims visiting the holy places.'[108] This extension would add 240 miles to the length of the Nile, and is further evidence of potential and controversial downstream uses of water. From the legal point of view, there may be a question of whether it requires consideration by the all basin states before inter-basin transfers are effected.

The Sudan

The Sudan has undertaken a major project in the Jonglei Canal which aims to drain the Sudd area of the Southern Sudan between Jonglei and Malakal. It is, therefore, entirely within Sudanese territory, but of major significance to Egypt because the net saving of water will be shared between the two countries under the 1959 Agreement. The hope is that this will reduce the loss of water through evaporation over the Sudd, open up a greater area in southern Sudan for agriculture, and release more water for irrigation in the northern Sudan and Egypt (see Chapter 12).

The idea is an old one, and has been a subject of engineering and ecological studies[109].(see also Chapter 5) However, its merits have been subject to controversy,[110] largely initiated by the hypotheses that the project would be an ecological catastrophe.[111]

The present projected scheme, which in any case has been halted by the Civil War in the southern Sudan, has been the subject of range and swamp ecology surveys, (Mefit-Babtie, 1983) and the predicted effects are described in some detail in Howell *et al.* (1988)[112].

Ethiopia

Ethiopia is considering increased utilization of Blue Nile and Sobat waters to an extent which Egypt might find threatening to its interests[113]. Reports on the issue are sketchy. According to an Egyptian newspaper in 1978:

> Egypt and the Sudan were studying with great interest feasibility studies being conducted by the USSR around Lake Tsana, where about 85 per cent of the Nile water originates. Egypt will not allow the exploitation of the Nile Waters for political goals, will not tolerate any pressure being brought to bear on it, or the fomenting of disputes between itself and its neighbours[114].

The Ethiopian Ministry of Foreign Affairs issued a series of terse and nonconciliatory responses directed largely to Egypt and in part, to the Sudan[115]. Their position was that 'Ethiopia has all the rights to exploit her natural resources'. Purportedly the statements also remind Egypt that, even though it receives 85 per cent of its Nile waters from Ethiopia, it has never shown friendship nor sought cooperation from Ethiopia. The Ethiopian statement points out that Egypt went ahead and built the Aswan Dam which has to depend on the Blue Nile waters, 'without even consulting Ethiopia.' [116] In the ultimate analysis, the situation illustrates a trend which will develop if basin states do not consult one another and develop a framework for cooperative utilization of the waters of an international river. This situation may have been mitigated by the discussions between the Ethiopian Head of State during his visit to Egypt in 1986 but the extent remains uncertain. However, a number of international conferences held between 1990 and 1992, among them the RGS/SOAS conference of May, 1990, have enabled representatives from upstream countries to project their particular points of view and there is now evidence of bi-lateral discussions between the various parties.

Tanzania

The Republic of Tanzania is understood to be planning two major development projects utilizing Lake Victoria Basin waters: one is to use the waters of the Kagera River as discussed above; and the other to abstract water from Lake Victoria itself for irrigation of the Vembere steppe in central Tanzania, which is outlined below.

Tanzania may have considered more than one approach to the utilization of waters of Lake Victoria, but one that stands out in history was narrated by H.E. Hurst from the Egyptian Ministry of Works, who went to Tanganyika in 1926 to ascertain if there was indeed such a plan for irrigation. He recounts the plan as follows:

> I found out that the Germans had, before the 1914-1918 War, a project to take water from Smith Sound, a long inlet at the south end of Lake Victoria, over the low country which separates the lake from the land sloping down towards Lake Eyassi. The water would have been used to irrigate arid land on the Vembere Steppe for the growing of cotton. The Scheme, which was not a government one, was to start on a small scale with a dam at Manyonga River to store its flood waters and irrigate a small experimental area. From this pilot project data would be built on the Manyonga, and hydro-electric stations at the dam would supply power to pump water from Lake Victoria. After passing through the turbine the water would irrigate land lower down and finally drain into Lake Eyassai.[117]

The area planned for irrigation in this project was 230,000 hectares, or 550,000 feddans. The extent to which this project has been considered seriously in modern Tanzania is not known. In critical commentary, Professor Rene Dumont wrote:

> The Smith Sound project, aiming to bring water at great cost from Lake Victoria to the south, will probably be worth studying towards the end of this century, to be finally carried out at the beginning of the next century. For the moment, the whole of small and medium-scale irrigation certainly has priority, especially in the spirit of the Arusha Declaration. I call attention to the Davidoff project from the era of Stalin, aiming to take into Central Asia water from the great Siberian rivers; it has been put off to a very distant date, very wisely[118].

As desert states that depend on the Nile waters, Egypt and the Sudan could have problems with the Smith Sound project, depending on the quantity of water to be extracted. At the time of his visit, Hurst thought the estimates to be about 82 cubic metres of water per second[119] and could make no appreciable difference to the Nile.'[120]

Kenya

The issue of first instance is that, although the level of Lake Victoria has had recorded variation in level over the years, an unusual trend began in 1961 and culminated in 1964 with a maximum rise of two and one half metres. This was an unprecedented rise[121] and the consequences in Kenya have been significant. There is loss around the lake of large tracts of land which have been inundated. Most of that land had been used for small-scale agricultural activities. The breeding grounds of some species of fish were submerged and resulting impact has been viewed as a possible contributor to the disappearance of some of the species, namely the *Tilapia esculenta* and *Protopterus*[122]. The increased flooding and swamps around the lake have provided breeding grounds for mosquitoes, creating a special health problem[123]. Finally, the raised level of the lake resulted in the submergence of pier facilities at Kisumu, Kendu Bay, Homa and Asembo Bays. Throughout the 1960s, temporary pierage facilities had to be deployed at each location until the East African Railways Corporation had the piers reconstructed in 1974. There may well be similar consequences felt in Tanzania and Uganda which remain unpublicized.

The theory that the control of outflow at Jinja is responsible for the increased lake level is strengthened by the background information on the construction of the dam which was to make the lake into a century storage head.[124] As evidenced in the background to the agreement discussed above, the dam was expected to produce an increase in the level of the lake to the margin projected. The agreement itself allowed for such a rise, and the conditions included provisions for compensation to injured parties. Therefore, arguments that the dam could not have caused the rise in the lake's level seem patently misleading.

The unprecedented rise in the level of Lake Victoria after 1961 was the result of the combined effects of the control works at Jinja and the heavy rains of the 1961-1964 period in the Lake Basin. A rise of up to 1.3. metres had been anticipated by those designing the Owen Falls dam; that the water level rose 2.5 metres reflects the impact of the unusually high precipitation in the early 1960s. Lake Victoria storage was increased to 170 km^3 causing flooding of the Lake Victoria shores of Uganda, Kenya and Tanzania, the release of very much higher than average volumes of water especially in the early 1960s, and even sustained higher than average flows through the 1970s with the consequence that the area of the Sudd doubled (See Chapter 12). The releases at the Owen Falls dam were consistent with the flooding in the Sudd and went some way to reduce the impact of storing over 30 times the normal annual recharge of Lake Victoria; in the period 1961-1964 releases averaged twice the normal level but this was apparently not sufficient to ameliorate the impact of the unusual weather events of the early 1960s.

It seems that the last word on the problem has not been said. The Hydrometeorological Survey team could not give a more complete answer to the question. Then, should Kenya and Tanzania find that report unsatisfactory, they could seek an agreement on a balanced formula for assessing the cause(s) of the unusual rise in the Lake Victoria. Some general observers report that the level has been rising since early 1978 and that the change already is noticeable at such popular spots as Hippo Point at Kisumu. The Ministry of Water Development is understood to be conducting studies to verify this state of affairs, and they may have their own explanations or a better hypothesis.

The establishment of the Lake Victoria Basin Development Authority to spearhead comprehensive development in the catchment area of Lake Victoria is a unique step. Through the working programmes of the Authority, Kenya has the means to consult with the other two riparian states. Programme planning might be risky unless such problems as an erratic rise in Lake levels are eliminated. On the other hand, increased use of the water of the rivers before they reach the lake may have an effect on lake levels. By the same token, if such utilization of water can make a difference in the lake level, then it could also have impact on the water flowing down the Nile. In this case, Egypt might want to discuss with Kenya the seasonality and quantity of water to be used on the Kenya side. This might be the case more particularly if Tanzania also decided to carry out the Smith Sound Project, because the combined impact of use by the major sources of Lake Victoria waters could make a significant difference to the water storage which Egypt has always coveted.

There is also the question of fishery resources in the lake. Although the popular view[125] is that there has been very little migration of fish in and out of Winam Gulf, conservation measures to maintain the proper resources balance might still be necessary. Absence of large-scale migratory patterns by the lake fish species does not mean that fish obey the territorial boundaries. Therefore, a basic consultative framework among the three littoral states will be needed if the Authority is to have effective long-term control of fish as an important resource.

It has been mooted in Kenya that, given adequate technology, Kenya should transfer Lake Victoria catchment water to the arid areas of the country for irrigation[126]. Perhaps the most appropriate location for that kind of experiment would be the Kerio Valley, for which a special development Authority has been established by the Kenya Parliament. The question of feasibility of such projects is an engineering one which, some observers say, is possible. Such an undertaking would use significant quantities of water if it were to be executed[127]. In large measure, the projects would be analogous to Tanzania's irrigation of the Vembere Steppes. It follows that Egypt would need to change its traditional position and opt for an agreement on a hydrological regime for the entire Nile Basin including Lake Victoria.

General

Some general developments in the international scene have had an impact on the use of internationally shared water resources. First, there are changes in the general political economy. No group of states demonstrated better that national natural resources are a powerful political weapon than the Arab States when they imposed an oil embargo against friends of Israel. This brings into question the new international economic order wherein states are called upon to cooperate in the management of resources to promote equitable development.

Secondly, the range of demands on water resources is increasing, and one of the most serious problems is that of pollution. As noted earlier, conservation of the resources of Lake Victoria must be approached on a lake-wide basis because pollution will not respect territorial boundaries. Municipal and industrial effluents discharged into one part of the lake in one of the three countries will have consequences for the other states. As shown by the experience of the North America's Great Lakes, Lake Victoria could become a cesspool.

Third, the applicable law on internationally shared water resources has been developing and is certainly more crystallized today than in 1960 when the last agreement on the Nile was signed. Therefore, it should be worthwhile for all of the basin states to take a fresh look at the local regime and begin working together on the formulation of a regional practice to meet local exigencies of the time.

Conclusion

Several conclusions have suggested themselves in this study. What seems clear throughout is the desirability of a framework for consultation and exchange of information on actual or intended projects involving utilization of the basin waters. One project to be accomplished within the framework could be actual hydrological and meteorological studies to ascertain basic or secondary facts and consequences of the use of such waters (see Chapter 7).

What the countries decide to call that framework is immaterial, so long as it involves all of the basin states and embraces the kinds of issue that have been apparent in the above analysis. It is recommended urgently that an agreement on a treaty creating a regulatory framework be reached, involving all the states of the Lake Victoria and Nile system. Such a framework would provide for the creation of development authorities to deal with development work for various parts of the basin, the latter category to include the Kagera Commission, and the Kenyan Lake Victoria Basin Development Authority. The disarray noticeable in the present treaty situation should not be allowed to continue.

Kenya, Tanzania and Uganda need to remember that, pursuant to the 1959 Agreement already discussed here, the Sudan and Egypt have undertaken to adopt a joint position in the event of any negotiations with third states. These two countries are better equipped in terms of hydrological and meteorological data, because they have worked at it since the 1950 agreement. They have been favoured, also in terms of access to the basic facts in the hydrometeorological survey carried out under the auspices of the World Meteorological Organization. There is very little advantage for the three East African countries in terms of qualified personnel to interpret technical information, as compared with Egypt and the Sudan.

The question is not one of renegotiation of the legal regime, but one of 'clean-slate' negotiation, because for the majority of the states within the drainage system there is no previously negotiated agreement which binds them. It is better to agree on such a framework while there is a propitious atmosphere than after a conflict of use has arisen among all or some of the basin states.

As a prerequisite to such a negotiation each of the basin states, especially the upper riparians, should work out a basin wide comprehensive Water Master Plan projected and phased up to, perhaps, fifty years. Within the scheme a careful assessment of available water and the possible uses for it, particularly for irrigation and industrial purposes, should be projected. Various aspects of possible inter-basin transfers within the riparian states should be assessed. And a determination made of the possible reserves.

It is pointless for the negotiations to start in the absence of such a Master Plan. A new agreement on the Nile should be long-sighted and realistic. For, indeed water will be increasingly important in the quest for stable agriculture and its place in viable economies for the African states.

Notes

1. Caponera, D.A. (1959) 'The Nile : Legal and Technical Aspects'. Mimeo paper of August 1958, being an English translation of an Italian version il Bachino Internationale del Nilo Consideration Giurridishe in XIV *La Communite Internationale* 45-46 (Jan.)
2. That was three years after the Sudan became independent . The Agreement on the Full Utilization of the Nile Waters was signed by the Sudan and

Egypt at Cairo on 8th November 1959. A most important feature was the Protocol Concerning the Establishment of the Permanent Joint Technical Commission for the implementation of that Agreement, which was signed by the two states at Cairo on 17th January 1960. See texts in United Nations Legislative Series, *Legislative Texts and Treaty Provisions Concerning the Utilization of International Rivers for Purposes Other than Navigation,* 143-49 UN Doc. ST/LEG/SER.B/12 (1962).

3. See comments in Okidi, C.O.(1988). 'The States and the Management of International Drainage Basins in Africa' in *Natural Resources Journal,* **Vol. 28** pp. 645-669 (Fall.).
4. Ongweny, G.S. (1979). 'Water Resources of Lake Victoria Drainage Basin in Kenya' in *Natural Resources and the Development of Lake Victoria Basin of Kenya* pp. 68-84, C.O. Okidi, (Ed). University of Nairobi. IDS/OP No. 34.
5. Kongere, P.C. 'Production and Socio-economic Aspects of Fisheries in Lake Victoria Drainage Basin in Kenya' in Okidi (1988), pp 407,410.
6. Ongweny, (1979).
7. Okidi, C.O. (1986). *Development and the environment in the Kagera Basin under the Rusumo Treaty.* (University of Nairobi, Discussion Paper No.284 Sept. The item was also covered widely in the popular press. See, for instance, *The Standard* (Nairobi, October 17th, 1978) p.8.
8. The Agreement for the construction of the Owen Falls Dam was reached through Exchange of Notes between Britain - the colonial administrator of Uganda - and Egypt. The construction started in May 1949.
9. *Report of the Hydrometeorological Survey of the Catchment of Lakes Victoria , Kyoga and Albert* (1974). (Burundi, Egypt, Kenya, Rwanda, Sudan, United Republic of Tanzania and Uganda), Meteorology and Hydrology of the Basin. Part II UNDP and WMO RAL 66-0250 Tech. Report No.1.
10. Garretson, A. (1967). 'The Nile Basin' in Garretson, A.H., Hayton, R.D., and Olmstead, C.J. (Eds). *The Law of International Drainage Basins.* 256-258. According to Garretson, of the 24 milliards (km^3) of water that flow downstream from Lake Albert (Mobutu Sese Seko) and the East African highlands, 12 milliards (km^3) are lost by evaporation in the Sudd area of southern Sudan (See also Chapter 12 p.).
11. Badr, G.M. (1959). 'The Nile Waters Question : Background and Recent Developments ', 15 *Revue Egyptienne de Droit International* 2. Badr estimates that the average 85% of total Nile discharge comes from Ethiopia. This estimate coincides with the figure given by Ethiopia. See also *Ethiopian Herald* (Addis Ababa) May 21st, 1978. Garretson draws attention to the fact that the Blue Nile supplies 90% of water passing Khartoum between April and September, but only 20% between January and March.
12. It seems that the first ever 'Agreement' on the Nile dealt with navigational uses of the river. It was expressed in the form of a unilateral declaration issued by the Viceroy of Egypt, under the Ottoman Empire on October 13th, 1841, granting foreigners the privilege of building ships for the

navigation of the Nile. See FAO *Systematic Index of International Water Resources Treaties, Declarations, Acts and Cases by Basins.* 45, 129, 135, 137, 146-7, 157-61. (FAO Legislative Study No 15, 1978)

13. UN Doc. ST/LEG/SER.B/12 (1963) pp.127-28
14. Hertslet, (1967). *The Map of Africa by Treaty.* **Vol.II No.100**, pp.432-42 (3rd Ed.)
15. UN. Doc ST/LEG/SER.8/12 (1963), *supra* Note 2 p.99
16. Hertslet,(1967). **Vol. II No.165**, pp.584-85.
17. 50 L.N.T.S. 282 (1926).
18. UN Doc. ST/LEG/SER.B/12 (1963), pp.100-107.
19. ibid. pp.108 -115
20. ibid. pp.143-48
21. ibid. pp.148-49
22. ibid. pp.127-28
23. Hertslet, (1967), pp.432-42
24. Caponera, *supra* Note 1, pp.13-14
25. See Articles 31-33 of the Vienna Convention on Treaties (1969)
26. See the relevant articles in UN Doc. ST/LEG/SER.B/12 (1963) *supra* Note 2, p.99
27. Hertslet, *supra* Note 23, pp. 436, 442
28. ibid.
29. 50 L.N.T.S. 282 (1926). In the exchange, the first Note (Dec 14, 1925) from Britain stated, inter alia 'I have therefore, the honour, under the instruction from His Majesty's Principal Secretary of State for Foreign Affairs, to request your Excellency's support and the assistance at Addis Ababa with the Abyssinian Government in order to obtain from them a concession for His Majesty's Government to construct a Barrage at Lake Tsana, together with the right to construct and maintain a motor for the passage of stores, personnel, etc, from the frontier of the Sudan to the Barrage ', (p.284). The Note added a *quid pro quo*: ' His Majesty's Government in turn are prepared to support the Italian Government in obtaining from the Abyssinian Government a concession to construct and run a railway from the frontier of Eritrea to the frontier of Italian Somaliland'. (p.285).
30. ibid. p.285
31. ibid. p.291
32. See discussions by Garretson , (1967). pp. 277-8
33. ibid., pp. 264 *et seq.*
34. ibid.
35. ibid., p. 268
36. ibid.
37. ibid. There were three members of the Commission , a nominee of the Indian Government, as chairman. A nominee from Cambridge University and a nominee of the U.S. Government. The US nominee, H.T. Cory submitted a separate report.
38. Batstone (1959), The Utilisation of Nile Waters, *8 int. and Comp. LQ.*, p.540

39. Garretson (1967), 264 *et. seq.*
40. ibid.
41. UN Doc. ST/LEG.SER.B/12 (1963), pp.100-7
42. ibid.
43. ibid.
44. ibid.
45. Garretson (1967), pp. 268; Caponera (1959), pp.10-11
46. Batstone (1959).
47. Text of Lloyd's Note is in UN Doc. ST/LEG/SER.B/12 (1963), p.107
48. Seaton, E.E and S.T. Maliti, (1973). *Tanzania Treaty Practice*, pp.90-91. Oxford, OUP.
49. ibid.
50. ibid.
51. ibid.
52. ibid. p.91
53. ibid. pp.148 -149
54. ibid.
55. ibid., Appendix V.
56. Badr, (1959).
57. Mutiti, M.A.B. *State Succession to Treaties in respect of Newly Independent African States* (East African Literature Bureau, 1976), p.23.
58. Hurst, H.E. (1952). *The Nile: A General Account of the River and the Utilization of its Waters* p.301. London, Constable.
59. ibid.
60. Hurst H.E.(1925, 1927). *Report on The Lake Plateau Basin of the Nile*, Cairo. Ministry of Public Works, Physical Department, Paper No 21.
61. Hurst (1952). *The Nile.* p.301.
62. ibid., p.302
63. ibid.
64. UN Doc. ST/LEG/SER.B/12 (1963), pp.108-9
65. ibid.
66. ibid., pp.111
67. ibid.
68. ibid.
69. ibid., pp.114-115
70. ibid., pp.108-109
71. ibid., pp.110-111
72. ibid., pp.114-115
73. ibid., pp.108-109
74. ibid.
75. ibid.
76. ibid. para (ii) of the Note from the Royal Egyptian Government.
77. As noted above, the last stage in the exchange of Notes constituting the Agreement came on January 5th, 1953. See text of both notes in UN Doc ST/LEG/SER.B/12 (1963) op. cit. pp.108-109
78. ibid. pp.112-113
79. ibid.

80. ibid.
81. ibid.
82. ibid. pp.143-148
83. ibid.
84. ibid.
85. ibid.
86. Badr (1959). p.20
87. ibid.
88. UN Doc ST/LEG/SER 12 (1963), pp.143-148
89. Batstone (1959), p.540
90. UN Doc ST/LEG/SER 12 (1963).
91. Article V of the 1959 Agreement.
92. UN Doc ST/LEG/SER 12 (1963). pp.148-149
93. ibid., pp.112-113
94. Hydromet Survey *supra,* Note, 9, Vol. 1 part 1, p.9
95. UN Doc ST/LEG/SER 12 (1963) op. cit. pp. 112-113
96. Seaton and Maliti (1973). p.91. Fahmy, (1977). *International Aspects of the River Nile.* (Conference paper) UN Doc E/CONF/TP 22, January 15, 1977, p.8
97. Seaton and Maliti (1977). p.92
98. ibid.
99. ibid.
100. Fahmy (1977).
101. ibid.
102. Hydromet Survey (1974).
103. Very little has so far been published on the Kagera Basin Organization. The information in this section is derived by the author's research published as *Development and the Environment in the Kagera Basin under the Rusoma Treaty.* (University of Nairobi, IDS/Discussion Paper No. 284. September, 1986); and Lwchabura, D.K. *'Cooperation in Management and Development of the Kagera River Basin'* (Kigali, May 1981)
104. At the time of writing the Agreement is pending registration by the UN Secretariat, for UNTS. A copy was obtained by the author from the KBO Secretariat.
105. Apart from the first preambular paragraphs of the original agreement, the other Articles affected by the amendments of 19th May 1981 were Articles 5, 6, 7, 9, 10, 13, 16, 18 and 20. Article 19 was repealed altogether. Thus, the last article was 21, followed by the testimonium.
106. See *The New York Times,* December 14th, 1975
107. ibid
108. *The New York Times* December 16th, 1979
109. See especially the five volume study *The Equatorial Nile Project and Its Effects in the Anglo-Egyptian Sudan: Being the Report of the Jonglei Investigation Team.* (Khartoum, 1954)
110. *The Weekly Review* (Nairobi), March 9th, 1979, pp. 26-27, May 5th, 1978 p. 2; May 12th, 1978, p.2; *African Business.* Nov. 1978, pp.14-16; *Earthscan Briefing,* Doc 8.

111. Mann, Oscar, (1977). *The Jonglei Canal: Environment and Social Aspects.* Nairobi: Environment Liaison Centre.
112. Howell, P., Lock, M. and Cobb, S. (1988) *The Jonglei Canal: Impact and Opportunity*. Cambridge University Press.
113. *Akhbar El Yom* (Cairo) May 13th, 1978
114. ibid.
115. *The Ethiopian Herald* (Addis Ababa) May 14th, 21st and June 2nd 1978.
116. ibid.
117. Hurst (1952). p.156 and Hurst, (1925/1927), pp.6-10
118. Dumont, R. (1969). T*anzania Agriculture after the Arusha declaration* (Dar es Salaam: Ministry of Economic Affairs and Development Planning) p.48.
119. Hurst (1925/1927). p.9
120. Hurst (1952). p.156
121. *Hydromet Survey* op. cit. pp.744-753
122. Welcome, 'The effects of Rapidly Changing Water Levels in Victoria upon Commercial Catches of Tilapias', in Obeng, L. (Ed) *Man-made Lakes: The Accra Symposium* 1969; and Kongere op. cit. and Odero' Fish Species Distribution and Abundance in Lake Victoria' in Okidi, *Natural Resources and the Development of Lake Victoria Basin.*
123. See strong contentions in the East African Legislative Assembly by Mr Orinda Sibuor and Mr Joseph Nyerere in June 1973, *The East African Standard* (NBI) June 23, 1973 p.5, and by Kenya's Minister for Health in *The East African Standard* (Nairobi) July 8th, 1976.
124 Edit. Comment: Dr. J.V. Sutcliffe writes: "The effect of the operation of the Owen Falls Dam on the rise in Lake Victoria during the exceptional rains of 1961-1964 has been examined by the WMO Hydrometerological Survey. The dam has operated on an 'agreed curve' with the aim that outflows follow the natural relationship between lake level and river flow; this agreed curve was derived from discharge measurements and extended by hydraulic modelling of the historical outfall at Ripon Falls. Kite (1981) has tabulated the cumulative effects of the operation of the dam on Lake Victoria levels and has shown that the effect was negligible." [Reference: Kite, G.W. (1981). Recent Changes in the Level of Lake Victoria, *Hydrol. Bull.*, **26**. 233-243].
125. Kongere (1988), pp. 407-413
126. The popular view has been expressed in the press. See *The Daily Nation,* (Nairobi) March 16th, 1979.
127. One expert opinion estimates that irrigable land in the Kenyan part of the Nile Basin is approximately 53,212 acres (22,000 ha) which would require 296.9 million cubic metres (0.29 cubic kilometres) of water annually, but that there may be additional areas that might require another 182 million cubic metres of water per year. There are other Kenyan estimates. See Dekke, 'A Note on the Nile' 8, *Water Resources Research,* No 4, pp.818, 827 (Aug 1972).

17

Principles and precedents in international law governing the sharing of Nile Waters.

S. AHMED

Introduction

The provisions and principles of international law, governing the utilization of the waters of international rivers, developed relatively recently, since the problems involving international rivers were primarily those concerning international navigation. The 1815 Vienna Congress ushered in the principle of the freedom of navigation in international rivers, the first such instance being in relation to the Danube.

The more recent application of scientific methods to the utilization of the waters of international rivers, the new trend towards the construction of dams, reservoirs, canals and the like, have shown that the problem of the proper utilization of the waters of international rivers is no less important than the problem of navigation; hence the increasing need for the formulation of precise rules governing the new problem.

As an international river, the Nile is naturally governed by the rules of international law on the administration and the uses of the waters of international rivers. Until the middle of the nineteenth century, the prevalent principle governing the uses of the waters of international rivers, i.e the Harmon doctrine, inferred that the absolute sovereignty of a state over its territory permitted that state a free hand in the exercise of its sovereignty over that portion of the international river which passed through its territory, without due consideration to the possible damage which this freedom might cause to the interests of other riparians. This doctrine held sway over the first part of the nineteenth century.

Fortunately, gone are the days of absolute sovereignty. With the progressive evolution of international law and the development of international institutions, more attention and more weight are now being given to the principle that 'rights' over international rivers should also be accompanied by certain 'obligations' towards other riparian states, obligations which limit and diminish a state's sovereignty over that portion of an international river passing through its territory.

By now, a majority of international lawyers and jurists confirm the rights of the other riparian states in international rivers, thus in effect discarding the Harmon doctrine. (UN Doc. 1952) A consensus of the opinions of

International law jurists and commentators holds that the rules of contemporary international law stipulate that a state cannot change the natural environment or topography of its territory in such a way as to constitute some damage or a significant change to the natural environment of a neighbouring state - sic utere tuo ut non alienum laedas - consequently no state should halt or divert the course of an international river; similarly a state should not exploit the waters of an international river in a manner which infringes upon the needs of another riparian, or prevents its proper utilization of its due share of its waters.

Article 38 (1) of the Statute of the International Court of Justice recognizes and accepts the significance of the contributions of jurists in the development and the enrichment of the rules of international law. The Article allows the Court to take into consideration the opinions of jurists as a secondary source in the determination of the rules of international law. Consequently, the findings and recommendations of such eminent legal bodies as the 'Institute de Droit International' and 'The International Law Association' carry a special significance in the development and the enrichment of international law.

Since 1910, the 'Institut de Droit International' has begun the study of the subject of the codification of the exploitation of the waters of international rivers in agriculture and industry, as well as a variety of other uses apart from navigation. In 1961 the study arrived at certain basic principles governing the rights and obligations which ought to be respected and accepted by the riparians of an international river. Among the most important of those principles are:

a. Cooperation in the utilization of the waters of that river.
b. Equality of distribution of its waters.
c. Due cooperation and consultation over proposed projects.
d. Adequate compensation for any possible damages befalling one of the riparians because of improper utilization by another.
e. The pacific settlement of disputes among riparian states, as an obligation arising from good neighbourliness.

On its part, the International Law Association, also after fifteen years of study, arrived at its Helsinki meeting in 1966 to a set of rules providing guidelines for the proper utilization and administration of international rivers, especially in those cases where there are no specific agreements or traditional norms of conduct among the riparian states. Those principles address themselves to the utilization of the waters of international rivers for irrigation, navigation, the transport of goods, and to the problems of pollution. The final provision of the Helsinki Rules stipulates the necessity of consultation among riparian states if any one of them wanted to begin construction works on the river, in order to avoid giving rise to disputes or misunderstanding. The same provisions explored ways and means to be followed for the solution of disputes.

The Helsinki rules define an international river by the newly-adopted term: an International Drainage Basin. The Rules distinguish between territories within the Basin which are entitled to priority benefits, and those outside the Basin of riparian states. The river basin's outside limits are delineated by chains of mountains, i.e. a watershed. The Rules, nevertheless, do not exclude the right of a riparian state to divert a part of its water allocation to its own territories outside the basin.

The most important Helsinki Rules comprise:

a. Equity of distribution is the governing factor among riparians.
b. Equity does not mean distribution by equal share, but by fair shares which can be decided by the following factors:

- the topography of the basin, in particular, the size of the river's drainage area in each riparian state;
- the climatic conditions affecting the basin in general;
- the precedents about past utilization of the waters of the basin, up to present-day usages,
- the economic and social needs of each basin state;
- the population factor;
- the comparative costs of alternative means of satisfying the economic and social needs of each basin state;
- the availability of other water resources to each basin state;
- the avoidance of undue waste and unnecessary damage to other riparian states.

If we apply the Helsinki definition of the 'International drainage basin' to the Nile river, north of Cairo, we find that this drainage basin extends eastward to the Sinai mountain ranges without any obstacles on the way. According to this principle the whole of this region therefore comes under the definition of the Nile Drainage Basin. The presence of the Suez Canal in between the Nile and the Sinai watersheds does not affect this, since the Canal is man-made and is not a natural barrier.

There is an obvious shortage in the number of juridical 'Court' opinions and decisions relative to international rivers, but the International Court of Arbitration and many other International jurists have made up for this shortage by the recourse to the decisions and findings of Federal Supreme Courts in federal states such as the United States, Switzerland, the German Federal Republic and the like.

One of the important cases submitted before international arbitration was the case of Lake Lanu situated between France and Spain. The International Court of Arbitration in its review of that case came out with certain relevant principles, namely:

a. the necessity of recognizing the right of sovereignty over its portion of the international river, of each riparian;
b. this right, however, should be subservient to all other international obligations of that riparian state;

c. there is no rule in international law which prohibits a riparian from the utilization of water-force to generate electricity, but in accordance with the rule of good faith the upper-river riparian should take into consideration, and on the same footing of equality, all the interests of all other riparian states;
d. the necessity of consultation and the exchange of all relevant information among the riparian states about any projected construction work on the international river.

Other Federal Courts in other Federal Governments have reached a consensus about the following principles:

a. International law limits the freedom of action of the riparian states of an international river; every one of them should avoid causing any detriment to other riparians;
b. equitable apportionment of the international river's waters;
c. due respect to acquired rights of the riparians;
d. the illegality of diverting the set course of an international river.

Water Commissions in the Indian Sub-Continent, in their review of inter-state water disputes, have added yet another important principle, namely that barren infertile lands have a priority over the waters of international rivers. Worthy of note, however, is the fact that the Helsinki Rules have not emphasized the priority status of barren lands even though they may be of primordial importance to many states.

International treaties and agreements on international rivers

A substantial number of international agreements among States sharing the waters of international rivers, as well as the studies conducted by the United Nations of international treaties dealing with international rivers constitute, in the aggregate, certain principles which point to the evolution of international law in this respect. These principles set aside the old theory of absolute sovereignty over an international river but accept and sanction the rights of all states sharing the waters of such rivers. They also guarantee to all riparians an equitable share in those waters, while imposing on them due respect to the rights of other neighbouring riparians. They also respect acquired rights, and prohibit the introduction of territorial changes infringing upon the status quo, without due agreement among all riparians.

Despite the fact that it should be difficult to establish a set of rules applicable to all international rivers, since the case of each river would have to be studied separately according to its specific circumstances, it follows from the above review that a consensus of the opinions of international jurists and the decisions of Federal Courts, as well as the existence of some basic rules of law governing the rights and obligations applicable to the riparians of an international river. Foremost among which are:

a. Equity of distribution and the utilization of the waters;

b. riparian states should refrain from either diverting the river course or constructing dams and reservoirs on the river which might reduce or affect the share of other riparians, without prior consultation and agreement;
c. due cooperation among riparian states in the improvement of the river's sources and flow and its utilization as a unit;
d. respect for the acquired rights of riparian states on the basis of the needs of each of them, and the degree of its dependence on the river's waters;
e. adequate compensation in case of damages.

The Organization of African Unity (OAU) Summit Conference at Algiers in 1968 came out with an Ecological Convention for the Preservation of Natural Resources. Article Five of the said Convention states that where surface or underground water resources are shared by two or more states, they shall act in consultation and if the need arises they should set up inter-state commissions to study and resolve problems arising from the joint development and conservation thereof. Article Sixteen also states that the contracting parties shall cooperate wherever any national measure is likely to affect the natural resources of any other contracting state.

International instruments governing the uses of the Nile waters
Following, in chronological order, is a review of the international instruments governing the uses and the sharing of Nile waters. They are eight in all, but it is important at the outset to shed some light and make a few points of clarification as to the signatories and the nature of those treaties.

The first six agreements, ending with the 1929 Agreement, have to do with the territorial status of the contracting parties. It is an agreed principle of international law that such territorial status agreements constitute an obligation and a limitation on the contracting parties' territory, unaffected by a change of sovereignty. The Vienna Convention of 1978 about State Succession and Treaties, has confirmed the above principle. Articles 11 and 12 of the said Convention stipulate that treaties dealing with the delineation of international boundaries or with territorial status are not affected by state succession. Such treaties remain valid and carry an obligation to the successor-state. They cannot be amended or abrogated except by the agreement of the signatories or in accordance with the measures stipulated by the Vienna Convention on the Law of Treaties of 1969.

The second observation to be made is about the signatories of the following treaties, for in many instances they were European colonial powers acting on behalf of an African colony or occupied country. Yet International Law recognizes the continuing validity of such instruments, in accordance with the law of state succession, and the territorial nature of the obligations arising from those treaties.

Moreover, the following treaties and instruments which govern and regulate the juridical status of an international river, the Nile, do not contain any exceptional or illegal principle. Rather, they merely confirm the principles already accepted by international jurisprudence and international norms, as well as the historical acquired rights which Egypt or some other country may have attained over many thousand years of dependence on the Nile as its sole life-giver:

1. The Protocol between Great Britain and Italy of 1891, for the demarcation of their respective spheres of influence in Eastern Africa. In its third article, the Protocol stipulates that Italy pledges not to construct on the Atbara river any irrigation work which might significantly affect the Atbara's flow into the Nile.
2. Treaties between Great Britain and Ethiopia; and between the first and Ethiopia and Italy, relative to the frontiers between the Anglo-Egyptian Sudan, Ethiopia and Eritrea, signed in Addis Ababa on May 15th 1902. In the third article of these Treaties, Emperor Menelek Second, the King of Kings of Ethiopia, engaged himself towards Great Britain not to construct or to allow to be constructed any work across the Blue Nile, Lake Tana or the Sobat River, which could arrest the flow from their waters into the Nile, except in agreement with the Government of Great Britain and the Government of the Anglo-Egyptian Sudan.
3. Agreement between Great Britain and the Congo Free State (now Zaire) signed in London on May 9th 1906 bringing modification to the Brussels Agreement of May 12th 1894. In its third article of the 1906 Agreement the Government of the Congo Free State undertook not to construct or allow to be constructed, any work on or near the Simliki or Isango rivers, which might reduce the volume of waters flowing into Lake Albert, except in agreement with the Government of the Anglo-Egyptian Sudan.
4. Exchange of Notes between the United Kingdom and Italy in December 1925, wherein the Italian Government recognizes the previously acquired 'hydrolic rights' [sic] of Egypt and the Sudan in the waters of the Blue and White Niles, and engage themselves towards the other contracting parties not to construct on the head-waters of the Blue Nile or the While Nile or their tributaries and effluents, any work which might substantially modify their flow into the main river.
   The Italian Government also took note that the Government of Britain has every intention of respecting the existing water rights of the population of the neighbouring territories which enter in the exclusive sphere of Italian economic influence. The Italian Government engaged itself that in so far as possible and was compatible with the paramount interests of Egypt and the Sudan, the 'scheme in contemplation' should be so framed and executed as to afford appropriate satisfaction to the economic needs of these populations.
5. Agreement between Egypt and Great Britain, (the latter on behalf of the Sudan, Kenya, Tanganyika, and Uganda) signed in 1929. This treaty stipulates that no work of any kind may be undertaken on the Nile, its tributaries or on the lakes which form its source, without Egypt's consent; and in particular if these works are related to irrigation or

power generation, or if they affect the volume of waters which reach Egypt, or in any other way be detrimental to Egypt. This treaty also stipulates that Egypt has the right to maintain supervision over the Nile from mouth to sources, conducting research; as well as over the implementation of any project which might prove beneficial to Egypt, and has the right to investigate feasibility.

6. Agreement between Great Britain (on behalf of Tanganyika) and Belgium (on behalf of Rwanda and Burundi) signed in London on 23rd November 1934, concerning the Kajera river, one of the tributaries of Lake Victoria. The first article stipulates that the contracting parties pledge to return to the river Kajera, before it reaches the common borders between Tanganyika, Rwanda and Burundi, whatever amounts of water which might be diverted for generation of power projects. Article four of the agreement allows for the diversion for industrial purposes of half of the volume of the river flow at its lowest season. Article six obliges the state which wishes to use the river's waters for irrigation purposes to notify the other contracting parties six months in advance in order to allow for possible objections to be raised and studied.
7. Exchange of notes between Egypt and Great Britain (on behalf of Uganda) in the period from July 1952 to January 1953, regarding Egypt's participation in the construction of the Owen Falls Dam for the generation of hydro-electric power in Uganda. It was agreed to heighten the Owen Falls Dam so as to raise the water level in Lake Victoria. Compensations were agreed upon for Uganda whose lands would be detrimentally affected by the rise of the water level in Lake Victoria. The rise would allow Egypt more water for irrigation; while the hydro-electric power generation would provide both Uganda and Kenya with more electricity.
8. Agreement between the Republic of the Sudan and the United Arab Republic (Egypt) signed on 8th November 1959, for the maximum utilization of the surplus waters by the two countries; and the utilization of the surplus waters resulting from the construction of the High Aswan Dam. The average annual flow of 84 billion cubic metres (km^3) of water was divided between the two signatories. Egypt's share would be 55.5 billion cubic metres per annum; the Sudan's would be 18.5 billion cubic metres per annum. Some 10 km^3 per annum was estimated to be lost in evaporation at the Aswan High Dam Lake (Lake Nasser).

The Egypto-Sudanese Nile Waters Agreement of 1959 has been regarded by some jurists as a model International Rivers' water-sharing agreement. It contains some 'advanced' ideas and principles governing cooperation and sharing of efforts and burdens relative to the international river. A joint Technical Commission is entrusted with the study of the various Nile projects. The principles of equity, compensation for damages and respect for acquired rights have all been included in the Agreement.

The Agreement has anticipated the possibility of other riparian states making further claims on larger shares of the Nile waters. Egypt and the Sudan have agreed that they should unify their stand vis-à-vis such future claims before entering - as one front - into negotiations with other claimants.

Thus, the obligations carried by the two parties according to the Agreement can be divided into two categories:

1. Obligations (Egyptian and/or Sudanese) vis-à-vis one another concerning, in general, cooperation and co-ordination in the fields of Nile waters sharing, utilization, administration, present and future, as well as legal obligations prohibiting unilateral actions or actions entailing any possible damage to the other partner.
2. Joint Egypto-Sudanese obligations towards other riparian states emanating from the Agreement's injunction to the two parties that they confront the other riparians as 'one unit' whenever the other riparians come forward with water claims.

Non-juridical precedents of Nile waters sharing: practical and diplomatic steps

Article Five of the Egypt-Sudanese Nile Waters Agreement stipulated the establishment of the 'Permanent Joint Technical Commission' mentioned above formed of an equal number of experts from each party. Since October 1961, the Commission has held meetings with representatives from Tanzania, Uganda, and Kenya, resulting in very fruitful exchanges of views on all technical aspects of Nile administration and the maximization of the benefits of water sharing. So much so that in 1967 it was decided to start a new Technical Commission made up of Egypt, the Sudan, Tanzania, Uganda, Kenya, Rwanda, Burundi and Zaire, to conduct hydrometeorological surveys of the catchments of lakes Victoria, Kioga and Albert. Ethiopia first participated as an observer, but has intimated a desire to attend in the future as a full member.

The objective of the above project was to collect and analyse hydrometeorological data of those catchments in order to study the water balance of the Upper Nile. This should assist the riparian states in the planning of water conservation and development and in providing the groundwork for future inter-governmental cooperation in the regulation and utilization of Nile waters.

The on-going project undertakes some specific interesting tasks:

a. Setting up additional data collecting stations - 24 hydrometeorological, 156 rainfall, including 6 rainfall recorders, 67 hydrological and 14 lake level recordings; and upgrading some of the existing stations in order to complete an adequate network from which basic data can be collected and analyzed;
b. establishing 7 small-index catchments for intensive studies of rainfall-runoff relationships for application to other parts of the catchment areas;
c. aerial photography and ground survey of those sections of the lake shore areas which are flat and which will be most subject to change due to

variations in levels of the lakes. Since only a few contours above and below lake waters level will be required, aerial photography will be used to prepare planimetric maps, and the contours will be properly located by topographic survey;

d. devising and proposing analytical procedures for the various parameters involved in the water balance of the lakes using data collecting from the new and existing stations, and from index catchments;
e. the training of the staff of the participating governments in hydrometeorological work.

The Nile Basin Commission

Cooperation in the area of hydrometeorological survey was a highly positive development in the field of cooperation among the States of the Nile Basin. As the Nile Basin constitutes a hydrological unit, providing a formidable base for fostering and reinforcing fruitful cooperation among the riparian states, the Egypto-Sudanese Technical Permanent Commission urged other riparians to join in a larger Nile Basin Commission. In 1977 a project was presented to the meeting of the Technical Committee on Hydrometeorological Survey of the Equatorial Lakes Plateau, which was held in Cairo in December 1977. Representatives of the riparian states approved the principle of establishing such an enlarged Commission. However, it was left to the governments concerned to decide on the political aspects involved in the establishment of such a far-reaching body.

Since then, an important development has taken place. The governments of Egypt, the Sudan, Uganda, Zaire, and the Central African Republic has been holding ministerial-level meetings with a view to coordinating their policies on a host of political and technical matters, including Nile waters, under what they have come to call the UNDUGU grouping. *Ndugu* is a Swahili word meaning brotherhood. Later on, Rwanda and Burundi decided to attend the Undugu meetings. The Undugu group has held already four ministerial-level annual sessions, and discussed a variety of political and technical matters, not excluding the benefits of Nile waters sharing and exploitation. Once Kenya, Tanzania and Ethiopia decide to join in these exploratory 'brotherhood' talks, the way may be open for the establishment of a Nile Basin Commission, and, hopefully, later on, of a Nile Basin Economic Community.

According to the African Summit (1980) Lagos Plan of Action, - which envisaged the establishment of an African Common Market as a final target - the stages of implementation of such a plan should include 'an African commitment to strengthen the existing regional economic communities, and establish other economic groupings in other regions of Africa, so as to cover the Continent as a whole; joint river-lake basin organizations should be established to promote inter-governmental cooperation in the development of shared water resources. Member states requiring such institutional arrangements should immediately start negotiations so that the new joint organizations can be brought into existence'.The UNDUGU grouping is just such an attempt.

A case study: The UNDUGU grouping of states

UNDUGU, is an unofficial African regional grouping created in 1983 in fulfilment of the resolution of the 16th OAU Summit of July 1979 at Monrovia, which called for self-reliance and African inter-dependence. It came also in fulfilment of the Lagos first African Economic Summit of April 1980 which specifically called for the creation of regional and sub-regional economic groupings as a first step towards inter-complementarity and the creation of an African Common Market by the year 2000.

The following is a summary of the UNDUGU grouping activities:

- The First Ministerial Session was held in November 1983 at Khartoum and was attended by Egypt, Sudan, Zaire, Uganda and the Central African Republic.
- The Second Ministerial Council was held at Kinshasa in September 1984.
- The Third was held in Cairo in August 1985.
- The Fourth was held in Kinshasa in May 1988.
- The Fifth was held in Cairo in May 1988.
- The Sixth and most recent Ministerial Council Meeting was held at Addis Ababa in February 1990, in conjunction with the Ministerial Council of OAU. The Sixth UNDUGU session was attended by Egypt, the Sudan, Zaire, Uganda, Tanzania, Rwanda, Burundi and the Central African Republic.

Members of the UNDUGU grouping now comprise Egypt, Sudan, Zaire, Uganda, Tanzania, Rwanda, Burundi and the Central African Republic. It still remains for Ethiopia and Kenya to join the UNDUGU grouping so that it covers all the Nile riparian states.

At the Fifth and Sixth sessions of UNDUGU, representatives of the following organizations attended as observers: ECA, UNDP, OAU, Association of African Highways, Kajera River Development and Management Organization, African Association for Communications and Telecommunications, African Association for Railways and the African Association for Pedagogical Sciences.

At the Fifth UNDUGU Ministerial Council Egypt submitted a review of its official contacts held with UNDP, at the instance of a previous UNDUGU decision, with a view to the UNDP undertaking an extensive technical and economic feasibility study of future cooperation among the members in the various economic, social, cultural and technical fields, and especially upon translating previous technical plans into concrete regional projects giving priority to the development of the members' infra-structure and in strengthening commercial exchanges among UNDUGU members in accordance with the Lagos Plan of Action.

Consequently, the UNDP sent two Missions during 1989 to visit the Nile Basin states to study the possibilities of enhancing cooperation among them. The first was a preliminary or exploratory mission, which visited some UNDUGU members during the one month period from 16th February to 18th

March 1989. The mission included Ambassador Paul Mark Henry and Mrs Tatiana Androsov and was in direct response to an invitation issued to UNDP by the then Minister of State for Foreign Affairs of Egypt, Dr Boutros Ghali, in the name of the UNDUGU Members, and in accordance with the Fifth UNDUGU Council's directive of November 1988. In its report, the UNDP mission laid the bases for a more comprehensive study of an economic and technical cooperative development Plan for long-term (25 years) infrastructural development among members in the fields of: highways, railways, rivers, air transport, energy, water resources, communications and telecommunications, and commercial exchanges.

The Second Mission was a fact-finding mission which visited the Nile Basin states during the period from 13th May to 23rd June 1989. It included nine experts from the UNDP under the chairmanship of a water resources engineer. The mission's task was based on the recommendations of the two Nile riparians' Ministerial Councils held at Bangkok in January 1986 and at Addis Ababa in January 1989 respectively, with a view to benefiting from the experience and achievements of the Mekong River Commission, for the future development of Nile resources. In its report, the fact-finding mission suggested a framework for regional cooperation among the Nile riparians, drawn from their own previous plans and projects. It also presented an evaluation of Nile water resources and the needs of the riparian states' populations, both in the long and medium terms. It also proposed a preliminary plan of action for the control of Nile resources.

While the preliminary (first) mission's report presented an account of the various fields of cooperation among the Nile riparians, the second mission, the fact-finding mission, concentrated, on the other hand, on water-resources; on the generation of electricity therefrom in view of the expected rise in demand both for water and electricity; on the expected rise in population and the necessity for food-security; on increasing arable lands by irrigation; and on the possibilities of developing fish resources in order to provide food proteins at the lowest possible cost by developing fish farms throughout the Nile Basin.

In short, the two missions' reports were complementary, and should be taken as a contribution to a comprehensive plan of cooperation for the long-term, while laying down at the same time a list of priorities of implementation, depending on future assessments of potential financing, both internationally and locally.

The proposed UNDP Development Project

The recommendations of the two UNDP missions for a comprehensive Plan of Action can therefore be summarized as follows:

1. The Plan of Action should be both comprehensive and long-term, so as to meet the developmental needs during the next 25 years, and not stop at improving the present infrastructure. It should aim at mobilizing additional financial resources and more investments in order to implement the projects of the comprehensive plan of action. The plan should also make allowances for the expected doubling of the populations in the Nile Basin states during the coming 25 years, the immigration flow from the countryside into the the urban centres, and finally the exigencies of the population explosion in terms of providing food supplies, goods and necessary services, as well as the protection of the

environment.

2. The necessity of the study of joining all the Nile Basin states by one electricity grid, making use of the projected extension of the lines of electricity generated at the Zaire water falls at Enga, extending them to Aswan(Egypt) and possibly extending the surplus from Aswan towards the North Africa States and the Middle East and even Europe.
3. Avoiding as much as possible, duplication in the planning and execution of projects, especially among neighbouring states, whenever the products of any one project suffice and meet the needs and the demand in the markets of the respective states, and suffice to export the surplus towards foreign markets. Priority should be given in the execution and enlargement of projects to already existing ones.
4. The fact finding mission estimated the needs of the Nile Basin populations of additional water resources by the year 2010 at about ten billion cubic metres (km^3) annually in order to meet the irrigation needs of Egypt(one million additional hectares needed) Rwanda, Burundi, Uganda, Kenya and Tanzania (400,000 additional hectares each). Surplus water resources should be divided between Egypt and the Sudan on the one hand, and the equatorial lake states surrounding Lake Victoria on the other, each side getting 50 per cent. The Mission found it impossible to increase the additional water resources needed by diminishing the amounts of wasted water through evaporation, especially in the southern Sudan, but did recognize the possibility of the improvement of water storage in the upper White Nile, which would necessitate a very close cooperation in water management among the Nile riparians in order to increase the flow of water into the White Nile north of Malakal.
5. The Mission estimated the energy productivity in the White Nile states in 1988 to be 4100 GWH. It also estimated the average yearly increase in energy production in all the Nile riparians at 3.5 per cent to 12 per cent or an additional 42,000 GWH per annum. The expected increase could be achieved through the development of water falls generated energy in a more balanced fashion, in addition to what Zaire could provide through the extension of the electricity grid from Enga to Aswan in Egypt and as a result of the improvement of water storage facilities and increasing water resources in the equatorial lakes region; and the joining of all riparian states by one and the same grid, in addition to the present incomplete grid joining Kenya, Uganda, Rwanda, Burundi and Zaire. Thus an element of stability could be introduced into the generation and the utilization of electricity.
6. Improvement of fisheries in general and fishing in the Nile Basin states especially in the Sudan, Rwanda and Burundi; improvement of fisheries in general in all other riparian states as a means of satisfying the needs of the expected increase in population from 114 millions in 1986 to 164 million by the year 2000, since fish is one of the cheapest sources of protein. The actual fish consumption ranges from 1 kilogramme per person a year in Rwanda, to 9 kilogrammes in Egypt. The UNDP Mission foresaw the possibility of increasing the average consumption per person in all the Nile riparian countries, with due regard to Egypt and its capabilities in this field.
7. The necessity of overcoming the negative effects of the increasing desertification owing to the clearing of forests in order to create more arable lands, and the effects of the use of chemicals in agriculture and industries, and the effects of industrial wastes, all of which have an adverse effect on the quality of life.
8. The necessity of preparing an 'executive program' from the development of water resources which would improve the flow and the utilization of water in the countries suffering from water shortages, and which would provide about 10 billion cubic metres (km^3) for the use of all the riparians. This programme should rely on the following means.
   a. Decreasing waste:
   Through the improvement of water management and making use of Egypt's acquired expertise; by the re-use of drainage water mixed with fresh water; and by decreasing the quantities of water lost into the sea; and by lessening wasted water in the upper reaches of the Nile; by the improvement of storage facilities; and by lessening wasted water due to excessive evaporation, all of which techniques should provide sufficient water for irrigation and for the extension of arable lands.
   b. Increasing water resources:
   Through the implementation of water preservation projects in the southern Sudan

especially through the completion of the Jonglei and the Machar canals (the latter on the Sobat), and through better control over the marshlands in the southern Sudan. (but see Chapter 12).

Finally the fact finding mission proposed that the 'executive programme' should include regional projects for the development of the Nile Basin, in addition to the projects already proposed by a technical group of experts held at Kampala 5-8 February 1989. Worthy of mention among suggested regional projects were:better management of the equatorial lakes; improvement of flood and drought forecasting systems; water fall generated electricity control and better evaluation; the creation of a data bank, and a new system of exchange of information between data banks among the riparian states; the creation of a regional centre for social studies; the setting up of a master plan for land survey and land reclamation; assessments of the possibilities of irrigation from the equatorial lakes; the control of wasted water; the development of fish farms, and an overall evaluation of transport and communications needs within the framework of an overall Nile Basin development plan.

The Mission estimated the requirements of financing the executive programme at 40-60 billion dollars over three to five year period. Two thirds of the credits should be financed through foreign aid and the rest by local currencies. The UNDUGU initiative has continued to be strongly promoted by Egypt as a framework for basin-wide cooperation. Egypt and the Sudan recognize the utility of creating staller agreements which will in turn enable the management and delivery of secure volumes of water both now and in the long-term. That two of the most significant upstream riparians, Ethiopia and Kenya, have so far not participated in the initiative, is an unfortunate impediment to the implementation of the principles being developed at the meetings of the Ministerial Council. But such is the need for the type of cooperation envisaged in the UNDUGU principles that it is appropriate that the agendas of subsequent Ministerial Councils will be expanded to include the issues which non-participants concede to be crucial to a sound basis for the basin-wide monitoring and management of water.

References

United Nations Document E/ECE/136, (1952)

Ministry of Foreign Affairs, Arab Republic of Egypt (1984). *Egypt and The Nile,* Ministry of Foreign Affairs, Cairo.

18

Law and The Nile River: emerging international rules and the *Sharīʻa*

C. MALLAT

Introduction

The law governing the Nile waters can be assessed against a background of several legal traditions, including the domestic laws of individual riparians, documented decisions of federal courts in countries like Australia, Germany or the United States, existing regional arrangements on international rivers, Nile-specific agreements between the basin states, as well as customary law, particularly among the peoples of the basin. There is a vast literature on the subject,[1] and the purpose of this chapter is not to review the subject overall but is much more modest in scope. Discussion will be limited to two traditions which could be of particular relevance to the subject of the Nile: the present reassessment and formulation of international watercourse systems law, and the Islamic legal tradition and some of its possible bearing on the law of water.

As for public international law, a survey of some recent developments is based on a comparison between the draft text established in the course of the 1980s by the International Law Commission, and the Restatement of the International Law Association in 1966, known as the Helsinki Rules. The factor-analysis approach, which is at the root of the international law experts' philosophy in both instances, will be assessed in the light of the Nile case. This is followed by an appraisal of some strands of the legal tradition offered by selected texts of the *sharīʻa*, with particular emphasis on their possible contribution to a comprehensive legal agreement on the management of the Nile resources. The study work is part of ongoing research in relatively uncharted terrain, which tries to throw some light on a delicate and complex issue. The situation in terms of international watercourse law is itself in a state of flux. In the circumstances of the scarcity of the Nile waters, the problem is the more daunting in the absence of specific rules, and of a comprehensive agreement on the present and future allotment of water use.

General uncertainty in international river law is exemplified by the mercurial nature of the Helsinki declaration of 1966, and a new scrutiny of the field by international lawyers is being conducted at present. Since the early 1970s, the International Law Commission has been working on codification of riparian law, and some of the results of its decade-long efforts are presented in the course of this paper.[2]

On the more specific rule of law relating to the Nile, three clear problems emerge, which add to the uncertain legacy of the framework provided by the Helsinki rules.

The first problem arises from the absence of a comprehensive text in common between the nine riparian states. The diversity of the various countries in terms of need, sheer interest, and mere geographical positions, is no doubt at the origins of the complexity of the issue taken as a whole, and so far, of the lack of success in offering an integrated legal approach.

The second problem is rooted in the historical span covered by few and far-between treaties, agreements, letters of intent, memoranda, and preparatory works, in a context of a colonial, then sovereign set of states. The earliest text mentioned in the literature goes back to 1891, with the Protocol Between Great Britain and Italy for the Demarcation of their Respective Spheres of Influence.[3] Since then, important texts have included the 'Exchange of notes in regard to the use of the waters of the River Nile for irrigation purposes between Egypt and the United Kingdom in 1929, and the revised Agreement Between the Sudan and Egypt for the full utilization of the Nile Waters' in 1959.[4] Fractional and ad hoc arrangements have tended to further cloud the overall picture of the legal rights over the use of the waters.

The third problem is also connected to the time factor, except that it is now 'projective'. Even very clear texts, such as the 1959 Egypt-Sudan agreement, seem to offer no more than a framework to a daunting problem constituted by climatic changes and the response of human interventions set to harness the flow of the waters, dramatically illustrated by such structures as the the Aswan dam and the Jonglei canal. At the heart of this problem is the difficulty of projecting into the future a number of interlocking data, such as the flow of the water, human intervention, and individual states' needs as regards the river Nile.

These three elements of the shifting equation -the absence of a comprehensive Treaty framework, historical legal fractionalization and mercurial projection interact in a way which seems to defy legal analysis, especially when the projection- must be made in conjunction with political and demographic unknowns in a tumultuous East African context.

It is not our purpose to offer ready answers to these questions. The chapter tries, rather, to look into some legal traditions that are relevant to water-sharing. Some of these traditions have been studied already with some care, in connection with international waterways, and we will consequently not dwell on them comprehensively. Less known perhaps are local practises of water laws in the Middle East. A study commissioned by the United Nations Food and Agriculture Organization in the 1950s has yielded interesting though incomplete results.[5]

Two other legal traditions may also offer some relevance: customary law and Islamic law. Howell has written extensively on the customary law in the Sudan,[6] and there may be very important arrangements specific to a given region of East Africa which can enlighten and enrich the debate on interstate watersharing. As for Islamic law, and despite a few remarks on the significance of the tradition by well-known *sharī'a* scholars,[7] there is in the twentieth century secondary literature an impressive dearth of material.[8] The classical law, however, is replete with the *sharī'a* contribution on water rights, and we propose to examine some of these classical texts more closely.

Public international law and domestic law operate in two separate spheres, and there is no intrinsic or consequential rapport between the two spheres.[9] In defence of the relevance of the 'domestic' tradition in the case of the Nile, including the one of interest here, that of Islamic law, it ought to be stressed that watersharing rules are a relatively unsystematic field, where the concern for a norm is high, but where models are insufficient. It is with the perspective of this insufficiency that one may turn to paradigms drawn from alien traditions - that is traditions alien to public international law. These traditions could offer challenging contrasts to such principles as those of international law, whether in the 1966 Helsinki Rules, or in the recent drafts of the International Law Commission.

In order to bring some focus to a vast and unwieldy field, it may be worth starting the analysis on a concept which encapsulates the heart of the Nile dilemma: the concept of 'natural and historical rights'. This concept appears in the Exchange of Notes in 1929 between the Sudan and Egypt, which established Egypt's 'natural and historical rights' to the Nile waters.[10]

Against this concept is the absolute sovereignty attitude of the upstream riparian state. This is the much decried Harmon doctrine which argued that water belonged to the natural entity in which it was located.[11] Arguably, in the case of the Nile, the 'natural and historical rights' doctrine and the Harmon doctrine constitute the two poles of the problem, with Egypt ultimately advocating the first doctrine to protect the Nile from being tampered with by upstream states (Ethiopia - and conceivably the Sudan and the Equatorial states), and the upstream states, conversely, finding their guiding principles in the Harmon doctrine. Obviously, the positions are not as sharp as they may have once been, as perhaps all the Nile states have now accommodated to the idea of some form of water sharing, which is a compromise between the two extreme doctrines. Still, the two doctrines offer neatly defined categories. The natural and historical rights doctrine (and conversely the absolute sovereignty doctrine of Harmon) can therefore serve as a counterpoint to the two legal traditions that are examined more closely: public international law, and Islamic law.

Public International law - from the International Law Association to the International Law Commission.

The United Nations International Law Commission has been involved since 1971 with the drafting of a comprehensive Code of Non-Navigational Uses of International Watercourses. A draft text has been adopted which covers three areas of prime importance:

The Scope of the Code

In the first place, the International Law Commission defined in articles 2-5 the scope of its work: the articles in the draft 'apply to uses of international watercourse[s] [systems] and of their waters for purposes other than navigation and ... measures of conservation' (art.2.1).[12] In Art.3, the meaning of a watercourse state is narrowed down.[13] This is followed by the relation of general principles of international watercourse law to more specific agreements 'concluded between two or more watercourse states... which apply and adjust the provisions of the present articles,' and which should not 'adversely affect, to an appreciable extent, the use by one or more other watercourse States of the waters of the international watercourse [system]'. Art.4.3 further establishes the necessity of consultation before such a specific agreement is concluded, with the right to participate and negotiate for any 'watercourse State whose use of an international watercourse [system] may be affected to an appreciable extent'.

The process of consultation

The process of consultation was precised in the articles adopted at the 40th meeting of the Commission, in 1988. They constitute Part III, Planned Measures (arts 11-21). Art.11 requires the 'exchange of information' over planned measures and Art.12 insists on 'timely notification' before the implementation of 'planned measures which may have an appreciable adverse effect'. From the time notification is made, a period of six months runs (Art.13), during which the notifying state 'shall not implement.. the planned measures without the consent of the notified states' (Art.14). If the notified state, within these six months, finds that the measures planned are inconsistent with the rules of Arts. 6 and 8 (See infra), it will provide the notifying state with a documented explanation of the appreciation of this inconsistency. In this case, a period of consultation will follow, during which, and for another period of six months, the notifying state will refrain from implementing the controversial measures. (Art.17).[14]

The remainder of the articles in Part III address more detailed questions related to absence of reply to notification (Art.16), the possibility for a state that may feel threatened by activities of another riparian in the case of absence of notification, to have recourse to information, to enter consultation with the state under similar procedures, and to benefit from the six-month grace period. (Art.18) There are also provisions in case of the implementation of urgent measures and for information vital to national security (Art.19), as well as ways to provide for indirect contacts between states which for whatever reason do not communicate directly (Art.21).

Watersharing Principles

Part II, on general principles, provides the most significant aspect of the regulation of international watercourse law.[15]

Article 6 establishes the principle of optimal utilization of an international watercourse system by all the riparian states. This 'right to utilize' is set against the necessity of 'an equitable and reasonable manner' in exercising it. Optimum utilization must be consistent with the adequate protection of the system, and must be achieved through interstate collaboration. This is the 'duty to cooperate', mentioned in article 6 as well as in article 9. It is developed in the consultation process of Part III mentioned earlier. Also significant is Art.8, which establishes the limits on the right to optimal use: 'Watercourse States shall utilize an international watercourse [system] in such a way as not to cause appreciable harm to other watercourse States.'[16]

On the one side therefore is the right of optimal use, with on the other side the necessity to avoid appreciable harm to a co-riparian. These two poles could be considered as a compromise solution between the Harmon doctrine and that of 'natural and historical rights' known to the Nile. The question is then how to appreciate the strength of each vis-a-vis the other. Indication on how the balance will be achieved can be seen in the factor-analysis approach of Article 7 of the International Law Commission draft, which was completed in 1987:

Article 7 - international Law Comission draft

Factors relevant to equitable and reasonable utilization

1. Utilization of an international watercourse [system] in an equitable and reasonable manner within the meaning of article 6 requires taking into account all relevant factors and circumstances, including:

(a) geographic, hydrographic, hydrological, climatic and other factors of a natural character;
(b) the social and economic needs of the watercourse States concerned;
(c) the effects of the use or uses of an international watercourse [system] in one watercourse State on other watercourse States.
(d) existing and potential uses of the international watercourse [system];
(e) conservation, protection, development and economy of use of the water resources of the international watercourse [system] and the cost of measures taken to that effect;
(f) the availability of alternatives, of corresponding value, to a particular planned or existing use.[17]

There does not seem to have been much controversy surrounding the adoption of these articles by the International Law Commission, save for disagreement on the concept of system, and for the exclusion of the concept of water as 'shared natural resource', as it appeared in the earlier texts of the Commission. Also, the word determining, which was to introduce the factors in article 7, was deleted to allow more flexibility in the interaction between the various factors.

With one exception, the codification of the International Law Commission does not seem to have broken in any significant way the legal pattern of international watercourses as defined two decades earlier in Helsinki. The present text offers some simplification of the 37 articles of Helsinki, and develops the mechanisms of regional consultation. As for watersharing principles, the present article 7 reproduces in the main the factor-analysis criterion adopted in Art. V of the Helsinki text.[18] The similarities and differences between present Article 7 and earlier Art.V can be seen in comparing the two texts:

Article V - Helsinki text

(1) What is a reasonable and equitable share within the meaning of Article IV[19] is to be determined in the light of all the relevant factors in each particular case.
(2) Relevant factors which are to be considered include, but are not limited

to:

(a) the geography of the basin, including in particular the extent of the drainage area in the territory of each basin Sate;
(b) the hydrology of the basin, including in particular the contribution of water by each basin State;
(c) the climate affecting the basin;
(d) the past utilization of the waters of the basin, including in particular existing utilization;
(e) the economic and social needs of each basin State;
(f) the population dependent on the waters of the basin in each basin State;
(g) the comparative costs of alternative means of satisfying the economic and social needs of each basin State;
(h) the availability of other resources;
(i) the avoidance of unnecessary waste in the utilization of waters of the basin;
(j) the practicability of compensation to one or more of the co-basin States as a means of adjusting conflicts among uses; and
(k) the degree to which the needs of a basin State may be satisfied, without causing substantial injury to a co-basin State;

(3) The weight to be given to each factor is to be determined by its importance in comparison with that of other relevant factors. In determining what is a reasonable and equitable share, all relevant factors are to be considered together and a conclusion reached on the basis of the whole.

Differences. The most significant difference between the Helsinki factor-analysis list and that of the International Law Commission draft can be seen to disfavour, in the case of the Nile, the Egyptian position. This is the factor mentioned in section (d) of Art.7: 'existing and *potential* uses of the international watercourse [system]'. This idea of 'potential uses' of the International Law Commission had been expressedly rejected in the Helsinki rules, which stated in Art.VII that 'a basin State may not be denied the present reasonable use of the waters of an international drainage basin to reserve for a co-basin State a future use of such waters.' In the Nile context, the Helsinki rules are a consecration of the status quo for Egypt, whereas the more flexible present approach of the International Law Commission suggests that the potential use (read here of Ethiopia and the other upstream states) must be taken into account along with the existing ones.

This may be a major departure from the previous rule, but it does not mean that under an impending agreement on the Nile, a state (e.g. Ethiopia) could *reserve* its right to the potential use of the waters. But in the analysis of factors, the former prevailing dimension of present over future use has now given way to an equal assessment of the use over time.

The factors of population (Art. V, section f, of the Helsinki declaration) and of the availability of other resources have also been reconsidered by the International Law Commission. Indeed the population factor has been deliberately ignored in the new text, and the absence of the 'factor population' appears to weaken the legal position of a more heavily populated Egypt.

In contrast, and in favour of Egypt and other downstream states, 'the availability of other resources' has now been replaced by the availability of *alternatives*. This is a slight distinction, which can be understood to be less specific to the rainfall element, which is almost non-existent for Egypt in contrast to the equatorial states and Ethiopia. 'Alternatives' is a word connoting human engineering, whereas 'resources' rings of more natural phenomena, such as rain. Thus 'the resource factor' weighs in the present balance more favourably on the side of the downstream states.

All in all, these differences notwithstanding, there is little dramatic departure from the Helsinki rules in the draft of the International Law Commission, and the factor-analysis approach has remained at the heart of the codification effort. This is not surprising, considering the variations in the rivers of the world and the difficulty of setting up a comprehensive and precise water allocation scheme in a single Code. Still, there is little doubt that the more factors enter on the balance, the more difficult it is to draw a legally precise picture of the rights of each state. The 1966 Helsinki declaration as well as the present text of the International Law Commission, will deserve to be complemented in each local application.

A Case study

A hypothetical case - which is strikingly reminiscent of the present Nile situation - was presented in the study of the International Law Association under the factor-analysis of Article V. Here is the simplified form of the hypothetical. Downstream State A, which uses water for irrigation purposes, and where several million people depend on agriculture for their survival is in dispute with upstream State B when the latter decides to set up a hydro-electric project on the shared river. The project obviously affects the irrigation needs of State A, but it does bring some conservation benefits to the whole system.

How would a judge or arbitrator consider the outcome of the factor-analysis of Article V?

The response of the International Law Association appears in its comment under this Article:

> The factor-analysis approach seeks primarily to determine whether (i) the various uses are compatible, (ii) any of the uses is essential to human life, (iii) the uses are socially and economically valuable, (iv) other resources are available, (v) any of the uses is existing under Art. VIII,[20] (vi) it is feasible to modify competing uses in order to accommodate all to some degree, (vii) financial contributions by one or more of the interested basin states for the construction of works could result in the accommodation of competing uses, (viii) the burden could be adjusted by the payment of compensation to one or more of the co-basin States, and (ix) overall efficiency of water utilization could be improved in order to increase the amount of available waters.[21]

These are therefore the general objectives set out in the Helsinki rules. In the case at hand, they mean a number of expert investigations on 'the incidental benefits of the new project', 'the availability of alternative sources for feeding the population in State A', and a host of other considerations such as: 'an existing reasonable use; dependence upon the waters; population; geographic, climatic and weather conditions, the existence of alternative sources of food supply; inefficient utilization; and the financial status of the respective co-basin states.'[22]

The Association concludes that all things considered, some accommodation and modification is required. Such would be for instance the abandonment of relatively wasteful practises of inundation in State A, against compensation by State B for the costs of developing a more modern irrigation system, or for State B possibly providing for the extra-food needed by the lower stream State. If however, the use of waters by State B is ultimately precluded or limited, then some compensation by State A for the continuation of the status quo may be required.

Would the illustration of the 1966 factor-analysis principles remain the same in the new proposals of the International Law Commission ?

The main difference, as suggested earlier, lay in the 'past use' factor. This has all but vanished from the new text, which puts the existing and potential element on a par. In its 1966 comment, the International Law Association had indeed emphasized 'the significant weight as a factor of an existing reasonable use' and derived a compensation scheme for the benefit of downstream State A in the hypothetical. It is doubtful that such compensation would be granted in a similar situation, if the new Article 7 were to be the measuring rod of river law. The development of international law has received considerable attention from the international community and its officials and the rules and statements devised over the past three decades have provided a useful framework for analyzing such cases as that of the Nile. Because, however, the problem of defining and monitoring the environmental variables, as well as the equally relevant socio-economic factors, is so complex, it is unlikely that the legal principles devised to date will be the basis of a speedy set of basin-wide agreements. In these uncertain circumstances it is likely that the search for appropriate principles and traditions upon which to base acceptable and appropriate water sharing agreements will be strengthened. One such tradition, affecting about half of the population of the Nile Basin is that of Islamic law, the *shari'a.*

Islamic water law

Ibn Manẓūr (d.711H./1311 A.D.), the most famous Arab lexicographer, mentions in his dictionary *Lisān al-'Arab* under the root "*sh r '* " that '*sharī'a* is the place from which one descends to water... and *sharī'a* in the acceptation of Arabs is the law of water (*shur'at al-mā'*) concerning the source which is regulated by people who drink, and allow others to drink, from'.[23] A later classical dictionary is more general: '*Ash-sharī'a*, writes Zubaydī, is the descent (*munḥadar*) of water. [The same word is also used] for what God has decreed (*sharra'a:* legislate, decree) for the people in terms of fasting, prayer, pilgrimage, marriage etc.... Some say it has been called *sharīʻa* by comparison with the *sharīʻa* of water in that the one who legislates, in truth and in all probablility, quenches [his thirst] and purifies himself, and I mean by quenching what some wise men have said: I used to drink and remained thirsty, but when I knew God I quenched my thirst without drinking'.[24]

The connection between *sharīʻa* as a generic term for Islamic law, and *sharīʻa* as the path as well as the law of water, is not a coincidence, and the centrality of water in Islam is obvious in the economic as well as ritualistic sense. What is more important however, is that the jurists -the exponents and expounders of the *sharīʻa*- did not fail to develop, in answer to this centrality, a highly sophisticated system of rules.[25]

One way of analyzing water law under the *sharīʻa* is to present it under two headings connected with the division of rights: property and servitudes.

On property and water categories

As for property, the principle, following a Prophetic *hadīth*,[26] is that 'Muslims are partners, *shurakā'*, in three [items]: water, grass and fire'. For water, this partnership entails that the right over it is held *in common* in two respects: the right of *shafa*, i.e. the right to 'lips' (from *shafa,* lip, i.e. the right deriving from thirst, the right to drink) for man and animal (and includes other amenities like to wash up for man), and the right of *shirb*, (to be differentiated from the word *shurb*, drinking), which for the jurists means 'the share in water for the irrigation of plants and trees', and sometimes more specifically 'the amount of time allotted for using the water' to such purposes.[27]

The property of water is distinguished from the servitudes which attach to water in that property, unlike servitudes, permits the sale of the water on which the property right is exercised.

In some texts,[28] four categories of waters are repertoriated: Sea water, great river basin waters, common water e.g. canals in a village, and 'containers'. Water is either free (*mubāḥ*) from appropriation and sale, as in the first two cases, or on the contrary, as in the two latter categories, subject to forms of appropriation. These categories notwithstanding, jurists, following the *hadīth* mentioned earlier, recommend that water be dispensed without reward, even when, as in the case of containers or vases, the water is clearly private property. The owner in this case cannot be compelled to provide the water freely except if the seeker's thirst is such as to create a danger on his life. In that exceptional case, the owner of water can be 'fought against'. There is here a distinction drawn between some jurists. If the water is found in a container, the fight against the owner must be achieved 'without weapons' (i.e. without shedding blood), and compensation must be provided: necessity here does not justify lack of compensation. If the water is found in a pond or a well, or a riverlet which is private property, full bloodied war (with weapons) is allowed. It is to be noted that in both cases, it is the surplus of potable water which is the object of the dispute. Water which is needed to fulfill the *shafa* right *of the original user* will be given him undisputedly.

Despite the Prophetic *hadīth* therefore, full property rights are recognized on water in the private category. The owner can keep the water to himself and prevent others from using it, except in the case of life-threatening necessity: the rule is that 'the owner can forbid the person entitled to *shafa* from entering his property if that person can find water nearer her. If the person entitled to *shafa* cannot find a nearer source, then the owner of the well can be required either to provide water to the person (*tukhrij al-mā' ilayhi*), or to allow her to come and get the water herself'.[29]

In the case of public waters such as those of 'great rivers', and in contrast with 'private waters' found in recipients or small canals, the possibility of private appropriation of water seems limited. But there is a *hadīth* which substantiates the 'sellable character' of water. This is the story of the well of Rūma, which is related, for instance, in the great Ḥanbalī compendium of Ibn Qudāmā (d.620/1223),[30] under two variants: in its simplest form, the Caliph 'Uthmān buys the well of Rūma from 'a Jew' upon the advice of the Prophet (viz. Prophetic *hadīth*: 'He who buys the well of Rūma and presents it to the Muslims will gain his way to Heaven'). This suggests that water - at least water in a well- can be sold. The interpretation of Ibn Qudāmā is that both the water and the well can be sold. They may be conceived of, and disposed of, separately. This 'individualistic' interpretation operates across the board of the water categories. In the case of water received in containers, the possibility of selling it is plain and not controversial. In the case of the 'public' category, viz. the great rivers' waters, the right of property can be exercised as soon as there is an 'application', or an exercise of human effort over the extraction and storage of water. 'As for running water', writes the Ḥanbalī jurist, 'if its source comes from a terrain with no owner as in the great rivers et al. it cannot be owned in any way even if a person goes on to the land to get the water, as in the case of a bird which goes onto a person's land, where it is permissible for any one to take it; but he will own the water only by preparing a container [receiver, *mustaqarr*] to retain it -- like a pond, or if a canal is dug to take advantage of the great river's water. In that case, as in the case of a [private] well, [the owner of the land who has transformed it] will have a preferred right to it.'[31] In the rendering of the story of the well of Rūma by Ibn Qudāmā, which is then elaborated upon in a somewhat more sophisticated manner,[32] this 'private property' option is clearly favoured. The right may not be over a commodity tradable by essence, it consecrates nonetheless for Ibn Qudāmā 'the possibility of dividing what includes a right but is not appropriated (*jawāz qismat mā fīhi haqq wa laysa bi-mamlūk*)'.[33]

The Ḥanbalī author is trying to combine here two conflicting traditions, the first of which renders the sale of water reprehensible,[34] and the second of which consecrates its validity. Ibn Qudāmā clearly leans in favour of the second opinion. Selling water that is being carried is allowed: 'this is how the use has been established in the various areas of the [Muslim] world, *'alā dhālika maḍat al-'āda fil-amṣār*'.[35]

* * *

Water property and use is addressed extensively in the Hanafī jurist Shams al-A'imma as-Sarakhsī's masterpiece, *al-Mabsūṭ*.[36] The issue of water is examined in a brief 'chapter on the river' in Volume 27, from a torts' perspective. Sarakhsī writes that if a well or a canal dug in someone's property flows over the neighbour's land and causes damage, there is no responsibility upon the original owner, 'except if he himself brought the water to overflow and knew that it would overflow onto the neighbour's land.'[37] Here the general Islamic rule of no-fault responsibility is inoperative without previous knowledge by the tort feasor.

But the issue of water is subject to a much more extensive treatment in Sarakhsī's long treatise. A chapter on irrigation (*kitāb ash-shirb*) is devoted to the question of water.

Sarakhsī starts the discussion with the definition of *shirb*: 'It is the share of water for land or otherwise'. The division (*qisma*) of water is an established tradition since the Prophet, he adds, 'it is a division which takes place upon the consideration of a right which is short of the right to property, as water in a river is no one's property'.[38]

Thus the specificity of water. The right to water use is not a right of property. Sarakhsī relates the Prophetic *hadīth* on the three common elements of partnership, and notes further that partnership involves both 'Muslims and non-Muslims (*kuffār*)'. For them all, 'the explanation of this partnership in water, which runs in villages and great rivers (*anhār 'iẓām*) like Jayḥūn and Sayḥūn, the Euphrates, the Tigris and the Nile',[39] is governed by 'the right of use (usufruct, *intifā'*) as in the right to enjoying the sun and breathing the air, inwhich Muslims and non-Muslims (*ghayruhum*) are equal... No person can prevent their use by another...'

After establishing the general principle, a distinction is made between different entitlements to water, depending upon whether the watercourse is specific (private, *khāṣṣ*), or public (*'āmm*) as in the great rivers. In the first case, limitations could be introduced on the right of use by privileging a given village over 'those outside' (*li-ghayr ahl al-qarya*): 'water which runs in a river which is specific to the people of a given village is subject to an opposable partnership which includes the [village dwellers'] self-satisfaction in terms of their drinking and the drinking of their household animals. They cannot prevent others from this [drinking], ...but people outside the village cannot water their palmtrees and plants from that river'.[40]

Apart from the public and specific rivers, there is the case of wells and ponds. Here also, there is no absolute right of property for the owner of the land where the wells and ponds lay. If a passer-by drinks from the well, the owner may not claim the water back from him. He could prevent him from entering his land by suggesting a nearby alternative where water flows outside owned property, but if this water is inaccessible, he is bound to allow the person to cross through to the well or pond, or, alternatively, he must bring the water out to him 'until satisfaction'.

The final 'type' of water, according to Sarakhsī, is that which may be found in private recipients like jugs or vases. In contrast to all that precedes, 'water is [here] property, and may [even] be sold.' However, the *hadīth* principle remains to the extent of 'a quasi-partnership' (*shubhat ash-sharika*). Were water to be stolen, the full retribution of the law does not apply: in such cases, the thief's hand will not be cut.[41]

On Servitudes

Water is also associated, in another legal register of the *sharī'a*, with the concept of the rights of servitude, *huqūq al-irtifāq*. Servitude is defined by the jurists as 'an objective right which is attached to realty in favour of another realty which is owned by a different person, independently from the identity of that person'.[42]

The exercise of the servitude rights (*usus* and *fructus*, as opposed to *abusus*, disposal, as *per* classical latin terminology) is limited by general, as well as specific, constraints. In that, it is significantly different from full property rights:

1 *La ḍarar wa la ḍirār*, that is no-fault injury,[43] is the rule in the exercise of the right. A person who irrigates his land in exercising the right of *shirb* cannot by so doing provoke harm to downhill or downstream neighbours.

2 Servitude can attach to both public and private property. For public property, like the great rivers, the fructus is established for all people without anyone's permission. Private property, in contrast, can allow usus and fructus only by permission of the owner.

Among the several special categories of servitudes recognized by the *sharī'a* under Zuḥailī's classification,[44] four attach to water:

1 the rights of irrigation, *shirb*.
2 drink and household rights, *shafa*.
3 access rights.
4 excess rights.

Excess rights and access (or flow) rights allow the owner of a land who needs either to get rid of damaging surplus water or to bring water otherwise inaccessible to his land, to use neighbouring plots to build diversion canals on them for that purpose. The exercise of these rights is subject again to the main tort principles under Islamic law: the principle of no fault-injury (*la ḍarar wa la ḍirār*); and the principle that prior rights must be respected, except in the case of damage, as 'injury can never be old.' (*ad-ḍarar la yakūn qadīman*).[45]

But the two most important servitudes are *shirb*, irrigation and *shafa*, drink and household use, which can be considered together. For private water, which would include in the present age 'the water in companies which are specialized in securing the needs of the city', the use as well as the disposal is restricted strictly to the owner.[46] He can sell it or dispose of it in whatever way he likes, subject to the exceptions mentioned earlier, necessity arising from dire thirst, but even then, compensation is due. Here there is no difference between *shirb* and *shafa*.

In the case of wells, sources and ponds, any person can extract the water (according to the partnership *hadīth* quoted earlier), whether the well, source or pond is private or held in common. Here only the right of *shafa*, and not the right of *shirb*, is part of the use.

The water of private waterways, similarly, is restricted to *shafa*, which can be exercised by anybody. Irrigating is possible only if the owner of the land on which the private water runs permits it.

Finally, the water of great rivers: here the principle of common control or ownership is absolute, and the distinction between *shafa* and *shirb* evaporates. Any person can use the water, by *shafa*, *shirb*, or by digging side canals and setting up tools to divert water to his land, subject to the general principle of *la ḍarar wa la ḍirār*. The right of irrigation is also subject to a number of other principles, most importantly:

a. the protection of the river coast, and the prevention of flooding other lands must be respected.
b. the right of irrigation allows the owner of a non-contiguous plot to construct a waterway which will allow the irrigation of his land. He must first build the waterway on available public property, but he may also, if public property is not available, force the passage on his neighbour.
c. irrigation rights can be inherited, and can be the subject of a will. There is some controversy as to the precise value of these rights, but custom has it that they may be inherited.
d. if water is to be distributed in common, as in the case of rivers, there is a set-up, based on equitable distribution, which will rule the rythm of irrigation. For this, prior use is one factor. There is here an important *hadīth* which is ascribed to the Prophet: 'The upstream irrigates before the downstream, up to the level of two ankles, and releases the remainder of the water,and so on , until all the needs are fulfilled, or until the water is exhausted'.[47]

* * *

Sarakhsī's treatise on water is not systematic by modern standards, but many of his remarks feed in the general appreciation of the servitude rights which are specific to water.

Sarakhsī finds difficulty in the concept of water appropriation and sale, and has a clear preference for the concept of partnership.[48] In a mention of a dispute over the irrigation of land between various riparians, he writes that 'if the original distribution scheme of a river in common is not known, ... irrigation is shared in proportion of the size of the riparians' lands'.[49] It is not here a matter of equality between individual riparians, but a direct relation of proportion to the size of the land, he precises. In any case, the upstream riparian is not allowed to block the river from flowing downstream, except by common agreement between all riparians. If work (such as a dam) is necessary on the common river for a more efficient distribution of water, 'each riparian profits from the river from beginning to end... The upstream people (*ahl al-a'lā*) are equal in that [i.e. the right to water use] to the downstream people (*ahl al-asfal*)'.[50] It is the duty of the state to finance such project from its coffers, and in the absence of funds, 'could force the Muslims' to finance the project 'for the good of the people'.

As for the ankle rule, which is also reported in *al-Mabsūṭ*, Sarakhsī introduces an important qualification by dismissing any attempt at deducing a scientifically measurable apportionment: 'If the *wādī* reaches two ankles, the upstream people cannot prevent its use downstream. What is meant here is the water in the *wādī* and *wādī* is the appellation of a place at the bottom of a mountain where water flows in and meets from all sides of the mountain to reach the place which people benefit from. His [The Prophet] saying 'if the *wādī* reaches two ankles [*ka'bayn*]' is not a compelling appreciation (*taqdīr lāzem*) of [the height] of two ankles, but a reference to a surplus of water.'[51]

In the following remarks, it is unclear in Sarakhsī's text whether, short of a surplus of water which must unequivocally be allowed to flow downstream, the priority of use is temporal or geographical. In one remark, it appears that whomever is the first to use the water will be allowed to have priority. In another, 'when water is scarce, the upstream people have a priority of use.'[52]

Whatever the case, it is clear that Sarakhsī gives particular importance to practise, and he refers to the 'later scholars of Balkh', to 'a practise known in Nasaf', and to irrigation schemes on the river Merv.[53]

* * *

The rich Islamic legal tradition remains effective in several Middle Eastern and Islamic countries, and water rights have been the subject of litigation and arbitration to date. Case law is unfortunately not systematically available, although more research would uncover, no doubt, the vast riches in the field. A recent and well-documented case involving water sharing in Morocco and decisions by the Chraa (*sharī'a*) tribunals will help illustrate this proposition.

A case study

The Moroccan case is complex procedurally. The dispute involved two tribes, the Bani Ourjane and the Ouloud Abdel Krim. Back in 1918, a conflict which had arisen over the use of River Shaqq al-Ard in Eastern Morocco had been settled among several tribes, with the Bani Orjanes receiving a share in half of the water from the river, which they used through canals or riverlets (*sāqiyas*, French *séguia*). These *sāqiyas* passed also through land owned by the Ouloud Abdel Krim, who also used the water. The case which came before the unique qāḍī court in 1946 was prompted by the Bani Orjanes seeking to establish their full entitlement to the water, and producing a 1918 registered document to the effect of their full right to the water now in competition with the Ouloud Abdel Krim's use.[54]

The first instance qāḍī found for the plaintiffs. Although the decision advocates 'the status quo which had prevailed between the parties for twenty-five years',[55] the classical legal treatise known as the *Tuḥfa*[56] is quoted to establish the right of property over an immovable (*immeuble*) 'held as of right for more than ten years'. On the basis of this decennal prescription, the Bani Orjanes were comforted in their claim to full right over the water.

Upon appeal by the Ouloud Abdel Krim, the Chraa tribunal reversed the qāḍī's decision on several procedural and legal grounds. The procedural grounds resulted essentially from evidence produced by the Ouloud Abdel Krim, which established their use of 1/6th of the disputed water since at least 1918. The Appeal court found the decennal prescription to be inoperative, or, alternatively, to be actually in favour of the defendants' claim. On the substantial level, the court also found for the Ouloud Abdel Krim:

> The simple use of water, however old it may be, does not confer a right of property...
> In matters of water not privately appropriated and consisting solely of river waters which are no one's property, the rule is that land is irrigated in decreasing order... In the words of Khalil,[57] 'rain water which flows in dead land can be retained for irrigation of upstream land, until it is submerged up to the ankle when it has been cultivated first.[58]

According to the appeal decision, two legal elements must be weighed in the appreciation of the right to water, which in any case cannot be one of full ownership: (1) the antecedence of use, and when it has been established, (2) the *hadīth* 'ankle rule' applies. The ankle rule stipulates that any excess of water, measured by 'submerging up to the ankle', will be allowed for the benefit of other users. In case the antecedence cannot be established, the statu quo must be respected. Otherwise the ankle rule operates.

At that stage of the case, the Ouloud Abdel Krim were therefore entitled to their claim of 1/6th share, provided that any surplus according to the ankle rule would be left for use by the Bani Orjanes.

That was not the end of the story. A full revision of the appeal decision was conducted on the instigation of the Moroccan Ruler's representative and after the production of new evidence. A commission of enquiry was dispatched to assess the situation on the ground. It concluded to the existence of several gardens and orchards belonging to the Bani Orjanes, which would be ruined in case the appeal decision was implemented. Meanwhile, the Bani Orjanes had produced a number of detailed documents which went back to the 19th century, and 'proved' their prior use of the water. The Appeal Chraa court then proceeded to revise its earlier decision and reversed it.

The final decision is based principally on the argument of prior use, as well as, to a lesser extent, on the fact that the expert commission and the 'Magistrat rapporteur' had concluded to the inevitable ruin of the previously irrigated orchards. The court 'rejected the appellants', Ouloud Abdel Krim, claim to the 1/6th of the water of the *sāqiya*... and recognized the right of the Orjane tribe over the *sāqiya* water as long as there is no surplus'.

'In case there is such evident surplus over the needs of prior use', the Court ended, 'the Ouloud Abdel Krim and all those who owned land which might be reached by the water would be entitled to use the excess...'[59]

It is noteworthy that there is no decisive upstream or downstream principle of priority. None of the three decisions reported gives exclusive weight to the geographical position of the riparian parties. Mention of upstream priority in the final decision is made in conjunction with the element of prior use, and the revised appeal judgment grants 'upstream land priority over downstream property'[60] but only in the particular context of the case, and only with the hypothesis that there is an excess which is available.

Conclusions

Whether in the Moroccan precedent or in the writings of classical scholars such as Ibn Qudāmā and Sarakhsī, the *sharī'a* offers rich but not totally consistent rules. Curiously, these tensions are reminiscent of the present status of international watercourse law. It is our contention that the input of the *sharī'a* and of other alternative legal traditions may enrich the present debate, not only because they are indigenous to a number of the Nile riparian states, but first and foremost as sophisticated branches of the family of world legal traditions. Putting forward the benefit of a closer examination of the legacy of the *sharī'a* in the apportionment of water might, in a Middle Eastern context, raise delicate political issues. But as a legal tradition established for more than a millennium, and because it is so rich a legacy on watersharing - in a region where water is almost by definition vital and scarce - , the *sharī'a* may be able to offer its riches to a situation which is, in international law, not yet settled.

Beyond general principles, which are adequately covered by the International Law Commission draft as well as the overarching principles of Islamic law, the quest for a norm must in the case of the Nile address several difficult areas, including the exact allotment of water, the assessment of injury, and the mechanisms of the factor-analysis.

On this last point, and against the limitations recently introduced by the International Law Commission on the evaluation of prior or existing use, both the *sharī'a* and the International Law Association seem to be in agreement that any change cannot be allowed to disregard or supersede established use. Beyond this important principle, the *sharī'a* offers an alluring scheme for a dismemberment of water rights on the basis of a distinction between property and servitudes, and a further breaking down of water rights along the lines of drinking, irrigation, and other, more modern industrial and electrical uses. In relation to to water, Sarakhsī coined in the eleventh century the concept of 'quasi-partnership', which encompasses in one word a vast agenda for intellectual enquiry.

To go back to the beginning of this contribution, and chart a medium course, in law, between the Harmon doctrine of absolute sovereignty (which grants the upstream states of the Nile complete control over the flow of water), and the 'natural and historical rights' converse doctrine (which privileges, along with the other tools forged over the century to protect downstream states, especially Egypt, in as much an absolute manner), there is no simple or straightforward answer in the emerging rules of international law. Whether under the factor-weighing rules spelled out in the Helsinki declaration, or the more recent balance between duty to cooperate and equitable utilization of articles 6 and 8 adopted in the draft of the International Law Commission in its 38th and 39th sessions, the lawyer remains on moving and uncertain legal ground.

It may well be that international law is unable to provide fully rigorous norms for the sharing of water resources. One is then left with limited legal choices. With the discredit cast on both the Harmon doctrine and the concept of sacrosanct natural and historical rights, the Helsinki declaration and the International Law Commission draft must be supplemented. One way which is suggested for the Nile is by means of the lessons of the *sharī'a* on water. There lies a rich and unique tradition which has been shaped over centuries. Similarly, and that may be a point for further investigation, a confluence of models may perhaps allow for a better approximation of rules of law which will be govern the Nile most adequately. The Nile is worth bringing together perspectives which have been traditionally opaque to each another.

---

1 See the extensive bibliography compiled by I. Kost, *Rights and Duties of Riparian States of International Rivers/ Droits et Obligations des Pays Riverains des Fleuves Internationaux,* The Hague, 1990.

2 On the work of the International Law Commission, see M. Solanes, "The International Law Commission and legal principles related to the non-navigational uses of the waters of international rivers", *Natural Resources Forum,* 11, 1987, pp.353-361. The United Nations General Assembly recommended in 1970 that the International Law Commission take up the study of the law of the non-navigational uses of international watercourses (G.A. Resolution 2669 (XXV) 8-12-70). Several reports have been produced by the successive rapporteurs, R.D. Kearney (A/CN.4/296, 1976), S.M. Schwebel (A/CN.4/320, 1979; A/CN.4/332, 1980; A/CN.4/348, 1981), J. Evensen (A/CN.4/367,1983; A/CN.4/381, 1984), and S.C. McCaffrey (A/CN.4/393, 1985; A/CN.4/399, 1986; A/CN.4/406, 1987; A/CN.4/412, 1988; A/CN.4/421, 1988). The latest reports are summarized in successive issues of the *American Journal of International Law,* see infra.

3 This Protocol was signed in Rome, April 15, 1891. The part relevant to the Nile affirms that "the Italian government shall undertake not to initiate any irrigation works on the Atbara [Ethiopia] which may alter the rate of flow of the Nile". The text of this agreement and of the major relevant texts can be found in Ministry of Foreign Affairs (Arab Republic of Egypt), *Egypt and the Nile,* Cairo, 1984. (1891 Protocol at p.38).

4 Nile Waters Agreement of 1929- Exchange of Notes Between the United Kingdom and the Egyptian Government in Regard to the Use of the Waters of the River Nile for Irrigation Purposes, Cairo, May 7, 1929. Ibid, at p.65; Agreement Between the Sudan and Egypt for the full Utilization of the Nile Waters, November 8, 1959. Ibid., p.69. There is a large legal literature on the Nile River agreements. See esp. B. Godana, *Africa's Shared Water Resources,* London, 1985; C.A. Pompe, "The Nile Waters Question", in *Symbolae Verzijl,* The Hague, 1958, pp.275-294; G.M. Badr, "Nile Waters Question", *Revue Egyptienne de Droit International,* 15, 1959, pp.94-117; A.M. Fahmy, "International River Law for Non-Navigable Rivers with Special Reference to the Nile", *Revue Egyptienne de Droit International,* 23, 1967, pp.39-62; A.H. Garretson, "The Nile Basin", in Garretson, Hayton and Olmstead, *The Law of International Drainage Basins,* Dobbs Ferry, 1967, pp.256-297. A recent dossier on the Nile has been compiled by the Egyptian *as-Siyasa ad-Dawliyya,* Cairo, April 1991, pp.114-179. Since his seminal *Hydropolitics of the Nile Valley,* New York, 1979, John Waterbury has published an excellent comprehensive survey of the agreements on the Nile as "Legal and Institutional Arrangements for Managing Water Resources in the Nile Basin", in R. Simon ed., *The Middle East and North Africa: Essays in Honor of J.C. Hurewitz,* New York and Oxford, 1990, pp.276-303. See also D. Whittington and K. Haynes, "Nile Water for Whom? Emerging Conflicts in Water Allocation for Agricultural Expansion in Egypt and Sudan", in P. Beaumont and K. McLachlan eds., *Agricultural Development in the Middle East,* London, 1985, pp.125-149; and the other chapters in this book, especially the articles of Ambassador S. Ahmad and C.H. Okidi.

5 D. Caponera, *Water Laws in Moslem Countries,* FAO Report, Rome, 1954.

6 See e.g. P.Howell, *A Manual of Nuer Law,* Oxford, 1954. (P.188 on collective water rights).

7 "C'est ainsi que les historiens ont relevé l'existence ou plutôt la survivance d'une loi sur le régime des eaux qui était appliquée dans certaines régions de la Mésopotamie et qui aurait été à l'origine de certaines lois occidentales". C. Chehata, *Droit Musulman,* Paris, 1970, p.11.

8 See J. Schacht, *An Introduction to Islamic Law,* Oxford, 1964, p.164-165; D.

Caponera, *Water Laws in Moslem Countries*, mentioned supra n.5; and J. Lapanne-Joinville, "Le Régime des Eaux en Droit Musulman (rite malékite)", *Revue Algérienne, Tunisienne et Marocaine de Législation et de Jurisprudence*, 1956, pp.12-78.

[9] "It would be incorrect to assume that tribunals have in practice adopted a mechanical system of borrowing from domestic law after a census of domestic systems... Moreover in some cases, for example the law relating to expropriation of private rights, reference to domestic law might give uncertain results and the choice of models might reveal ideological predilections". I. Brownlie, *Principles of Public International Law*, Oxford, 1973, p.16-17; "Mais la méthodologie du droit international ne se limite pas à l'étude de l'influence -par transposition ou par osmose- du droit commun. Etant une science distincte, le droit des gens postule nécessairement une technique juridique particulière, qui se différencie des texhniques de droit interne". C. Rousseau, *Droit International Public*, Vol.1, Paris, 1970, p.52. Strictly speaking, of course, Islamic law is not municipal law, and its relevance to public international law therefore less problematic.

[10] Letter of Lord LLoyd, UK High Commissioner to M. Mahmood Pacha, Egyptian Chairman of the Council of Ministers, 7 May 1929: "In conclusion, I would like to remind your Excellency that Her Majesty's Government in the United Kingdom has already recognized the natural and historical right of Egypt to the waters of the Nile", in *Egypt and the Nile*, op. cit. p.68.

[11] See e.g. K. Krakau, *Die Harmon Doktrin*, Hamburg, 1966.

[12] System is in brackets in the original because of a disagreement at the International Law Commission over the proper appellation of the water basin. This is why a final decision on the definition of Article 1 was postponed until the completion of the full text. The Commission had started work on the non-navigational uses of international watercourses in 1971. The 1983 report, submitted by then Rapporteur Jens Evensen provided: "Article 1- An 'international watercourse system' is a watercourse system ordinarily consisting of fresh water components, situated in two or more system States. ..." Quote in "The Thirty-Fifth Session of the International Law Commission", *American Journal of International Law*, 78, 1984, p.475-476. On the work of the Commission, see the reports quoted in this chapter, and the article of the present Rapporteur, Professor Stephen McCaffrey, "The law of international watercourses: some recent developments and unanswered questions", *Denver Journal of International Law*, 17, 1989, pp.505-526.

[13] "For the purposes of the present articles, a watercourse State is a State in whose territory part of an international watercourse [system] is situated". "The Thirty-Ninth Session of the International Law Commission", *American Journal of International Law*, 82, 1988, p.148.

[14] On the process of consultation, see "The Fortieth Session of the International Law Commission", *American Journal of International Law*, 83, 1989, pp. 160-166. Concern for the process of consultation is also significant in view of the similar requirements by the World Bank before it gives its agreement for the financing of an international waterway project. Its Operational Directive, issued in September 1989, emphasises this dimension in its introduction: "Basic Policy: 1. Projects on international waterways require special handling as they may affect relations not only between the Bank and its borrowers but also between states, whether members of the Bank or not. The Bank recognizes that the cooperation and good will of riparians is essential to the most efficient utilization and exploitation of international waterways for development purposes. The Bank, therefore, attaches the utmost importance to riparians entering into appropriate agreements or arrangements for the efficient utilization of the entire waterway system or any part of it, and stands ready to assist in achieving this end. In cases where differences remain unresolved, the Bank, prior to financing the project, will normally urge the state proposing the project to offer to negotiate in good faith with other riparians to reach appropriate agreements or arrangements." *The World Bank Operational Manual*, September 1989. Text kindly provided to the author by Mr Raj Krishna, International Law Adviser with the World Bank. See also his article, "Legal Regime of the Nile River Basin" in J. Starr ed., *Politics of Scarcity, Water in the Middle East*, Boulder, 1988.

[15] "The Thirty-Ninth Session of the International Law Commission", *American Journal of International Law*, 82, 1988, p.149-150.

16 "The Fortieth Session of the International Law Commission", *American Journal of International Law,* 83, 1989, p.161.

17 There should be added that the subsequent meetings of the Commission have made some headway on remaining issues, especially in relation to pollution control.

18 The Helsinki rules, along with detailed comments, were published by the International Law Association as a separate booklet. *Helsinki Rules on the Uses of Waters of International Rivers,* London, 1967. (Hereinafter *Helsinki rules, Comment)*

19 Article IV, in turn, refers to "the key principle of international law in this area" (*Helsinki rules, Comment,* p.9): "Each basin State is entitled, within its territory, to a reasonable and equitable share in the beneficial uses of the waters of an international drainage basin."

20 Article VIII.1: "An existing reasonable use may continue in operation unless the factors justifying its continuance are outweighed by other factors leading to the conclusion that it be modified or terminated so as to accommodate a competing incompatible use". As suggested, there is no equivalent to this article in the International Law Commission recent text.

21 *Helsinki rules, Comment,* p.11-12.

22 Id.

23 Ibn Manẓūr, *Lisān al-'Arab,* Beirut, 1959, Vol. 3, p.175.

24 Az-Zubaydī (d.1205/1790), *Tāj al-'Arūs,* N.d., Benghazi, Vol.5, pp.394ff. See also Schacht, s.v. *sharī'a,* in *Encyclopaedia of Islam* (1st ed. Leiden, 1916-): 'the road to the watering place, the clear path to be followed, the path which the believer has to tread, the religion of Islam, as a technical term, the canon law of Islam.'

25 This is obviously not a treatise on water law in Islam, and we have reduced our study to presenting a few introductory remarks on an understudied field.

26 The *hadīths* are the sayings attributed to the Prophet Muḥammad. With the Qur'ān, they constitute the two original textual sources for Islamic law.

27 Wahbe az-Zuḥailī, *al-Fiqh al-Islamī wa adillatuhu,* Damascus, 8 Vols, 1984, Vol. 4, p.450; Vol. 5, p.592. Zuḥailī is a prominent contemporary jurist from Syria.

28 We have selected here texts from various schools (*madhhab)* of Islamic law. Islamic law is often presented under categories relating to the four "official" madhhabs of Sunnī Islam, the Ḥanafīs, Ḥanbalīs, Mālikīs and Shāfi'īs. More important than this formal distinction is the geographical and historical setting of individual scholars' work, especially in the case of water law. See examples infra in our treatment of the Central Asian 'Ḥanafī' scholar Sarakhsī, and in the case study in Moroccan 'Mālikī' courts. The treatment of water is generally found in a chapter on irrigation (*shurb)* within the "book of land revival" (*kitāb ihyā' al-mawwāt)* ; see e.g. Ibn Nujaim (16th century Levantine Sunnī jurist), *al-Baḥr ar-Rā'eq Sharḥ Kanz ad-Daqā'eq,* with the commentaries of Ibn 'Ābidīn (d.1252/1836). Vol.8, pp.238-246 and 'Abd al-Ḥakīm al-Afghānī, *Kashf al-Haqā'eq Sharḥ Kanz ad-Daqā'eq,* and the commentaries of 'Abdallāh Ibn Mas'ūd, Cairo, 1318-1322, Vol.2, p.243-245.

29 Zuḥailī, *al-Fiqh al-Islamī,* Vol.4, p.452.

30 Ibn Qudāmā, *Al-Mughnī wash-Sharḥ al-Kabīr,* Beirut, 1984, Vol.4, p.217-218; See also *Mu'jam al-Mughnī fil-Fiqh al-Ḥanbalī,* (This is a 2 Volume index to Ibn Qudāmā's extensive *Mughnī),* Beirut, n.d., Vol. 2, p. 920.

31 Ibn Qudāmā, *Al-Mughnī,* Vol.4, p.216.

32 The case is more complicated in *al-Mughnī:* 'Uthmān is said to have bought only half of the water source. He then started distributing the water to the Muslims, and by dumping the value of the other half which was being sold by the Jew, forced him to sell it to him. The example is used by the Ḥanbalī jurist to illustrate the flexibility of the right to ownership of water despite the original ban on selling it.

33 Ibn Qudāmā, *Al-Mughnī,* Vol.4, p.218.

34 To the *hadīth* earlier cited on the partnership in water is added an even clearer *hadīth*: The Prophet "has forbidden to sell the surplus of water (*nahā 'an bay' faḍl al-mā'*)". Zuḥailī, *al-Fiqh al-Islamī,* Vol.4, p.454. The practise, of course, is different, whether as

illustrated in Ibn Qudāmā's interpretation, or in the well-known trade of water carriers in history. See e.g. André Raymond, "Les porteurs d'eau du Caire", *Bulletin de l'Institut Français d'Archéologie Orientale*, 57, 1958, pp.183-202. But see a qualified version of the *hadīth* note 46 infra.

35 Ibn Qudāmā, *Al-Mughnī*, Vol.4, p.218.

36 Muḥammad as-Sarakhsī (d.483/1090), known as Shams al-A'imma, *al-Mabsūṭ*, 30 Vols, Cairo, 1906-1912.

37 Sarakhsī, *al-Mabsūṭ*, Vol. 27, p.22-23.

38 Sarakhsī, *al-Mabsūṭ*, Vol. 23, p.161.

39 Id., p.164. Jayḥūn and Sayḥūn are Central Asia's Amu-daria (Oxus) and Syr-daria rivers. See in the classical geography literature in André Miquel, *La Géographie Humaine du Monde Musulman*, 3 Vols, Paris, 1973-1980.

40 Sarakhsī, *al-Mabsūṭ*, Vol. 23, p.164.

41 Ibid., p.164-165.

42 Zuḥailī, *al-Fiqh al-Islamī*, Vol.5, p.588.

43 On this important principle, see W. Zuḥailī, *Naẓariyyat ad-Ḍamān fil-Fiqh al-Islamī*, Damascus, 1970; E. Tyan, *Le Système de Responsabilité Délictuelle en Droit Musulman*, Paris, 1926.

44 Zuḥailī, *al-Fiqh al-Islamī*, Vol.5, pp. 588-610.

45 See the *Majallat al-Aḥkām al-'adliyya*, the 19th century Ottoman Code of Obligations, where these two general principles are reproduced as Art.19 (*la ḍarar wa la ḍirār*), and Art.7 (*ad-ḍarar la yakūn qadīman*). Salīm Bāz, *Sharḥ Aḥkām al-Majalla*, Beirut, 3d ed. 1923, p.22, 29.

46 *Apud* the *hadīth*: "nahā 'an bay' al-mā' illā mā ḥumila minhu (the Prophet has forbidden the sale of water, except for what has been carried)", see Zuḥailī, *al-Fiqh al-Islamī*, Vol.5, p.593.

47 This is known as the ankle rule. As an apparently measuring device, it is important for water law, see infra. There are also provisions for *karī*, i.e. the construction and repair of rivers for the use of water. Zuḥailī, *al-Fiqh al-Islamī*, Vol.5, p.600-601.

48 Sarakhsī, *al-Mabsūṭ*, Vol. 23, p.172.

49 Id., p.173.

50 Id., p.174.

51 Id., p.164.

52 Id.

53 Respectively Id. p.170. p.171, and p.178.

54 The case is reported in French by J. Lapanne-Joinville, "Arrêt du Tribunal d'Appel du Chraa sur une Question d'Eau", *Revue Algérienne, Tunisienne et Marocaine de Législation et de Jurisprudence*, 1956, pp.79-90. [Hereinafter Arrêt]

55 Decision of the qāḍī, 23 Dec. 1946, in Arrêt, p.82.

56 Ibn 'Āsim (Mālikī, d.829/1427), *Tuḥfat al-Aḥkām*, transl. L. Bercher, Algiers, 1958.

57 Khalīl Ibn Iṣḥāq (d.767/1367), author of *al-Mukhtaṣar*, a widely used compendium of Mālikī law.

58 Decision of the Chraa Appeal Tribunal, 22 Dec. 1947, in Arrêt, p.84.

59 Revised Decision of the Chraa Appeal Tribunal, 14 Mar. 1949, in Arrêt, pp.89-90.

60 Id., p.90.

19

Developing policies for harmonised Nile waters development and management

J.A. ALLAN

Nile Basin water allocation and management

The review of the status of the Nile resource, of the planning options of the major users and of economic, political and legal issues associated with the allocation and use of Nile water in the preceding chapters has shown that environmental and development factors have begun to interact in various parts of the Basin which have environmental, economic and political outcomes which require careful management.

The geography of Nile water has been shown to be significant in that the location of individual sites within the Basin and of individual political units determine whether the sites or countries have a potential to be water exporting or water importing. In the past when technologies for controlling water were not very effective, the advantage lay with the downstream riparians. The advantage of the downstream location was enhanced by the terrain of the lower Nile valley which is flat, slows down the flow of the river and has the capacity to accumulate the eroded silt from Ethiopia. Until the second half of the twentieth century the lower Nile countries therefore had sufficient land, sufficient water and the necessary institutions to develop Nile waters to support enduring rural and urban livelihoods.

By the second half of the twentieth century the development context had begun to change. The environmental advantage still appeared to lie with the downstream countries but the major user of water, Egypt, began to find that the volume of water available was no longer adequate for its sectoral needs. Even the construction in the 1960s of the High Dam at Aswan provided only a very brief respite from the spectre of seriously economically embarrassing water scarcity. Since the early 1970s Egypt has been living beyond its water means and has solved its economic crisis by the political device of subordinating its regional position to the United States in return for the provision of the means to obtain commodities to fill its food gap. Only the industrialised world could by the 1980s and 1990s, and for the foreseeable future, reliably supply the volumes of food required annually by Egypt.

The geographical position of the upstream riparians has not changed in physical terms, unless the less than average rains, especially of past two decades were to prove to be the rule. But even if they had, it would be the potential change in the capacity to mobilise resources to store and utilise water in the national territory of Ethiopia which would be the major change. The development of water by Ethiopia for agriculture and industry would in past decades have appeared to be an economically destabilising act by Egypt. In the seriously water scarce 1990s such initiatives would appear at first sight to be even more destabilising. But since Egypt has already adjusted to even greater hydrological difficulties the impact of Ethiopian water developments should not be exaggerated.

Egypt has already coped with the problem of much greater magnitude than the reduction of water from the Ethiopian tributaries through the use of quantities of water, as much as 12 km^3 annually on the proposed Ethiopian irrigation projects. Such use would reduce the volume in the Nile annually by between six and eight billion cubic metres - the precise volume depending on soil properties at the irrigation sites. By the early 1990s Egypt was coping with a water deficit of over ten billion cubic metres annually and rising. Egypt has shown one way in which the inevitable water deficits of the Nile Basin can be dealt with. Not all the countries will take the course adopted by Egypt and in any event such a course may not be available to other riparians in future as the United States may not have the economic strength or will to take on additonal burdens on the scale of Egypt. The Sudan will certainly 'run out' of Nile water in a period of ten to twenty years, and Ethiopia could very quickly fully develop an internationally acceptable volume of Nile water. The East African countries are unlikely to face problems of water supply partly because they enjoy adequate rains at least in part of their national areas and the volume of water supplied to the north via the White Nile is much smaller than that of the Ethiopian tributaries. Also the reduction in evaporation in the Sudd marshes through canalisation, if this is ever completed or expanded, will alleviate the impact of any diminution of flow caused by abstractions in the East African Highlands.

Sound principles of water allocation and management are based on economic efficiency, ecological sustainablity, the provision of an adequacy of safe water, and the arrangement of equitable access to water. The principles are occasionally recognised in the speeches and writings of officials and politicians, but for the moment they have no place in the actual management of water not in the development of national and international water allocation and management policies. There follows a concluding discussion of the conflicting obligations of the downstream and upstream riparians.

Egypt has already given in by substituting for water and this substitution has been achieved in two ways - economically or politically. The economic strategy requires that Egypt diversifies the sources of national income so that a rate of increase in the GDP is generated which accelerates faster than the burgeoning population. Industrial development must be a major component of this diversification and income generation. The record in this area has been useful to date in that Egypt has more than kept pace with demographically induced demand.

Egypt has also taken a political option in order to meet its water and food deficits by accepting massive assistance from the United States in the form of food aid and other financial support. In order to have a political relationship which would enable this support Egypt shifted from a position of being pro-

Soviet Union prior to 1973, to reaching an accord with Israel at Camp David five years later and at various points in the 1980s and the 1990s following a carefully balanced pro-United States line in regional affairs such as during the Gulf Crisis and War of 1990-91.

Water resources and water management in the Sudan

The construction of major dams in the Sudan began later than in Egypt although irrigation by pumping from the Nile with modern systems began as early as 1906. Success with the raising of cotton in the first decades of the twentieth century stimulated the Sudan Government to design and construct the Sennar dam in 1925 partially to control the Blue Nile so that 300,000 feddans (125,000 hectares) of the Gezira area could be irrigated (Gaitskell, 1952). An alternative interpretation of the economic and social impact of the Sennar/Roseires/Gezira scheme has been argued by Barnett (1974) which shows that the benefits were eroded by inefficiencies in engineering and administrative systems.

The Sudan's water development options are more diverse than those of Egypt. While the water storage and power capacity of the Sudan's present structures to allocate and manage water are small (Chapters 3 and 9) the potential for hydropower generation are considerable (Chapter 9) and the Sudan's hydrological resources are potentially substantial.

The waters of the Sudan's flood region comprise a vast potential water resource and the only major source of water which could be engineered to become accessible to extend irrigation and other water uses in this segment of the Nile's hydrological system. It is argued that *in situ* development would not be appropriate because of the combination of adverse climatic and soil conditions (Chapter 12), and the utilisation of Sudd water is not likely to be used outside the south of the Sudan as originally anticipated in the original Jonglei Scheme. Much has changed since the concept of the Jonglei Canal was given substance by the start of construction in the 1970s. At that time the argument that agricultural economies further north would be enhanced by the availability of additional water appealed to the governments in both Khartoum and Cairo. However by 1992, after almost a decade of stalled construction and political disarray the attitude of the international community to these wetlands has changed dramatically. It will be extremely difficult to restart the construction of the Jonglei Canal for technical and political reasons; it will almost certainly be impossible to do so because no international or bilateral agency will finance the perceived impairment of one of Africa's major wetlands.

Water development in the Sudan, as everywhere, has to be seen in the context of a number of conflicting policy priorities - economic, conservation and environmental, as well as social and political. The Government and its officials have to take into account issues such as productivity of water, the fair use of water in terms of current users in the south of the Sudan and potential users in the north, as well as the sustainable and safe use of water. These sometimes incompatible goals have complicated the evaluation of schemes to use the water according the the 1959 Nile Waters Agreement, which agreed that joint investment would be followed by the equal benefit to the joint investors, Egypt and the Sudan, in terms of the allocation of water.

Almost three quarters of Stage I of the Canal has been constructed but it is unlikely to be completed in the foreseeable future. The important principle highlighted by the situation in the southern Sudan is that the interests of those living in areas with a water surplus should not be underestimated, or worse, ignored. Governments in both Cairo and Khartoum will have to offer some tangible reciprocation if southern attitudes are to be changed to favour the export of the Sudd water.

The issue of evaporation is a crucial one from many points of view and especially significant with respect to the consideration of the water resources of the Sudd and similar wetlands in the south. It is relevant to the overall economy of water management projects and their sustainability both economically and ecologically. The estimates of evaporation from the Sudan's southern swamps provide some very high figures, amounting to half the annual flow. As water becomes more scarce with respect to growing demand there will be increasing pressure to minimise evaporation from all surface storage, whether natural or man-made, by constructing storage capacity in cooler environments in the south of the Basin including Uganda. In such arguments Ethiopia is also in a strong position.

The Nile in Ethiopia: an undeveloped power and irrigation potential

Ethiopia has a very large water resource potential within its national area. The development of these resources has been impeded for decades first by agreements made by colonial powers and then by conflict and political instability. As the Ethiopian Blue Nile and other tributaries contribute over 80 per cent of the currently utilisable water in the Sudan and Egypt, the potential of Ethiopian geographical advantages preoccupy the major downstream users.

Political strife has affected Ethiopia in terms of national security and has especially affected its economic development. Ethiopia has been unable to develop its huge power potential (US Department of the Interior, 1964) and to irrigate from its Nile tributaries in its remote north western region. The construction of the necessary civil and irrigation works will require a decade or more of effort and investment. In order to bring them about Ethiopia's economic situation and its economic and political relations externally will have to be transformed if the developments are to take place with the approval of the downstream states and international funding bodies. If the downstream riparians are to feel that they have achieved some reciprocal benefit, then it will be necessary for the states involved to devise a framework for evaluating regional water budgets and the benefits and disbenefits of upstream development in both economic and resource security terms. (Whittington and Haynes, 1985)

The Ethiopian authorities have not yet decided on their strategy for the development of water for agriculture. At the outset it is recognised by the Government (Abate, 1991, p.62) that water is not a free commodity and that a charging system would have to be installed when water is used in large quantities in the agricultural sector in future. It is also recognised that there are different approaches to the use of water and that the high-technology approach for water distribution is expensive, requiring between US$10,000 to US$15,000 (per hectare) to develop. Small scale irrigation projects require much less finance and much more quickly to achieve a pay-back on investment. And because increases in yield of such staples as wheat resulting from the allocation of irrigation water to agriculture are unlikely to provide economic returns, nor contributions to the national food economy of strategic importance, the Ethiopian authorities are unlikely to devote scarce investment resources to large scale irrigation projects where both initial and running costs are high. The small scale irrigation project is likely to be the preferred development strategy. Moreover Ethiopia's agricultural and irrigation strategies are likely to emphasise crops which would benefit from the optimum growing conditions in the high temperature regions of the North-Western lowlands and the cultivation of sugar cane is being considered as an economical option (Abate, 1991).

The Ethiopian Government has estimated that over the next half century it will require US$ 60 billion for irrigation and US$ 19 billion for the hydropower schemes. Ethiopia's gross domestic product was US$ 5.8 billion in 1988 and as such the rate at which it can devote finance to the development of its water resources will be slow. Overseas development assistance could accelerate the rate of development and there are strong arguments to favour this approach in that if the period of exporting power could be brought forward, the economic returns on investments would also be enjoyed earlier, to the advantage of the Ethiopian economy.

The construction of Ethiopia's dams will take place during an era when notions of sustainability, environmental impact and environmental rehabilitation have gained currency and have become integral to the planning and implementation of projects. These priorities complicate the inter-sectoral and international relations in both the pre-construction and the implementation phases. The Ethiopian Government is explicit in its statements that its future development of water will take such issues into account. Ethiopia is also explicit in its expectation that there will be a re-examination of the basis of international water agreements and that new arrangements will evolve and new agreements will be reached which do not necessarily conform with agreements and precedents from the colonial era.

Investment requirements

The essential investment to mobilise new environmental monitoring systems and in due course water management structures can only come from outside the region. This means that international agencies and major national donors such as those in the United States, the EC and Japan have a potentially very large role and must be taken into account as part of the economic, political and even legal scenarios One of the problems in the mobilisation of the finance will be the agreement of assumptions on how the water will be allocated as the result of the construction of new water management structures. Sums of tens of billions of dollars will be needed; in other words sums far exceeding the annual GDPs of all except Egypt. But even Egypt lacks the surpluses from which to allocate the massive investment resources required

Investment will be differently targeted in the upstream states compared with those downstream The latter have already invested heavily and further investment will be much more in ameliorating measures designed to reduce the effects of intensive irrigated farming as well as to enable the further intensification of land and water use. Upstream investment, especially in Ethiopia, will lead to more dramatic changes in production and productivity and the governments of upstream riparians are likely to resist the attempts of international financial interests to constrain their use of water according to principles which maximise the effectiveness of the investment basin-wide. This despite assurances to the contrary. (Abate, 1991 supplement, p.16)

Conclusions

Dams and other engineering structures including especially those works which have an impact on drainage and evaporation are integral to existing and future basin-wide use of water in the Nile catchment. The construction of new works will only take place after there is a much more widespread recognition amongst Nile riparians that there is mutual interest in managing water comprehensively. The preconditions for the adoption of such an approach are first that the partners must feel that their interests will be taken into account; in other words that they can expect new developments to provide water security in terms of supply and environmental impact, as well as better economic returns to water in some parts of the basin. Secondly they must contribute to, and recognise the benefits of, installing a reliable basin-wide hydro-meteorological monitoring system. The success of future negotiations on water allocation depends on the reliability of such data. Thirdly the international agencies and relevant bilateral donors must earn a place in the evaluation, implementation and water utilisation phases of Nile water planning and management. They can play an especially important technical role in the creation of the environmental monitoring systems. Fourthly the governments of the Nile Basin countries must begin to recognise the need to take a reciprocal approach to the development of water through gaining of a share of the fruits of investment including sharing in investment. Of these the second requirement is of most immediate importance and potential donors must be made to realise that this is a priority for financial support.

References

Abate, Z., (1991). Planned national water policy: a proposed case for Ethiopia. Paper presented at a World Bank International Workshop on Comprehensive Water Resources Management Policies. The World Bank, Washington DC.

Abate, Z., (1991), summary. Planned national water policy: a proposed case for Ethiopia. Paper presented at a World Bank International Workshop on Comprehensive Water Resources Management Policies. The World Bank, Washington DC.

Allan, J.A., (1981). The High Dam is a success story. *Geographical Magazine*, Vol LII, 6, March (1981). pp. 393-396.
Allan, J.A., (1991). *Nile Basin water planning*. Paper given at the FAO Symposium on the Nile held in the University of Bologna, March 1991.
Barnett, T., (1974). *The Gezira Scheme: an illusion of development*. Cass, London
Egyptian Government. (1921). *Nile Control*, Vol. **1**, edited by Sir Murdoch MacDonald, Cairo, Government Press.
FAO, (1987). *Irrigation and water resources potential for Africa*. Rome, FAO.
Gaitskell, A., (1952). *Gezira: a story of development in Sudan*. London, Faber and Faber.
Hurst, H.E., Black, R.P., and Simaika, Y.I., (1946). *The Future Conservation of the Nile. The Nile Basin*, **VII**. Cairo.
Hurst, H.E., (1959). *The Nile*. London, Constable.
Little, T., (1965). *High Dam at Aswan*. London, Methuen.
McGregor, R.M., (1945). 'The Upper Nile Irrigation Projects'. University of Durham, W.N. Allan Papers.
Morrice, H.A.W. and Allan, W.N., (1958). *Report on the Nile Valley Plan*. 2 vols, Khartoum, Sudan Government.
Nasser, President G.A., (1960). *The Year Book of the UAR for 1960*. Cairo, Government of the UAR.
Permanent Joint Technical Commission, (1989). *Statistical Information*. Cairo, Ministry of Public Works and Water Resources, and Khartoum, the Ministry of Irrigation.
UNDP, (1989). *Nile Basin integrated development: fact finding mission report*. RAF/86/003-RAB/86014.
United Nations, (1962). *UN Document ST/LEG/SER.B/12*. United Nations, New York. The Nile Waters Agreement was signed in Cairo on 8 November 1959; the Permanent Joint Technical Commission was set up by agreement in Cairo on 17 January 1960. The texts have been recorded in the UN Legislative Series.
US Department of the Interior, (1964). *The Land and Water Resources of the Nile Basin: Ethiopia*. 17 vols., Washington DC.
Whittington, D. and Haynes, K.E, (1980). Valuing water in the agricultural environment of Egypt: some estimation and policy considerations, *Regional Science Perspectives*. Vol 10, pp. 109-126.
Whittington, D. and Haynes, K.E., (1985). Nile Water for whom? Emerging conflicts in water allocation for agricultural expansion in Egypt and Sudan. in Beaumont, P. and McLachlan, K.S., *Agricultural development in the Middle East*. John Wiley, London, pp. 125-149.
WMO/UNDP, (1982). *Hydrometeorological survey of the catchments of Lake Victoria, Kyoga and Mobutu Sese Seko, Project findings and recommendations*. WMO, Geneva.

Index

nb - page numbers in *italics* refer to figures and tables